BARRON'S E-Z STATISTICS

Douglas Downing, Ph.D.
Seattle Pacific University

Jeffrey Clark, Ph.D.
Elon University

Better Grades or Your Money Back!

As a leader in educational publishing, Barron's has helped millions of students reach their academic goals. Our E-Z series of books is designed to help students master a variety of subjects. We are so confident that completing all the review material and exercises in this book will help you, that if your grades don't improve within 30 days, we will give you a full refund.

To qualify for a refund, simply return the book within 90 days of purchase and include your store receipt. Refunds will not include sales tax or postage. Offer available only to U.S. residents. Void where prohibited. Send books to **Barron's Educational Series, Inc., Attn: Customer Service** at the address on this page.

ABOUT THE AUTHORS

Douglas Downing teaches economics and statistics at Seattle Pacific University. He is the author of several books in the *E–Z* series: *Algebra, Trigonometry, Calculus,* and *Java Programming*, along with *Dictionary of Mathematics Terms* and *Dictionary of Computer and Internet Terms* (with Michael and Melody Covington). He holds a Ph.D. degree in economics from Yale University.

Jeffrey Clark teaches mathematics at Elon University. He is the author of *Business Statistics* and *Quantitative Methods* (both with Douglas Downing for the Barron's *Business Review* series). He holds a Ph.D. degree in mathematics from Yale University.

ACKNOWLEDGMENTS

Our thanks go to Mark Yoshimi for his help with the probability experiments, to Liz Ashburn for providing experimental data, and to our students.

All inquiries should be addressed to:
Barron's Educational Series, Inc.
250 Wireless Boulevard
Hauppauge, New York 11788
www.barronseduc.com

Library of Congress Control Number 2008936750

ISBN-13: 978-0-7641-3978-9
ISBN-10: 0-7641-3978-9

PRINTED IN THE UNITED STATES OF AMERICA
9 8 7 6 5 4 3

CONTENTS

optional sections that require the use of more advanced mathematics

*optional sections that require the use of more advanced mathematics

INTRODUCTION

Statistics is the study of how to acquire meaningful information by analyzing data, such as a list of numbers. Sometimes this data comes from planned experiments in controlled environments where it is possible to vary the factors that we are concerned with. When it is not possible to do experiments, then we must rely on empirical data from observations of the uncontrolled "real" world. In either case, this book teaches you how statistical methods can be used to analyze the situation.

You're probably wondering whether statistics is a hard subject. The answer to that question is "yes and no." Many of the important ideas in statistics can be understood even by someone with only a slight mathematical background. However, there are also parts of statistics that are hard. Many require the use of advanced mathematical methods such as calculus. For this reason, many of the important results in statistics have remained inaccessible to the general reader. The purpose of this book is to make these important results understandable by everyone. Many books attempt to simplify the material by presenting you with all the formulas that you need to know, making it possible for you to plug in numbers mindlessly and crank out results without understanding what you are doing. If you follow that method, you will sooner or later run into a situation where your lack of understanding could get you into trouble.

In this book we explain why everything works. However, the subjects that require the most mathematics are left as exercises or as optional mathematical sections. (You can peek in the back of the book if you wish, since the answers to most of the exercises are given there.) A star marks the exercises that require advanced math, and the optional mathematical sections are clearly labeled with warnings to prevent nonmathematicians from inadvertently straying into them. However, if you know calculus you will find that these sections of the book provide valuable explanations of the underlying theory.

You will need to know a little bit of algebra to use this book; for example, you will need to know what exponents are. We will use some set notation, but we will develop all of that from scratch. We also use some Greek letters. You can get to know them by checking the "Cast of Greek Characters" section at the beginning of the book. (We use Greek letters for two reasons: (1) Everybody else does it; and (2) it's easier to keep track of which symbol stands for what when more symbols are available.)

To understand statistics it is necessary to understand probability. Probability, the study of chance phenomena, is also an interesting subject in its own right. There are many fun aspects of probability, since it includes the study of dice, cards, and related games. There are also many important practical applications of probability, and the important concepts of statistical inference are based on the ideas of probability. The first few chapters cover the ideas of probability that you will need to know.

In both probability and statistics you will face the need to apply long and complicated formulas. That is the kind of work that is best left to a computer, so this book takes advantage of the fact that many people today have access to a small computer. You can still learn from this book even if a computer is not available to you, but it will be much easier for you to come up with numerical answers if you do have a computer. Answers expressed as numbers are far more satisfying than answers expressed as complicated formulas.

There are several types of computer software you can use to perform statistical calculations. Many popular spreadsheet programs have some built-in statistical capabilities, and there are other programs that are specifically designed for statistics. Appendix 2 gives more information on the use of computers. In the book we provide several examples using the popular Microsoft spreadsheet Excel.

Best wishes as you embark on the challenging yet fascinating journey of learning statistics.

Cast of Greek Characters (and other symbols)

Here are the main Greek letters that you will have to know:

μ (lowercase mu), pronounced *mew*; μ stands for the mean of a random variable.

σ (lowercase sigma); σ stands for the standard deviation of a random variable.

σ^2 (sigma squared) represents variance.

χ (lowercase chi), pronounced like *kite*, but without the *t* at the end.

χ^2 (chi squared) represents an important random variable distribution.

π (lowercase pi), pronounced *pie*; π represents a special number about equal to 3.14159 that is used extensively in mathematics.

Φ (capital phi), pronounced *fie*; Φ is used for the cumulative distribution function for a standard normal random variable.

Σ (capital sigma); Σ represents summation. In other words, ΣX means "Add up all the values of X."

These Greek letters also appear:

λ (lowercase lambda); λ is used for the parameter in the Poisson and exponential distributions.

Δ (capital delta); Δ represents "change in."

ψ (lowercase psi), pronounced *sigh*; ψ represents a moment generating function.

ε (lowercase epsilon); ε represents a small positive number.

Γ (capital gamma); Γ represents the gamma function.

Other symbols that will be used follow:

$\cup$ union

$\cap$ intersection

$|$ given that

$\sqrt{n}$ square root

$|\ |$ absolute value (for example, $|-3|$ is 3)

e 2.71828 …

$\emptyset$ empty set

$>$ greater than

$<$ less than

$\int_a^b f(x)dx$ area under the curve $f(x)$ from $x = a$ to $x = b$

1

Introduction to Statistics

Your mission is to study the population. Find out how tall they are, who they'll vote for, and which medicines will cure them. However, you confront a big problem: it is very hard to check everyone in the population. You are considering investigating a sample selected from the population, and you hope that the results you find for the sample will be representative of the entire population. We must turn to the study of statistics to see if this will work.

There are many times when we encounter statistics in our lives. In this book you will learn how to understand and apply some of the most common statistical methods. Here are two examples of the types of problems we will investigate:

- How can a poll based on a sample of about 1,000 people predict the results of an election with about 100 million voters? If the election is not close, then typically the results can be predicted accurately before the election (although there are some notable examples of wrong predictions from 1936 and 1948 that provide warnings about how not to conduct polls). Also note that no poll can predict an election if the vote is very close, as in the U.S. presidential election of 2000. The success of polls in predicting elections does provide some assurance that polls can be trusted in other cases where the poll results are never verified by checking the population, as is done in an election.

- Doctors study the relation between certain chemicals (such as nicotine) and the diseases they might cause. However, not everyone exposed to the chemicals will get the disease. For example, you may be able to think of a smoker who has lived a long time without ever suffering any serious disease, and you may also be able to think of a nonsmoker who has become ill. This does not mean that there is no connection between the chemicals and the disease. Instead, it illustrates the dangers of using anecdotes as evidence. An anecdote is a story about one particular incident. If you search long enough you can probably find an anecdote to illustrate almost anything. However, in order to see if there really is a connection between two things, such as a chemical and a disease, it is necessary to obtain a large amount of data covering many cases. Once the data have been collected, then you need to use statistical analysis to determine if you have found a significant relationship or not. We will discuss how to test this type of hypothesis.

The word *statistics* has two different but related meanings. In the most common usage, statistics means "a collection of numerical data." For example, we could look at the statistics that show the populations of cities in a state or that describe the performance of a baseball team.

The word *statistics* also refers to the branch of mathematics that deals with the analysis of statistical data. There are two key branches of statistics: descriptive statistics and inferential statistics. *Descriptive statistics* is the process of obtaining meaningful information from sets of numbers that are often too large to deal with directly. A large pile of numbers that have not been summarized is called *raw data*. Even though the raw data contains a lot of information, it is not very meaningful because people have a limited capacity to absorb the information. In order to convey meaning, it is usually necessary to summarize the data. You are already familiar with some concepts from descriptive statistics, such as the use of the average (also known as the *mean*) to indicate the typical value in a group of numbers. Descriptive statistics will be introduced in Chapter 4.

Statistical inference refers to the process of obtaining information about a larger group from the study of a smaller group. The complete group you are interested in is called the *population*. For example, we might be interested in the entire population of the United States or in the population of a particular city. In statistics, the word population does not have to refer to people; for example, the population we are interested in might consist of all the fish in a lake, or it might consist of all ice cream cartons produced at an ice cream factory.

The group of items chosen from the population is called the *sample*. Examples include

- population: the 31 flavors of ice cream at a 31-flavor ice cream store
  sample: the 5 flavors you have tested to determine whether this store sells good ice cream
- population: all voters in the United States
  sample: the 1,000 people in a poll that attempts to predict the result of an election
- population: all people in the United States who have television sets
  sample: the people who are surveyed by the Nielsen television rating firm
- population: all households in the United States
  sample: the 50,000 households interviewed in the Census Bureau's monthly Current Population Survey, which is used to estimate employment and unemployment.

In each case, you can save on limited resources by checking a sample instead of the population. There are also other reasons why it is sometimes necessary to do this. In some cases, the process of investigating the item destroys it. For example, if you are tasting the ice cream coming off the assembly line or checking the amount of stress that a post can withstand without breaking, then testing the entire population of items would leave you without any items. It is also possible that data collected from a sample will be measured more accurately than data from a population because the smaller size of the sample makes it possible to be more careful in training interviewers, entering data, and processing answers.

How should the sample be chosen? If the sample is unrepresentative of the population, then our estimates based on the sample will be wrong. There is no way we can guarantee that the sample will be representative, but it turns out that the best plan is to choose the sample completely at random. Imagine that we have a list of the names of everyone in the population and we inscribe them all on marbles. We put all the marbles in a giant drum that we can spin so that they will be mixed thoroughly. Then we draw some marbles to determine who will be included in the sample. In practice, the sample will most likely be chosen by a computer programmed to make a random selection, but the concept is the same.

Selecting the sample randomly does guarantee that there will be no systematic bias in the sample. Any other selection method carries the risk of a biased sample. If you choose your sample by investigating only people who live in a particular town, your sample will be biased because people in different locations have different characteristics. If your sample consists only of people who subscribe to a particular magazine, your sample will be biased because those subscribers have their own characteristics that don't match those of the population. If your sample consists only of people with phones, it will be systematically biased by not including people without phones (although this is not as big a problem now as it once was). If your sample consists only of people who answer their home phone in the afternoon, it will be biased by systematically excluding people who are not home during the afternoon.

To choose the sample randomly, you need to make sure that every member of the population has an equal chance of being selected and that all possible samples are equally likely to be chosen. (In practice, it is hard to meet this condition exactly, but you need to try to approximate these goals.) If you choose the sample randomly, there is a danger that you might have bad luck and that the sample is unrepresentative. If your population consists of a drum containing 60 percent blue marbles but your sample has 70 percent red marbles, you will fail to accurately picture the population. Do you need to lose sleep over this possibility? What we can do is calculate the probability of having bad luck that would make your sample unrepresentative. In order to do this we need to understand *probability*.

We will introduce the study of probability by concentrating on a familiar random processes: tossing a coin (see Chapter 2). Then we will develop the connection between probability and statistical hypothesis testing in Chapter 3. Chapter 4 will cover the use of descriptive statistics such as the mean, median, and standard deviation to summarize the properties of data. Chapters 5 to 7 provide more detailed development of probability concepts, and Chapters 8 to 15 apply the study of probability to random variable distributions. Chapters 16 to 22 cover essential topics in statistical inference: confidence intervals, hypothesis testing, polling, nonparametric methods, and regression. Finally, Chapter 23 includes some warnings to help you avoid fallacies that can trap you if you're not careful about applying statistical methods.

2

Introduction to Probability

If you flip a coin, the result will be either heads or tails. It would be very useful if you could predict exactly how the coin will turn up when it is flipped (especially if you are the captain of a football team meeting the referee for the opening coin toss). However, it is practically impossible to do this. You could in theory calculate the path of the coin by using the laws of mechanics, if you knew the initial force that was applied to it and the other forces acting on it, and thus determine which side would be up when the coin comes to a stop. This would be very hard. As the coin is spinning, the difference between the rates of spin that would cause it to end up heads are imperceptibly different from the rates of spin that would cause it to end up tails. For practical purposes, a coin toss is a good example of a random event: there is no accurate way to predict what the result will be.

If you toss a coin more than once, though, you can make some interesting predictions about what will happen. This is a general feature of random events, and it is the key idea that makes statistical inference possible. You can't say what will happen in one occurrence, but if you watch the same random event happen many times, you will be able to predict some patterns. For example, it would be very difficult to accurately guess the height of one randomly selected person, but it would be much easier to guess the average height of 100 randomly selected people. If you toss a coin many times, you intuitively should sense that you would be unlikely to get all heads or all tails and that the most likely possibilities will be somewhere near half heads and half tails. At the opening of the play *Rosencrantz and Guildenstern Are Dead*, Guildenstern tosses 92 heads in a row, thereby losing a lot of money to Rosencrantz. Intuitively you know that this is extremely unlikely to occur.

If you flip a coin twice, there are four possible outcomes: (HH, HT, TH, TT, where H = heads and T = tails) (see Figure 2–1).

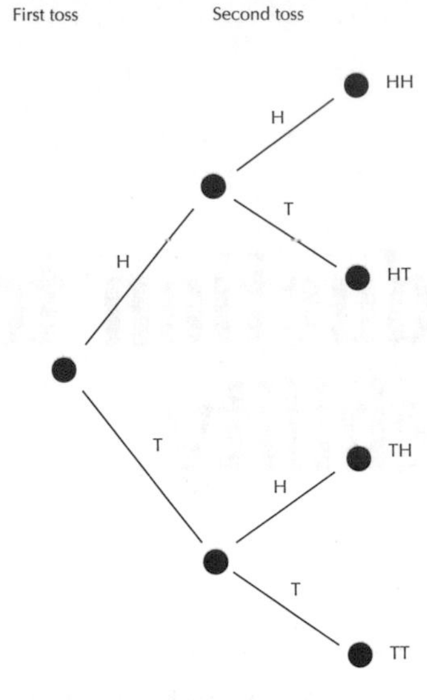

FIGURE 2–1

Assuming that the coin is a fair coin, not biased toward heads or tails, each of the 4 outcomes is equally likely. The chance of two heads is only 1/4 since only 1 of the 4 outcomes has two heads (HH). The chance of two tails is also 1/4 (one outcome: TT). If you had to guess the number of heads that will appear, then guess 1: since there are 2 outcomes (HT and TH) with one head, the probability of getting one head is therefore 1/2. In general, when all outcomes are equally likely, then the probability of a particular event is equal to the number of outcomes corresponding to that event divided by the total number of outcomes.

If we toss the coin 3 times, there are 8 outcomes (see Figure 2–2). This type of tree diagram can help to make sure that all possible outcomes are listed.

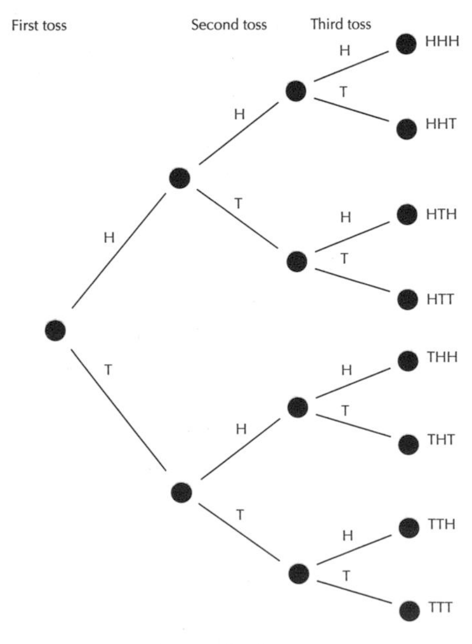

FIGURE 2–2

HHH, HHT, HTH, HTT, THH, THT, TTH, TTT

We can then make a table of all the outcomes:

Event You're Interested in	Outcomes That Lead to That Event	Number of Outcomes	Probability of Event
0 heads	TTT	1	1/8 = .125
1 head	HTT, THT, TTH	3	3/8 = .375
2 heads	HHT, HTH, THH	3	3/8 = .375
3 heads	HHH	1	1/8 = .125

There are 16 outcomes if we toss the coin 4 times:

HHHH, HHHT, HHTH, HHTT, HTHH, HTHT, HTTH, HTTT,
THHH, THHT, THTH, THTT, TTHH, TTHT, TTTH, TTTT

It becomes challenging to list all the outcomes as the number of tosses increases. The best way to do this systematically so as to not miss any outcomes is to proceed in the same way that an odometer works. Consider an odometer reading

0020358

The far right digit advances first:

0020359

Once that digit reaches the end (9), it resets to zero, but at the same time that it resets, the digit one space to the left advances by 1:

0020360

Then the far right digit repeats the pattern. The digit that is now 3 will not advance until much later. When the odometer finally reaches

0020399

the next step resets the far right digit to 0 while also advancing the second-from-the-right digit. This digit also gets set to 0, which means that the next digit to the left needs to advance by 1:

0020400

For coins, there are only two possibilities for each digit: H and T. Start at

HHHH

Then advance the far right digit to T:

HHHT

Next, the far right digit resets to H while also advancing the next digit to the left to T:

HHTH

Advance the far right digit to T again:

HHTT

Next, the far right digit resets to H, which will also cause the next digit left to reset to H, which will cause the far left digit to advance to T:

HTHH

Continue following this pattern. Once we have all the outcomes, we have to group them according to the number of heads, and then we can set up the table of probabilities:

Event You're Interested in	Outcomes That Lead to That Event	Number of Outcomes	Probability of Event
0 heads	TTTT	1	1/16 = .0625
1 head	HTTT, THTT, TTHT, TTTH	4	4/16 = .2500
2 heads	HHTT, HTHT, HTTH, THHT, THTH, TTHH	6	6/16 = .3750
3 heads	HHHT, HHTH, HTHH, THHH	4	4/16 = .2500
4 heads	HHHH	1	1/16 = .0625

We can begin to see the patterns. Each time we add another toss, the number of outcomes doubles. Therefore,

$$\text{number of outcomes with } n \text{ tosses} = 2^n$$

The number of outcomes increases rapidly, so that we will have nightmares if we keep trying to list them all. For example, if there are 10 tosses, there will be $2^{10} =$ 1,024 possible outcomes. Our only hope is to work out a way to count how many outcomes have a specific number of heads without having to list them all.

Let's see if we can determine how many outcomes have 2 heads out of 4 tosses.

There are 4 possible slots where we could put an H:

H _ _ _ ; _ **H** _ _ ; _ _ **H** _ ; _ _ _ **H**

For each of these 4 choices, there are 3 slots remaining where we can put the other H. That leads us to think there are $4 \times 3 = 12$ possible ways of tossing two heads out of the four tosses:

3 outcomes with a head in the first slot:

H H _ _ ; **H** _ H _ ; **H** _ _ H

3 outcomes with a head in the second slot:

H **H** _ _ ; _ **H** H _ ; _ **H** _ H

3 outcomes with a head in the third slot:

H _ **H** _ ; _ H **H** _ ; _ _ **H** H

3 outcomes with a head in the last slot:

H _ _ **H** ; _ H _ **H** ; _ _ H **H**

However, we can quickly see that something is wrong with this approach. The outcome H H _ _ is listed twice. In fact, each possible outcome is listed twice.
Therefore, there are only 6 distinct outcomes with 2 heads out of 4 tosses:

3 outcomes with a head in the first slot:

H H _ _ ; H _ H _ ; H _ _ H

2 outcomes with a head in the second slot and not already listed:

_ H H _ ; _ H _ H

1 outcome with a head in the third slot and not already listed:

_ _ H H

Note that these six outcomes are listed in the previous table.
We'll work out the case of 5 tosses by first listing all $2^5 = 32$ outcomes, using the odometer principle:

HHHHH, HHHHT, HHHTH, *HHHTT*, HHTHH, *HHTHT, HHTTH*, **HHTTT**
HTHHH, *HTHHT, HTHTH*, **HTHTT**, *HTTHH*, **HTTHT, HTTTH**, HTTTT
THHHH, *THHHT, THHTH*, **THHTT**, *THTHH*, **THTHT, THTTH**, THTTT
TTHHH, **TTHHT, TTHTH**, TTHTT, **TTTHH**, TTTHT, TTTTH, TTTTT

Then we will group them according to the number of heads and find the probabilities. In the list above, outcomes with two heads are shown in boldface type, and outcomes with three heads are shown in italics to facilitate grouping them in the following table:

Event You're Interested in	Outcomes That Lead to That Event	Number of Outcomes	Probability of Event
0 heads	TTTTT	1	1/32 = 0.0313
1 head	HTTTT, THTTT, TTHTT, TTTHT, TTTTH	5	5/32 = 0.1563
2 heads	HHTTT, HTHTT, HTTHT, HTTTH, THHTT, THTHT, THTTH, TTHHT, TTHTH, TTTHH	10	10/32 = 0.3125
3 heads	HHHTT, HHTHT, HHTTH, HTHHT, HTHTH, HTTHH, THHHT, THHTH, THTHH, TTHHH	10	10/32 = 0.3125
4 heads	HHHHT, HHHTH, HHTHH, HTHHH, THHHH	5	5/32 = 0.1563
5 heads	HHHHH	1	1/32 = 0.0313

The hard part is deriving the general formula that shows the number of outcomes with k heads out of n tosses. We'll show you what the formula looks like here. See if you can see the pattern. (We won't try to explain why this pattern works until Chapter 6; see page 56).

Five tosses:

$$\text{outcomes with 0 heads out of 5 tosses} = 1$$

$$\text{outcomes with 1 head out of 5 tosses} = \frac{5}{1} = 5$$

$$\text{outcomes with 2 heads out of 5 tosses} = \frac{5 \times 4}{2 \times 1} = 10$$

$$\text{outcomes with 3 heads out of 5 tosses} = \frac{5 \times 4 \times 3}{3 \times 2 \times 1} = 10$$

$$\text{outcomes with 4 heads out of 5 tosses} = \frac{5 \times 4 \times 3 \times 2}{4 \times 3 \times 2 \times 1} = 5$$

$$\text{outcomes with 5 heads out of 5 tosses} = \frac{5 \times 4 \times 3 \times 2 \times 1}{5 \times 4 \times 3 \times 2 \times 1} = 1$$

Six tosses:

$$\text{outcomes with 0 heads out of 6 tosses} = 1$$

$$\text{outcomes with 1 head out of 6 tosses} = \frac{6}{1} = 6$$

$$\text{outcomes with 2 heads out of 6 tosses} = \frac{6 \times 5}{2 \times 1} = 15$$

$$\text{outcomes with 3 heads out of 6 tosses} = \frac{6 \times 5 \times 4}{3 \times 2 \times 1} = 20$$

$$\text{outcomes with 4 heads out of 6 tosses} = \frac{6 \times 5 \times 4 \times 3}{4 \times 3 \times 2 \times 1} = 15$$

$$\text{outcomes with 5 heads out of 6 tosses} = \frac{6 \times 5 \times 4 \times 3 \times 2}{5 \times 4 \times 3 \times 2 \times 1} = 6$$

$$\text{outcomes with 6 heads out of 6 tosses} = \frac{6 \times 5 \times 4 \times 3 \times 2 \times 1}{6 \times 5 \times 4 \times 3 \times 2 \times 1} = 1$$

The formula for coin tossing is built into Excel:

=BINOMDIST(A1, B1, 0.5, false)

calculates the probability of A1 heads out of B1 tosses. We can use this formula to create graphs to show us how the probabilities change as the number of tosses increases. See Figures 2–3 and 2–4. Note that the probabilities are highest near the middle. Also note that the probabilities are symmetric (since the probability of getting k heads is the same as the probability of getting k tails, for any k). Further, note that the probabilities tend to follow a bell-shaped curve as the number of tosses increases. This curve is known as the *normal distribution* and will play a crucial role later in the book.

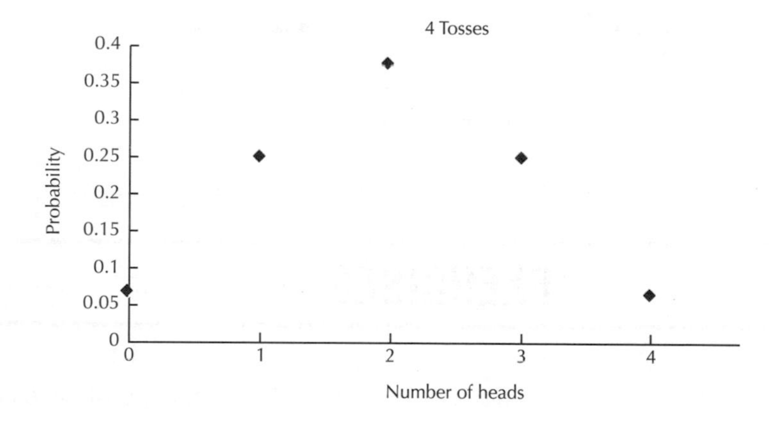

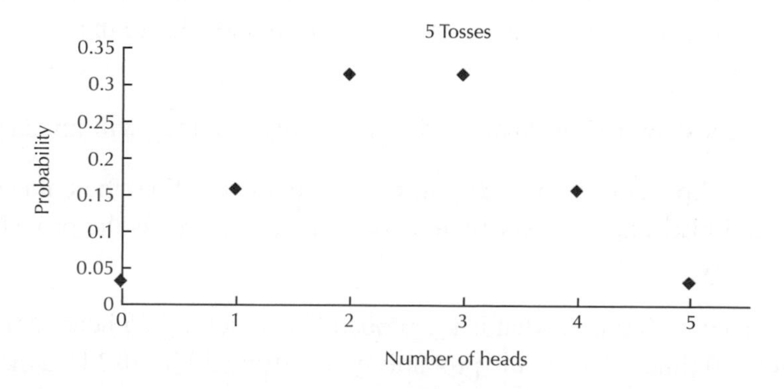

FIGURE 2–3

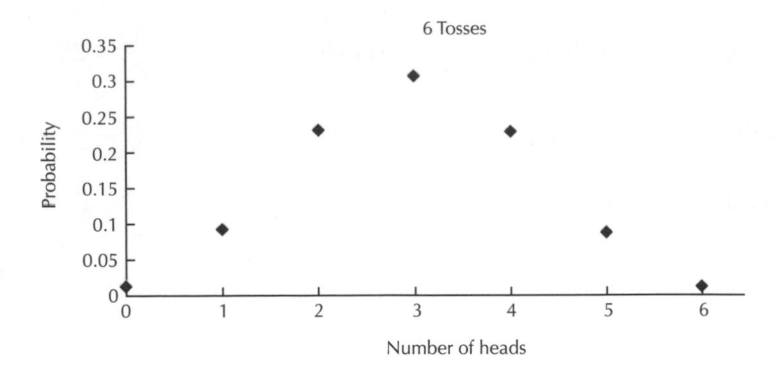

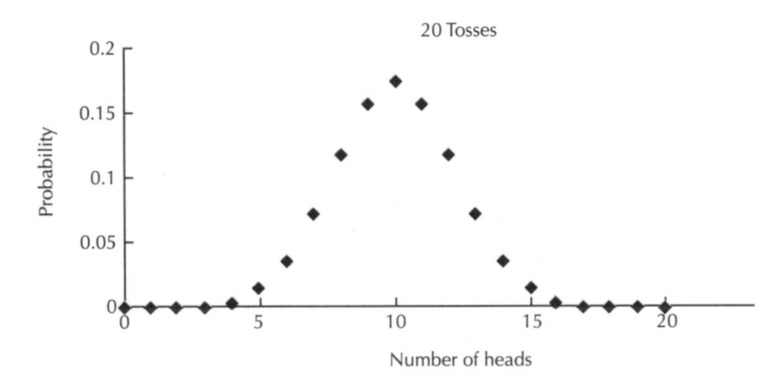

FIGURE 2–4

EXERCISES

1. Toss a coin 50 times and keep track of the results. Is the number of heads that appear close to the number that is predicted?
2. Toss five coins and write down the number of heads that appear. Repeat this process 50 times and then compare the observed frequencies to the probabilities.
3. Suppose you toss two coins. What is the probability that they will match?
4. Suppose you flip a coin 5 times. If you win a dollar every time the coin comes up heads and lose a dollar every time it comes up tails, what is the probability that you will win 3 dollars?
5. If you flip a coin 50 times, what is the probability of getting 25 heads? If you flip a coin 100 times, what is the probability of getting 50 heads? Use the BINOMDIST function in Excel.

3

Introduction to Hypothesis Testing

In the last chapter we investigated a probability problem: If you toss a coin a certain number of times, how many heads are likely to appear? During that calculation we assumed that the coin was a fair coin—that is, during any particular flip there was a 50 percent chance of flipping heads and a 50 percent chance of flipping tails. Now we will consider an even more perplexing problem: How can we tell whether or not the coin is really fair? You will especially need to know the answer to this question if you are considering playing a coin-flipping game with an unkempt-looking stranger in a strange town. To put it formally, if we let p stand for the probability that the coin will come up heads, how do we know that $p = 1/2$?

In our coin example we should first, of course, make an obvious check. If the coin has two heads, then $p = 1$; if it has two tails, then $p = 0$. Once we've done that, though, it is very difficult to tell just by looking at the coin whether it is fair or not. Intuitively, we can't think of any reason why it might be more likely to come up heads rather than tails (or vice versa), but it might be unbalanced in such a way as to make one outcome more likely than the other. If we flip the coin once, we won't have a clue as to whether or not it is fair. However, if we flip the coin many times, then we will start getting some information that we can use to estimate how fair it is.

Problems of this sort are called *hypothesis testing* problems. First we decide on the hypothesis we want to test. In this case our hypothesis is that $p = 1/2$. The hypothesis that is to be tested is often called the *null hypothesis*, symbolized by H_0. The only other possibility is that the null hypothesis is wrong. The hypothesis that says, "The null hypothesis is wrong" is called the *alternative hypothesis*, symbolized by H_a or H_1. In our case, the alternative hypothesis is that the coin is not fair, ($p \neq 1/2$). We know that either the null hypothesis or the alternative hypothesis must be true, since they are the only two possibilities. The question is, do we accept the null hypothesis and say that the coin is fair, or do we reject the null hypothesis and say that the coin is unfair?

It is clear intuitively that we should flip the coin many times; let n be the number of flips. Then, if the number of heads that appears is close to $n/2$, we should accept the hypothesis that the coin is fair. If the number of heads is very far from $n/2$, we should reject the hypothesis that the coin is fair. For example, if we flip the coin 100 times and come up with 44 heads, it seems quite likely that the coin is fair. However, if we come up with only 10 heads in 100 flips, we can almost surely say that something fishy is going on and that the coin is not fair.

Therefore, our test procedure will work like this: We will pick a number c. If the number of heads (h) is between $(n/2) - c$ and $(n/2) + c$, we will accept the null hypothesis and say that the coin is fair; otherwise we will say that the coin is not fair. We can call the region from $(n/2) - c$ to $(n/2) + c$ the *zone of acceptance*. If h is not in the zone of acceptance, we will be very critical of the hypothesis and will reject it. Therefore, the zone for which the hypothesis will be rejected is called the *critical region*.

The main problem now is, how far away from $n/2$ can we let the number of heads get before we say that the coin is unfair—that is, how big should we make c?

We would like to make the right judgment about our null hypothesis. There are two ways we can be right: we can accept the hypothesis when it is true, or we can reject it when it is false. However, that means that there are also two ways we can be wrong: we might reject the hypothesis when it is really true, or we might accept the hypothesis when it is really false. The first kind of mistake can be called error type 1 and the second kind can be called error type 2. Table 3–1 shows the possibilities.

TABLE 3–1

	Accept Hypothesis	Reject Hypothesis
Hypothesis is true	Right	Type 1 Error (we want to avoid this)
Hypothesis is false	Type 2 error	Right

If we choose a large value for c, then we will have a wide zone of acceptance and we will be more likely to accept the hypothesis than we would with a small value of c. That means that there is less chance of committing a type 1 error—that is, we're not likely to reject the hypothesis if it is really true. However, it we make the zone of acceptance large, we are increasing the risk that we will accept the hypothesis even if it is really false, which means that we would commit a type 2 error.

The other strategy is to choose a narrow zone of acceptance. If we do that, it is unlikely that we will commit a type 2 error (we're not likely to accept the hypothesis if it is really false), but we stand a much greater chance of committing a type 1 error (rejecting the hypothesis when it is really true). There is an inherent trade-off involved in hypothesis testing. We usually can't devise a single test procedure that will minimize the chances of committing both types of error.

Often we will be more worried about the possibility of incorrectly rejecting the hypothesis, so we will be more careful about avoiding type 1 errors. If we're going to have the courage to tell the unkempt stranger that we think his coin is unfair, then we want to make almost certain that we're right. (Otherwise he might get nasty.) In scientific work, if we decide to accept the hypothesis, then we're likely to keep on searching for more evidence to see if we can make a convincing case for it. If we decide to reject the hypothesis, that means we're really convinced that the hypothesis is false, so we can stop.

What is often done in statistics is to set an upper limit to the probability of committing a type 1 error. Usually this limit is set at either 10 percent or 5 percent. At first it can be confusing to remember the difference between type 1 and type 2 errors. Just remember that our number one priority is to avoid errors of type 1, which we do by being very polite and making sure that we don't reject the hypothesis unless we're pretty sure it's wrong. If we decide on a 10 percent test, that means we want to make sure there is only a 10 percent chance that our test procedure will say that the coin is unfair when it is really fair.

Once we've decided this, we need to figure out how wide to make the zone of acceptance. Start by making the assumption that the null hypothesis is true (so the coin is fair). Then we can calculate the probabilities as in the previous chapter. We'll toss the coin 50 times, so set up this table in Excel:

	A	B
1	Number of heads	Probability
2	0	=BINOMDIST(A2, 50, 0.5, FALSE)
3	1	

Enter the numbers 0 to 50 in column A. Then type the formula

=BINOMDIST(A2, 50, 0.5, FALSE)

into cell B2. When you copy that formula down the column, you'll have the table of probabilities. You can create the graph from this table (see Figure 3–1). Most of these probabilities are close to zero, so to save space we'll show only the values from 19 to 31:

Number of Heads	Probability
19	.02701
20	.04186
21	.05980
22	.07883
23	.09596
24	.10796
25	.11228
26	.10796
27	.09596
28	.07883
29	.05980
30	.04186
31	.02701

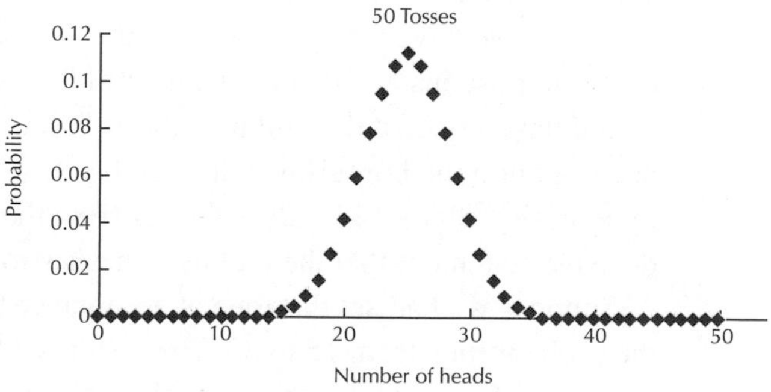

FIGURE 3–1

If we add up all these probabilities, we'll find that the probability is .93509 that there will be between 19 and 31 heads out of 50 tosses of a fair coin. This means that if the null hypothesis is true, we will have a .93509 probability of being right if we set the zone of acceptance from 19 to 31. However, that means we also have a $1 - .93509 = .06491$ probability of being wrong—which means committing a type 1 error by saying the null hypothesis is false when it is really true. Since this probability is greater than .05, we probably could not sleep well at night by risking this much chance of a type 1 error. The remedy is that we will have to widen our zone of acceptance.

The probability of 18 heads is .01603. By symmetry this is also the probability of 32 heads, so the probability that the number of heads will be between 18 and 32 is

$$.01603 + .93509 + .01603 = .96716$$

With the zone of acceptance set from 18 to 32, the probability of a type 1 error is $1 - .96716 = .03284$. We will be able to sleep more easily now that the probability of type 1 error is less than 5 percent. (Note: Sometimes probabilities are expressed as decimals, such as .05; other times they are expressed as percents, such as 5 percent. You need to be familiar with both.)

However, widening the zone of acceptance carries the opposite kind of risk: it becomes more likely that we will accept the null hypothesis when it is really false—that is, commit a type 2 error. There is not a single number that gives the probability of a type 2 error since the null hypothesis can be false in a variety of different ways. In our case, the null hypothesis will be false if the probability of heads (p) is equal to .51, or if $p = .52$, or if $p = .53$, etc. In general, a larger sample will give you a more powerful test that will have a better chance of distinguishing whether the null hypothesis is true or not. However, the benefit of a better test comes at a cost: it will be more expensive for you to take a larger sample (or toss the coin more times).

Suppose we tossed the coin 50 times and only 14 heads appeared. This would make us suspicious. We had previously decided that the zone of acceptance would be the range 18 to 32, so with 14 heads we would clearly reject the null hypothesis. Since this zone of acceptance provides for a probability of type 1 error equal to .03284, we will say that we can reject the null hypothesis at the 3.28 percent level of significance. However, we have a nagging feeling that we can make a stronger statement in this case. Suppose instead that the number of heads had been 17. In that case, we also would have rejected the null hypothesis at the 3.28 percent level, but we would have been right on the borderline (since 18 heads would have caused us to accept the null hypothesis). Since 14 heads is further from the boundary, we can make a more decisive statement that the null hypothesis strongly deserves to be rejected in this case.

Suppose we had set the zone of acceptance to be 15 to 35. Use Excel to add up all the probabilities from 15 to 35. The result is .99740. This means that if we had used this zone of acceptance, the probability of type 1 error would be $1 - .99740 = .00260$. This value is so small that we would have little worry about the chance of a type 1 error with the zone of acceptance 15 to 35. Since the appearance of 14 heads would cause us to reject the null hypothesis even at the very low 0.26 percent level of significance (.00260), we can be very confident that the null hypothesis has been very decisively rejected. The value .0026 is called the *p-value* of this test. The *p*-value tells you how low could you let the chance of type 1 error get and still reject the null hypothesis. If the *p*-value is below .05 (5 percent), then you will reject the null hypothesis at the 5 percent significance level. However, if the *p*-value is .049, then you realize that the test is right on the borderline, and you should do further research to determine if rejecting the null hypothesis is the correct decision. On the other hand, if the *p*-value is very small, then you can be very sure that rejecting the null hypothesis is the right decision.

For example, suppose you have developed a new drug and you are conducting a hypothesis test where the null hypothesis states that the drug has no effect on the disease it supposedly cures. You are hoping that the null hypothesis will be decisively rejected, so you would be overjoyed to see a low *p*-value for your test. (However, don't let your hopes cloud your judgment; you still need to make sure that the experiment is conducted as objectively as possible.)

Hypothesis testing will be developed more completely in Chapter 18.

EXERCISES

Use a spreadsheet to answer these questions:

1. What is the probability that you will get 44 heads in 100 tosses of a fair coin?

2. What is the probability that you will get 10 heads in 100 tosses of a fair coin?

3. Suppose you flip a coin 100 times and you want to test the hypothesis that the coin is fair, making sure that there is less than a 5 percent chance of erroneously rejecting the fair coin hypothesis. How wide should the zone of acceptance be? How wide should the zone be if you flip the coin 5 times? How wide should the zone be if you flip the coin 10 times?

4. Suppose a friend has tossed two dice, one blue and one red. You are told the total of the two numbers appearing on the dice but not the individual numbers. You want to test the hypothesis that the blue die shows 3, with only a 5 percent chance of a type 1 error. What should your testing procedure be?

5. (This one might take a while.) Flip a coin 20 times. Decide whether to accept or reject the hypothesis that the coin is fair, based on what we did in this chapter. Now repeat this procedure 100 times. How many times did you accept the hypothesis?

State the null hypothesis and the alternative hypothesis in these cases:

6. You want to see if cars made on Monday have more defects than cars made on other days.

7. You want to see if football teams that mostly run the football win more on the average than teams that mostly throw the football.

8. You want to see if people who drink coffee without caffeine are healthier than people who drink regular coffee.

4

Descriptive Statistics

At first, many people think statistics is a dull subject that is not very much fun. However, think about how much *more* dull work you would have to do if you didn't know statistics. If you've just been given a large pile of numbers, you have no hope of understanding them unless you can figure out some way to summarize them. And that's what statistics is all about.

Mean and Median

Suppose you have traveled across the country and crossed these seven suspension bridges:

Bridge	Length
Verrazano-Narrows (New York)	4,260
Bronx-Whitestone (New York)	2,300
George Washington (New York)	3,500
Mackinac (Michigan)	3,800
Tacoma Narrows (Washington)	2,800
Golden Gate (San Francisco)	4,200
Transbay (San Francisco)	2,310

(The length column gives the length of the longest span of the bridge, in feet.)

When faced with a list of numbers such as this, it is human nature to want to summarize them. One way to do this is to calculate the average. To do that, we add up the lengths of the bridges and then divide by the number of bridges. We find the average is 23,170/7 = 3,310 ft.

The average is also called the *mean*, symbolized by the Greek letter mu (μ), or by putting a bar over the quantity. If there are n numbers in the list (symbolized by $x_1, x_2, \ldots, x_n$) the average is given by this formula:

$$\mu = \overline{x} = \frac{x_1 + x_2 + \ldots + x_n}{n}$$

(For now we will use μ and $\overline{x}$ interchangeably to represent the average; later we will see that it is common to use μ to represent the mean of a population and $\overline{x}$ for the mean of a sample.)

The sum of x_1 to x_n can also be written with summation notation:

$$x_1 + x_2 + \ldots + x_n = \sum_{i=1}^{n} x_i$$

The symbol Σ, the Greek capital letter sigma, stands for summation; the designation "$i = 1$" at the bottom means to start adding where i equals 1; the "n" at the top means to stop adding where i equals n. The expression written to the right of the sigma (x_i in this case) tells you what to add.

The average is an example of a *statistic*. A statistic is a single number that summarizes the properties of a larger group of numbers. We will discuss some other important statistics later.

Another way of summarizing a list of numbers is to look at the number that is in the middle (called the *median*). To do that, we need to put the numbers in order:

Bridge	Length
Verrazano-Narrows (New York)	4,260
Golden Gate (San Francisco)	4,200
Mackinac (Michigan)	3,800
George Washington (New York)	3,500
Tacoma Narrows (Washington)	2,800
Transbay (San Francisco)	2,310
Bronx-Whitestone (New York)	2,300

Since 7 is an odd number, the fourth item is the middle item (three are higher and three are lower). Therefore, the median bridge length is 3,500 ft.

In general, you will be able to find an exact middle item for any list with an odd number of items. If the list has an even number of items, you need to find the two items closest to the middle—then the median is halfway between those two elements. For example, the median of the list 10, 30, 42, 50, 65, 84 is found halfway between 42 and 50: $(42 + 50)/2 = 46$.

Often the value of the median is close to the average, as with the bridge example. However, if a list contains one number that is unusually large, then the average will be pulled up above the median. If one number is unusually low, then the average will be pulled down below the median. The median is not sensitive to changes in the values at the ends of the list.

For example, suppose you cross four short bridges and then cross the Golden Gate Bridge. The lengths of the five bridges are 200, 210, 220, 230, 4,200. The mean length of these five bridges is

$$\frac{200 + 210 + 220 + 230 + 4{,}200}{5} = \frac{5{,}060}{5} = 1{,}012.000$$

Because the Golden Gate Bridge is so much longer than the other bridges, the mean is pulled up to be much larger than the typical value. The median value of 220 is a better indicator of the typical bridge length you crossed on this particular trip.

In Excel, the function AVERAGE(*range*) will find the average of the values in the given range, and MEDIAN(*range*) will find the median.

Variance

There are also times when we would like to measure the degree to which the numbers in a list are spread out. That is, we would like to know if the numbers are all close to the mean, or if they are all far away from the mean. The first thing we can do is calculate the distance between each length and the mean bridge length $\bar{x}$ (3,310):

Bridge	Length (x_i)	($x_i - \bar{x}$)
Verrazano-Narrows (New York)	4,260	950
Golden Gate (San Francisco)	4,200	890
Mackinac (Michigan)	3,800	490
George Washington (New York)	3,500	190
Tacoma Narrows (Washington)	2,800	−510
Transbay (San Francisco)	2,310	−1000
Bronx-Whitestone (New York)	2,300	−1010

We will use x_i to represent the length of bridge i; then $(x_i - \bar{x})$ is the difference between x_i and the mean.

Now we need to figure out how to use these differences to measure the amount of spread. We can't simply add up the differences, because then the positive numbers and negative numbers cancel out, giving zero. Before adding up the distances, we need to turn them all into positive numbers by squaring them:

Bridge	Length (x_i)	($x_i - \bar{x}$)	($x_i - \bar{x})^2$
Verrazano-Narrows (New York)	4,260	950	902,500
Golden Gate (San Francisco)	4,200	890	792,100
Mackinac (Michigan)	3,800	490	240,100
George Washington (New York)	3,500	190	36,100
Tacoma Narrows (Washington)	2,800	–510	260,100
Transbay (San Francisco)	2,310	–1000	1,000,000
Bronx-Whitestone (New York)	2,300	–1010	1,020,100
Total	**23,170**	**0**	**4,251,000**
Average	**3,310**	**0**	**607,285.7**

After squaring all the distances, add them up (giving 4,251,000) and then divide by 7 to find the average: 607,285.7. This figure, the average of the squared distances from the mean, is our measurement of the spread of a list. It is called the *variance*.

In general, the variance of a list of numbers $x_1, x_2, x_3, \ldots x_n$ is symbolized by Var(x), and is given by this formula:

$$\mathrm{Var}(x) = \frac{(x_1 - \bar{x})^2 + (x_2 - \bar{x})^2 + (x_3 - \bar{x})^2 + \ldots + (x_n - \bar{x})^2}{n}$$

This can be written using summation notation:

$$\mathrm{Var}(x) = \sigma^2 = \frac{\sum_{i=1}^{n} (x_i - \bar{x})^2}{n}$$

The variance is also represented by the symbol σ^2, which is read as "sigma squared." The symbol σ is the lowercase version of the Greek letter sigma. Make sure you don't confuse this with the uppercase sigma (Σ) used for summation notation.

The square root of the variance is called the *standard deviation*, and is represented by σ. For the bridge example, we have $\sigma = \sqrt{607{,}285.7} = 779.3$.

One advantage of the standard deviation is that it is measured in the same units as the original data. For example, the standard deviation of the bridge length is measured in feet. The variance is measured in units that are the square of the units used to measure the original data, which makes it harder to interpret. However, there are times when the variation is more convenient to work with. For the rest of the book you must be prepared to work with either the variance or the standard deviation. Be sure to remember the fundamental rule connecting them: the standard deviation σ is always the square root of the variance.

In practice, it is slightly easier to calculate the variance using this formula:

$$\mathrm{Var}(x) = \overline{x^2} - \bar{x}^2$$

The quantity $\overline{x^2}$ is found by squaring each value of x, and then finding the average of those squares:

Bridge	Length (x_i)	x_i^2
Verrazano-Narrows (New York)	4,260	18,147,600
Golden Gate (San Francisco)	4,200	17,640,000
Mackinac (Michigan)	3,800	14,440,000
George Washington (New York)	3,500	12,250,000
Tacoma Narrows (Washington)	2,800	7,840,000
Transbay (San Francisco)	2,310	5,336,100
Bronx-Whitestone (New York)	2,300	5,290,000
Total	**23,170**	**80,943,700**
Average	**3,310**	**11,563,385.7**

Therefore, $\overline{x^2} = 11{,}563{,}385.7$. We have already found $\bar{x} = 3{,}310$, so $\bar{x}^2 = 3{,}310^2 = 10{,}956{,}100$. The variance is $11{,}563{,}385.7 - 3{,}310^2 = 607{,}285.7$, the same answer that was found using the longer formula.

The above formulas for variance and standard deviation apply when we have data taken from a population. When investigating data from a sample that is being used to estimate the properties of a population, the variance is often calculated from a slightly different formula:

$$\text{sample variance} = s_2^2 = \frac{\sum_{i=1}^{n}(x_i - \bar{x})^2}{n - 1}$$

The only difference is that $n - 1$ appears in the denominator of the sample variance. By contrast, n appears in the denominator of the population variance. We will discuss the difference between the population variance and the sample variance more fully in Chapter 17.

In Excel, the functions VARP(*range*) and STDEVP(*range*) calculate the variance and standard deviation of population data (using the formulas with n in the denominator). The functions VAR(*range*) and STDEV(*range*) calculate the variance and standard deviation of sample data (using $n - 1$ in the denominator).

Frequency Histograms

Suppose that you are interested in the height and weight of the human body, perhaps so you can give a general description of the human race to a friendly extraterrestrial. Suppose further that you have a wonderful computer called the Brain that can give you the height and weight of every person. How would you give the information to the extraterrestrial?

You probably wouldn't take the time to give him the 6 billion or so heights and weights. These numbers, called *raw data*, often give us more information than we really need. The extraterrestrial really doesn't need the raw data (the height and weight of every human being). Instead, you would organize the information and tell the extraterrestrial how many people weigh between 50 and 55 kilograms, how many between 55 and 60 kilograms, and so on. Giving the number of people in various height and weight categories is enough.

Let's look at another set of raw data. Suppose a history class consisting of 30 people has the following scores on a one-hundred-point test:

53	87	76	73	62	99
78	93	82	69	65	93
92	92	78	82	89	65
63	49	88	87	94	73
85	77	98	59	93	82

We can learn more about how the scores are distributed if we group the scores into classes like this:

Interval	Number of Scores in Interval
41–50	1
51–60	2
61–70	5
71–80	6
81–90	8
91–100	8

To display these data in even clearer form we can draw a *frequency histogram*. Figure 4–1 shows a frequency histogram for these data.

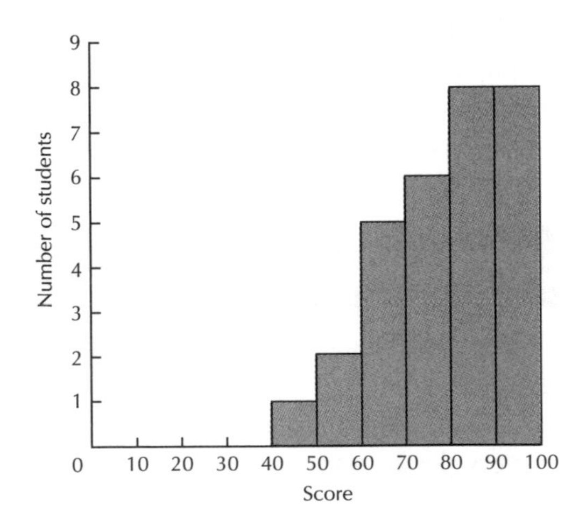

FIGURE 4–1

The height of each bar is the number of people whose scores are in that interval.

An interesting statistic is the *mode*. The mode (or modes) is (or are) the data that occur most frequently. In the previous sample, there are two modes, 82 and 93. Each mode occurs three times. None of the other scores occurs three or more times.

The Excel function FREQUENCY will help you generate the frequency histogram. Enter the data for the 30 scores in the range A1:A30. In cells B1 to B6 enter the values at the top of each interval: 50, 60, 70, 80, 90, and 100. Then highlight the range C1:C7 and enter this formula: =FREQUENCY(A1:A30, B1:B6). After entering the formula, type three keys simultaneously: Control, Shift, and Enter. By highlighting the range C1:C7 first, then entering the formula, and then pressing Control-Shift-Enter, you will enter the FREQUENCY function as an array formula. The result in columns B and C will be the frequency table:

B	C
50	1
60	2
70	5
80	6
90	8
100	8
	0

The final 0 in column C indicates that there are 0 values greater than 100. With the frequency numbers it is easy to create a column graph in Excel to illustrate the distribution.

Now we've introduced some important statistics, but we still have to figure out when to use which statistic, and we have to determine what the statistics mean. For example, if we've gathered data on two different groups of rats that received different treatment, how can we tell how much difference (if any) the treatment made? If we have surveyed the opinions of a sample of people, how can we tell whether or not their opinions are representative of the population as a whole? If we have looked at the relation between national income and consumer spending in the past, do we think we can reliably predict the relationship between these two quantities in the future?

We'll cover all of these questions later. First we need to develop the main concepts of probability. Probability is an amusing subject in itself. Historically, probability concepts were developed to help understand games of chance, and we will talk a lot about games. Probability also has important applications in fields such as insurance and decision-making.

DESCRIPTIVE STATISTICS SUMMARY

$$\text{mean} = \mu = \bar{x} = \frac{x_1 + x_2 + \ldots + x_n}{n} = \frac{\sum_{i=1}^{n} x_i}{n}$$

If a list contains an odd number of numbers, the median is the middle number; if the list contains an even number of numbers, then the median is midway between the two numbers closest to the middle (in both cases, the list must be arranged in order first).

$$\text{population variance} = \text{Var}(x) = \sigma^2$$

$$\sigma^2 = \frac{(x_1 - \bar{x})^2 + (x_2 - \bar{x})^2 + (x_3 - \bar{x})^2 + \ldots + (x_n - \bar{x})^2}{n}$$

$$= \frac{\sum_{i=1}^{n}(x_i - \bar{x})^2}{n}$$

$$= \overline{x^2} - \bar{x}^2$$

$$\text{standard deviation} = \sqrt{\text{Var}(x)} = \sigma = \sqrt{\frac{\sum_{i=1}^{n}(x_i - \bar{x})^2}{n}}$$

$$\text{sample variance} = s_2^2 = \frac{\sum_{i=1}^{n}(x_i - \bar{x})^2}{n - 1}$$

NOTE TO CHAPTER 4

In this chapter we considered only cases in which the classes were of equal width. You can draw a histogram even if the classes are of unequal width by drawing the diagram so that the area of each rectangle is proportional to the number of numbers in its interval.

EXERCISES

1. The temperatures in degrees Celsius each day over a three-week period were as follows: 17, 18, 20, 22, 21, 19, 16, 15, 18, 20, 21, 21, 22, 21, 19, 20, 19, 17, 16, 16, 17. Compute the mean, median, and mode of these raw data, and, using two-degree intervals starting with 15–16 draw a frequency diagram.

2. Calculate the same statistics as in Exercise 1 using all of the positive numbers that are perfect squares and that are less than 1,000. (For the frequency diagram use these intervals: 1–100, 101–200, etc.

3. Calculate the mean, median, and standard deviation for these numbers: 5, 10, 6, 11, 0, 0, 0, 10.

4. Change the 11 in the previous exercise to 100 and calculate the new answers.

5. Here are the scores on an English paper for the same 30 students discussed in this chapter:

58	88	79	76	66	99
80	94	84	72	69	94
93	92	80	84	90	69
67	54	90	89	95	76
86	80	98	63	94	84

Here are some chemistry scores for the same students:

76	94	88	86	81	99
89	96	91	84	82	96
96	96	89	91	94	82
81	74	94	93	97	86
92	88	99	79	96	91

Calculate the mean, median, and standard deviation for the two sets of scores.

6. In 2006, the mean U.S. household income was $66,570, while the median income was $48,201. What do you think is the significance of the fact that the mean is greater than the median?
(Source: http://www.census.gov/hhes/www/income/income06.html.)

7. Calculate the mean, median, and standard deviation for these lists:

(a) 0, 1, 2, 3, 4, 5, 6, 7, 8
(b) 0, 1, 4, 9, 16, 25, 36, 49, 64
(c) 0, 1, 2, 3, 7
(d) 0, 2, 4, 6, 14
(e) –1, 0, 1, 0
(f) 0 –1, 0, 1

8. Make a list of some of your friends and relatives, and then draw a frequency diagram showing the number of people whose first names start with a particular letter. Which letters seem to be the most popular for starting names?

9. The following table gives the area of the 50 U.S. states in square miles. Determine the mean, median, and standard deviation for each region, and for the entire United States.

East North Central	*Area*
Illinois	56,400
Indiana	36,291
Michigan	58,216
Ohio	41,222
Wisconsin	56,154

East South Central	*Area*
Alabama	51,609
Kentucky	40,395
Mississippi	47,716
Tennessee	42,244

Mountain	*Area*
Arizona	113,909
Colorado	104,247
Idaho	83,557
Montana	147,138
Nevada	110,540
New Mexico	121,666
Utah	84,916
Wyoming	97,914

Middle Atlantic	*Area*
New Jersey	7,836
New York	49,576
Pennsylvania	45,333

New England	*Area*
Connecticut	5,009
Maine	33,215
Massachusetts	8,257
New Hampshire	9,304
Rhode Island	1,214
Vermont	9,609

Pacific	*Area*
Alaska	589,757
California	158,693
Hawaii	6,450
Oregon	96,981
Washington	68,192

South Atlantic	*Area*
Delaware	2,057
Florida	58,560
Georgia	58,876
Maryland	10,577
North Carolina	52,586
South Carolina	31,055
Virginia	40,817
West Virginia	24,181

West North Central	*Area*
Iowa	56,290
Kansas	82,264
Minnesota	84,068
Missouri	69,686
Nebraska	77,227
North Dakota	70,665
South Dakota	77,047

West South Central	*Area*
Arkansas	53,104
Louisiana	48,523
Oklahoma	69,919
Texas	267,338

5

Definition of Probability

Interpretations of Probability

What exactly is probability? That is a tricky (and sometimes controversial) question.

Consider the statement, "If we flip a coin, the probability is 1/2 that the result will be heads." It is a difficult philosophical question to determine exactly what this statement means. According to the *relative frequency* view of probability, this statement means that the number of heads will be close to 1/2 of the total tosses if you toss the coin a large number of times. (We will show that this is indeed the case, but does that mean that the probability is nothing more than the relative frequency?)

There are some events for which the relative-frequency interpretation is difficult. The weather report often says, "There is a 20 percent chance of rain today." However, we can't have today repeat itself 100 times to see if it rains 20 of those times.

The *subjective* view of probability states that the probability is an estimate of what an individual *thinks* is the likelihood that an event will happen. In that case two individuals might estimate the probability differently. The subjective view makes it possible to talk meaningfully about the probabilities of a wider class of events, but the probabilities become more intangible because we can't objectively specify what the probabilities are.

We'll assume that we know the meaning of the statement, "The probability is 1/2 that a coin toss will result in heads." If this is true we can calculate what the probabilities will be for any number of coin flips.

Random Events

We can state some characteristics of a *random event* in which each outcome is equally likely. Examples of such random events are the toss of a single coin, the roll of a fair die, and the selection of a single card from a well-shuffled deck. The main characteristic of a random event is that there is no way to predict the outcome that is any better than any other method.

For an example of a random event, let us consider tossing a coin again. If you're captain of a football team, you'll face this problem: How do you predict the outcome of a single toss? As we'll see, there is no accurate way to do this, even if you've spent a lifetime studying probability.

Not only is there no best method; there is also no worst method. Suppose that you want to guess wrong as often as possible. This is just as hard as guessing right, since if you could guess wrong on purpose you could also guess right by calling each toss the opposite way.

Probability Spaces

Now we will develop the formal method of determining the probabilities associated with a random experiment. First, we'll make a list of all the possible results of our experiment. We're going to use the technical name *set* for this list. *Set* is just a formal name for a collection of objects. Some examples of sets follow:

$$\{\text{Washington, Oregon, California, Alaska, Hawaii}\}$$
$$\{1, 2, 3, 4, 5, 6\}$$

Sets can be defined by either of two methods. We can use the listing method, as we just did. That just means to write down all of the members of the set. The set members are surrounded by braces: { }. Then it's clear what is in the set and what is not.

Sets can also be defined by stating a rule that makes it clear what is in the set. For example, the sets above could have been defined by these rules:

- the set of states that touch the Pacific Ocean
- the set of all possible results of tossing one die

It should be noted that there need not be a rule that fits the set; {Bill Clinton, New York Giants, 4} is also a set (although one that is not likely to come up in practice).

With small sets, the listing method or the rule method will work equally well. However, with large sets the rule method works much better. For example, it would be very difficult to list all of the members of these sets:

- the set of all whole numbers from 1 to 1 million
- the set of all possible results of flipping a coin 12 times

We are especially interested in sets that contain all of the possible results of an experiment. The set of all possible results is called the *probability space* (or sometimes the *sample space*). Mathematicians like to use the term *space* for this type of set. Here are some examples of probability spaces:

- experiment: flip coin one time
  probability space: {H, T}
- experiment: flip coin three times
  probability space: {HHH, HHT, HTH, HTT, THH, THT, TTH, TTT}
- experiment: roll one die
  probability space: {1, 2, 3, 4, 5, 6}
- experiment: roll two dice
  probability space:

{(1,1) (1,2) (1,3) (1,4) (1,5) (1,6)
(2,1) (2,2) (2,3) (2,4) (2,5) (2,6)
(3,1) (3,2) (3,3) (3,4) (3,5) (3,6)
(4,1) (4,2) (4,3) (4,4) (4,5) (4,6)
(5,1) (5,2) (5,3) (5,4) (5,5) (5,6)
(6,1) (6,2) (6,3) (6,4) (6,5) (6,6)}

- experiment: draw one card from a deck of 52 cards
  probability space:

{ace hearts, ace diamonds, ace spaces, ace clubs,
two hearts, two diamonds, two spades, two clubs, etc.}

(There will be 52 elements if we list them all.)

In order to save us the bother of writing "probability space" each time, we'll use the letter S to stand for a probability space. (S is short for space. Or, if you like Greek letters, the letter Ω (omega) is often used to represent probability spaces.)

The important thing is that every single possible result of the experiment must be included in the probability space. We'll call each possible result an *outcome*. We'll use a small s to represent the total number of outcomes in the probability space S. In the examples above, the first probability space has two possible outcomes and the others have 8, 6, 36, and 52 outcomes, respectively. For now, we'll assume that each outcome is equally likely. (This is called the *classical approach* to probability.) Then, since there are s outcomes, the probability of any one outcome is $1/s$.

Now we'll use a probability space to solve a practical problem. Suppose that we're playing Monopoly, and that we need to roll a 7. As we saw earlier, there are 36 possible outcomes for the experiment of rolling two dice. We want to get a 7, so it doesn't make any difference to us if we roll (1,6) or (2,5) or (3,4) or (4,3) or (5,2) or (6,1). We can put all of these outcomes together in a set. We'll often use capital letters to stand for sets, so we may as well call this set A:

$$A = \{(1,6), (2,5), (3,4), (4,3), (5,2), (6,1)\}$$

We can also define set A by the rule method: Set A is the set of all possible outcomes from rolling two dice such that the sum of the numbers on the two dice is 7.

Set A contains 6 outcomes. If we roll the dice, we have an equal chance of getting any one of the 36 outcomes. Since 6 of these outcomes give us what we want (a total of 7 on the dice), the probability of getting a 7 is 6/36 = 1/6 (see Figure 5–1).

(1,1)	(1,2)	(1,3)	(1,4)	(1,5)	**(1,6)**
(2,1)	(2,2)	(2,3)	(2,4)	**(2,5)**	(2,6)
(3,1)	(3,2)	(3,3)	**(3,4)**	(3,5)	(3,6)
(4,1)	(4,2)	**(4,3)**	(4,4)	(4,5)	(4,6)
(5,1)	**(5,2)**	(5,3)	(5,4)	(5,5)	(5,6)
(6,1)	(6,2)	(6,3)	(6,4)	(6,5)	(6,6)

FIGURE 5–1

Set A is an example of what we will call an *event*. An event is a set that consists of a group of outcomes. In our case, (1,6), (2,5), (3,4), (4,3), (5,2), (6,1) are outcomes, and the set {(1,6), (2,5), (3,4), (4,3), (5,2), (6,1)} is an event.

An event can also be defined this way: An event is a *subset* of the probability space. A subset is a set that contains some (or possibly all) members of another set. If set A is a subset of set S, that means that every outcome in set A is contained in set S. Here are some examples of subsets:

- set: all people living in New York State
  subset: all people living in New York City
- set: all natural numbers less than 10: $\{1, 2, 3, 4, 5, 6, 7, 8, 9\}$
  subset: all odd numbers less than 10
- set: all possible outcomes from flipping a coin three times
  subset: all possible outcomes of three flips with one head: {HTT, THT, TTH}

Probability of an Event

Now we're ready to give a formal definition for the probability of an event A. First, count the number of outcomes in A. Call that number $N(A)$. (Read this "N of A," which is short for "number of outcomes in A.") Remember that s is the total number of outcomes in the probability space S. Then we can define the probability:

$$\text{Probability that event } A \text{ will occur is } \frac{N(A)}{s}.$$

In other words, count the number of outcomes that give you the event A, and then divide by the total number of possible outcomes.

It would save a lot of writing to have a shorter way to write "probability that event A occurs." So we will use Pr to stand for probability, and write it like this:

$\Pr(A)$ means "probability that event A will occur."

This gives the following result:

$$\Pr(A) = \frac{N(A)}{s}$$

The probability of an event can also be written as $P(A)$.

EXAMPLE What is the probability of getting one head if we toss a coin three times?

SOLUTION

In this case, there are $2^3 = 8$ possible outcomes, so $s = 8$. There are three outcomes that give one head (HTT, THT, TTH), so if A is the event of getting one head, then $N(A) = 3$. Therefore, $\Pr(A) = 3/8$.

EXAMPLE What is the probability of getting a total of 5 if we roll two dice?

SOLUTION

There are 36 possible outcomes, so $s = 36$. Let B be the event of getting a 5 on the dice, so B contains 4 outcomes:

$$B = \{(1,4), (2,3), (3,2), (4,1)\}$$

Then $N(B) = 4$, so $\Pr(B) = 4/36 = 1/9$ (see Figure 5–2). (So your chances of getting a 5 are worse than your chances of getting a 7.)

(1,1)	(1,2)	(1,3)	(1,4)	(1,5)	(1,6)
(2,1)	(2,2)	(2,3)	(2,4)	(2,5)	(2,6)
(3,1)	(3,2)	(3,3)	(3,4)	(3,5)	(3,6)
(4,1)	(4,2)	(4,3)	(4,4)	(4,5)	(4,6)
(5,1)	(5,2)	(5,3)	(5,4)	(5,5)	(5,6)
(6,1)	(6,2)	(6,3)	(6,4)	(6,5)	(6,6)

FIGURE 5–2

EXAMPLE What is the probability of drawing an ace if we draw one card from a deck of cards?

SOLUTION

There are 52 possible outcomes, so $s = 52$. Let C be the event of drawing an ace, so C contains four outcomes:

$$\{\ A \text{ hearts}, A \text{ diamonds}, A \text{ clubs}, A \text{ spades}\ \}$$

Then $N(C) = 4$, so $\Pr(C) = 4/52 = 1/13$.

There are two special events that are interesting. Let us consider the probability that the event S occurs. Remember that S contains all of the possible outcomes of the experiment. Using the formula gives:

$$\Pr(S) = \frac{N(S)}{s} = \frac{s}{s} = 1$$

That result should be obvious. All it says is, "The probability is 100 percent that the result will be one of the possible results." (Just ask yourself: What is the probability that the result will *not* be one of the possible results?)

Another possible event is the set that contains *no* outcomes. You should be able to convince yourself that this set has zero probability of occurring. This set is often called the empty set, or the null set, because it doesn't have anything in it:

$$\Pr(\text{empty set}) = 0$$

The empty set is often symbolized by a 0 with a slash, /, through it, like this: $\emptyset$. So we can say:

$$\Pr(\emptyset) = 0$$

Now we can develop another important result that is fortunately very obvious. Many times we will want to know the probability that an event will *not* happen. For example, suppose we're interested in the probability that we will *not* get a 7. It should be clear that if there is a 1/6 chance of getting a 7, then there is a 5/6 chance of not getting a 7. We can demonstrate that. Let A be the event of not getting a 7. Then A contains 30 outcomes:

{(1,1)	(1,2)	(1,3)	(1,4)	(1,5)	
(2,1)	(2,2)	(2,3)	(2,4)		(2,6)
(3,1)	(3,2)	(3,3)		(3,5)	(3,6)
(4,1)	(4,2)		(4,4)	(4,5)	(4,6)
(5,1)		(5,3)	(5,4)	(5,5)	(5,6)
	(6,2)	(6,3)	(6,4)	(6,5)	(6,6)}

Therefore $N(A) = 30$, so $\Pr(A) = 30/36 = 5/6$. In general, if p is the probability that a particular event will occur, then $1 - p$ is the probability that the event will not occur.

We will give a special name to the set that contains all of the outcomes that are not in set A. It is called the *complement* of A, written A^c. (The little c stands for complement. Read the symbol as "A complement.")

Then:

$$\Pr(A^c) = 1 - \Pr(A)$$

This result is important for calculations, since sometimes it is easier to calculate the probability that an event will not occur than it is to calculate the probability that it will occur.

Here are some examples of complements:

- If the experiment consists of flipping a coin, and A is the event of getting a head, then A^c is the event of getting a tail.
- If the experiment consists of flipping four coins, and B is the event of getting four heads, then B^c is the event of getting at least one tail.
- If the total set is the set of all 52 cards in a deck of cards, and R is the event of drawing a red card, then R^c is the event of drawing a black card.

Probability of a Union (*A* or *B*)

Now suppose we need to know the probability that we will get either a 5 or a 7 on two dice. It often happens in probability that we need to find out the probability that either one of two events will occur. Let's say that A is the event of getting a 7. Then A contains these outcomes:

$$\{(1,6), (2,5), (3,4), (4,3), (5,2), (6,1)\}$$

$$\Pr(A) = 6/36$$

If B is the event of getting a 5, then B contains these outcomes:

$$\{(1,4), (2,3), (3,2), (4,1)\}$$

$$\Pr(B) = 4/36$$

Let's say that C is the event of getting either a 5 *or* a 7. Then C contains these outcomes:

$$\{(1,6), (2,5), (3,4), (4,3), (5,2), (6,1), (1,4), (2,3), (3,2), (4,1)\}$$

Set C contains 10 outcomes, so $\Pr(C) = 10/36$.

There is a special name for the set that contains all of the elements that are in either or both of two other sets. It is called the *union* of the other two sets. In this case, C is the union of set A and set B. You can think of it this way: set A and set B get together and join forces to form a union, and the result is set C. In mathematics the word *union* is symbolized by a little symbol that looks like a letter *u*: ∪. Therefore, we can write "C is the union of A and B" as

$$C = A \text{ union } B$$

or

$$C = A \cup B.$$

Here are some examples of unions:

- If A is the set of even numbers and B is the set of odd numbers, then $A \cup B$ is the set of all natural numbers.
- If V is the set of vowels and C is the set of consonants, then $V \cup C$ is the set of all letters.
- If A is the set {AH, KH, QH, JH, 10H}, and B is the set {QH, JH, 10H, 9H, 8H} then $A \cup B$ is {AH, KH, QH, JH, 10H, 9H, 8H}.

There is another interesting fact that we can learn from the dice example. We said that (probability of getting a 7) = $\Pr(A) = 6/36$, (probability of getting a 5) = $\Pr(B) = 4/36$, and (probability of getting a 5 or a 7) = $\Pr(A \text{ or } B) = \Pr(A \cup B) = 10/36$. It looks as though you could just add the probabilities for the two events to get the probability that either one of them will occur, since $10/36 = 6/36 + 4/36$. This amazingly simple rule will work a lot of the time:

$$\Pr(A \text{ or } B) = \Pr(A \cup B) = \Pr(A) + \Pr(B)$$

However, we can quickly find a counterexample where our proposed rule does not work. Toss two dice. Let S_1 be the event of getting a 6 on the first die and S_2 be the event of getting a 6 on the second die. We would like to know the probability of getting at least one 6 on the two dice—in other words, $\Pr(S_1 \cup S_2)$. If we add together the two probabilities, we find:

$$\Pr(S_1) + \Pr(S_2) = 1/6 + 1/6 = 2/6 = 12/36$$

However, from Figure 5–3 we can see that the probability of getting at least one 6 is 11/36. Why can't we add the probabilities in this case?

(1,1)	(1,2)	(1,3)	(1,4)	(1,5)	(1,6)
(2,1)	(2,2)	(2,3)	(2,4)	(2,5)	(2,6)
(3,1)	(3,2)	(3,3)	(3,4)	(3,5)	(3,6)
(4,1)	(4,2)	(4,3)	(4,4)	(4,5)	(4,6)
(5,1)	(5,2)	(5,3)	(5,4)	(5,5)	(5,6)
(6,1)	(6,2)	(6,3)	(6,4)	(6,5)	(6,6)

FIGURE 5–3

Further analysis of Figure 5–3 reveals the problem. If we add the probabilities, we are double-counting the one outcome that has a 6 on both dice. Therefore, we have to add a condition to our rule: If events A and B can never happen together (that is, they are *mutually exclusive*), then the following is true.

$$\Pr(A \cup B) = \Pr(A) + \Pr(B)$$

Another name for mutually exclusive events is *disjoint* events.

Probability of an Intersection (*A* and *B*)

Now we will consider an example in which two events can happen together (that is, they are not mutually exclusive). Suppose you are selecting one card from a deck and you want to know the probability that you will get a red face card. Let's say the F is the event of getting a face card. Then F contains these outcomes:

$$\{\text{JH}, \quad \text{JD}, \quad \text{JC}, \quad \text{JS}, \\ \text{QH}, \quad \text{QD}, \quad \text{QC}, \quad \text{QS}, \\ \text{KH}, \quad \text{KD}, \quad \text{KC}, \quad \text{KS}\}$$

(Here we are using J to stand for jack, Q = queen, K = king, H = hearts, D = diamonds, C = clubs, S = spades.)

Since $N(F) = 12$, the probability of getting a face is 12/52.

Let R be the event of getting a red card. So R contains these outcomes:

$$\{\text{AH}, \text{2H}, \text{3H}, \text{4H}, \text{5H}, \text{6H}, \text{7H}, \text{8H}, \text{9H}, \text{10H}, \text{JH}, \text{QH}, \text{KH}, \\ \text{AD}, \text{2D}, \text{3D}, \text{4D}, \text{5D}, \text{6D}, \text{7D}, \text{8D}, \text{9D}, \text{10D}, \text{JD}, \text{QD}, \text{KD}\}$$

Set R contains 26 outcomes, so $\Pr(R) = 26/52 = 1/2$.

Let C be the event that both event F and event R occur—in other words, C is the event that you get a card that is both a red card and a face card. Then C contains these 6 outcomes:

$$\{\text{JD}, \text{JH}, \text{QD}, \text{QH}, \text{KD}, \text{KH}\}$$

Therefore, $\Pr(C) = \Pr(F \text{ and } R) = N(C)/s = 6/52$ (see Figure 5–4).

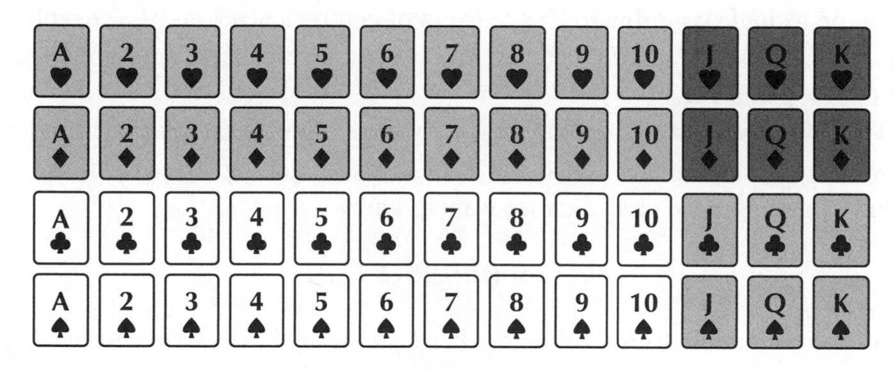

FIGURE 5–4

There is a special name for the set that contains all of the elements that are in both of two other sets. It is called the *intersection*. For example, in this case set C is the intersection of set F and set R, because it contains those outcomes that are in both set F and in set R. The symbol of intersection is the symbol for union turned upside down: $\cap$. Now we can write:

$C =$ set that contains the elements in both A and B;
$C = A$ intersects B; or
$C = A \cap B$

Here are some examples of intersections:

- If V is the set of vowels and C is the set of consonants, then $V \cap C$ is {y}.
- If A is the set of hands with five cards in sequence, and B is the set of hands with five cards of the same suit, then $A \cap B$ is the set of all straight flushes.
- If you flip a coin twice, and A is the event of getting a head on the first toss and B is the event of getting a head on the second toss, then $A \cap B$ is {HH}.

There is no general formula for the probability of the intersection of two events, although if the two events are *independent*, we can find the probability that they both occur by multiplying:

$$\Pr(A \cap B) = \Pr(A \text{ and } B) = \Pr(A) \times \Pr(B)$$

Two events are independent if they do not affect each other. This idea will be developed more fully in Chapter 7. For now, we'll have to count the numbers of outcomes in the intersection and calculate the probability directly from that.

If two events can't happen together (in other words, they are mutually exclusive events) then their intersection is the empty set. For example, we can't get both a head and a tail on a single coin flip, so if $H =$ event of getting a head, and $T =$ event of getting a tail, then $\Pr(H \text{ and } T) = \Pr(H \cap T) = 0$. In general, if A and B are disjoint, then $A \cap B = \emptyset$ and $\Pr(A \cap B) = 0$.

Now let us consider the possibility that we will get either a face card *or* a red card when we draw a card from the deck. If F is the event of getting a face card, and R is the event of getting a red card, then we want to know

$$\Pr(F \text{ or } R) = \Pr(F \cup R).$$

We can't use the simple formula $\Pr(F) + \Pr(R)$, because these two events can happen together. We can make a list of all the outcomes in $F \cup R$:

{AH, 2H, 3H, 4H, 5H, 6H, 7H, 8H, 9H, 10H
JH, QH, KH,
JC, QC, KC,
AD, 2D, 3D, 4D, 5D, 6D, 7D, 8D, 9D, 10D,
JD, QD, KD,
JS, QS, KS}

Altogether there are 32 outcomes, so

$$\Pr(F \text{ or } C) = \Pr(F \cup C) = 32/52 \qquad \text{(see Figure 5–4).}$$

Now we can figure out a general formula for the probability of $A \cup B$. We know that

$$\begin{aligned} N(A) &= (\text{number of outcomes in } A \text{ but not in } B) \\ &\quad + (\text{number of outcomes in } A \text{ and } B) \end{aligned}$$

$$\begin{aligned} N(B) &= (\text{number of outcomes in } B \text{ but not in } A) \\ &\quad + (\text{number of outcomes in } A \text{ and } B) \end{aligned}$$

If we add together $N(A) + N(B)$, we get

$$\begin{aligned} N(A) + N(B) &= (\text{number of outcomes in } A \text{ but not in } B) \\ &\quad + (\text{number of outcomes in } B \text{ but not in } A) \\ &\quad + 2 \times (\text{number of outcomes in } A \text{ and } B) \end{aligned}$$

But we know that the number of outcomes in $A \cup B$ is

$$\begin{aligned} N(A \cup B) &= (\text{number of outcomes in } A \text{ and not in } B) \\ &\quad + (\text{number of outcomes in } B \text{ and not in } A) \\ &\quad + (\text{number of outcomes in both } A \text{ and } B) \end{aligned}$$

When we just add $N(A)$ and $N(B)$, we're counting the outcomes in $(A \cap B)$ *twice*. So in order to get the number of outcomes in $A \cup B$, we can subtract $N(A \cap B)$, like this:

$$N(A \cup B) = N(A) + N(B) - N(A \cap B)$$

Therefore,

$$\Pr(A \text{ or } B) = \Pr(A) + \Pr(B) - \Pr(A \text{ and } B)$$

or, written mathematically:

$$\Pr(A \cup B) = \Pr(A) + \Pr(B) - \Pr(A \cap B)$$

This formula is good for any two events, whether or not they are mutually exclusive. Notice that, if A and B are mutually exclusive, then $\Pr(A \text{ and } B) = 0$, so we get the same formula that we had before: $\Pr(A \text{ or } B) = \Pr(A) + \Pr(B)$.

EXAMPLE Suppose, in the middle of a backgammon game, you want to know the probability that you will get a total of 8 or doubles on the dice.

SOLUTION

Let's say that $E1$ is the event of getting an 8. Then:

$$E1 = \{(2,6), (3,5), (4,4), (5,3), (6,2)\}; \quad \Pr(E1) = 5/36$$

Let $E2$ be the event of getting a double. Then:

$$E2 = \{(1,1), (2,2), (3,3), (4,4), (5,5), (6,6)\}; \quad \Pr(E2) = 6/36$$

These two events are not mutually exclusive, since you can get both an 8 and a double (if the dice turn up (4,4)). Then $(E1 \cap E2)$ is the event of getting (4,4), which has a probability of 1/36. So now we can use our formula:

$$\begin{aligned} \Pr(\text{getting an 8 or a double}) &= \Pr(E1 \text{ or } E2) \\ &= \Pr(E1) + \Pr(E2) - \Pr(E1 \text{ and } E2) \\ &= 5/36 + 6/36 - 1/36 \\ &= 10/36 \end{aligned}$$

EXAMPLE Use the formula to find the probability that you will get at least one 6 when you roll two dice.

SOLUTION

We'll call S_1 the event of getting a 6 on the first die and S_2 the event of getting a 6 on the second die. Then the event that you will get at least one 6 is $S_1 \cup S_2$. We know that $\Pr(S_1) = \Pr(S_2) = 1/6$. Since $S_1 \cap S_2$ is the event of getting 6's on both dice, we know that the probability of that happening is 1/36. Therefore, we can use the formula

$$\begin{aligned} \Pr(S_1 \cup S_2) &= \Pr(S_1) + \Pr(S_2) - \Pr(S_1 \text{ and } S_2) \\ &= 1/6 + 1/6 - 1/36 \\ &= 11/36 \end{aligned}$$

So your chances of getting at least one 6 are 11/36 (see Figure 5–3).

Axioms of Probability

We've now discussed everything we need in order to develop the formal mathematical model of probability. Formal mathematics works like this: First, make some assumptions (which are called *axioms* or *postulates*). Then, use the postulates to prove *theorems*.

Start with a probability space S. We will adopt these axioms:

$$\text{Axiom 1: } \Pr(S) = 1$$

This axiom just says that we are absolutely positive that the result of an experiment will be one of the possible results.

Now, let X be any event. (In other words, X is any subset of the probability space S.)

$$\text{Axiom 2: } \Pr(X) \geq 0$$

This axiom just states the obvious fact that there is no such thing as a negative probability.

Axiom 3: if A and B are any two mutually exclusive events, then

$$\Pr(A \cup B) = \Pr(A) + \Pr(B)$$

Now we can use these axioms to prove theorems. (The proofs will be left as exercises if you're interested.)

Theorem 1: $\Pr(X) \leq 1$, for any event X
Theorem 2: $\Pr(A^c) = 1 - \Pr(A)$
Theorem 3: $\Pr(\emptyset) = 0$
Theorem 4: $\Pr(A \cup B \cup C) = \Pr(A) + \Pr(B) + \Pr(C)$ if A, B, and C are any three mutually exclusive events
Theorem 5: $\Pr(A \cup B) = \Pr(A) + \Pr(B) - \Pr(A \cap B)$

This framework allows us to generalize our results further. So far we have only considered experiments where we could count the number of outcomes and each outcome was equally likely. We can still do probability calculations even if the outcomes are not all equally likely. All we need to do is assign a probability to each outcome, and then make sure that all of the probabilities add up to 1.

Continuous Probability Spaces

We will also want to consider random experiments in which we can't count the number of outcomes. The probability spaces we have discussed up to now are called *discrete* probability spaces. Now we'll look at an example of a *continuous* probability space. For example, suppose we throw a dart at the dart board in Figure 5–5. (We won't count throws that don't hit the dart board, so the probability is 1 that any throw that does count will hit the dart board.) Suppose we have no idea where on the dart board the dart will land, one place being as likely as another. Then we can regard the dart throw as a random experiment, but we can't count all of the outcomes. For example, the position of the dart might end up being the point 1 inch down and 1 inch to the left, but it could also be at the point 1.5 inches to the left, or 1.25 inches left, or 1.00000002 inches left, etc. (assuming that we can measure the point where the tip of the dart hits with infinite accuracy). In this case we can't calculate the probability that the dart will hit a particular point, but we can calculate the probability that it will hit within a particular area. For example, the probability that we will hit the left half of the diagram (shown as the shaded region in Figure 5–5) is 1/2. In general, if s is the area of the whole dart board and a is the area of a particular region A, then the probability that the dart will hit the area A is a/s. In the case of continuous probability spaces, the events we're interested in can be represented as areas, and we'll often draw these dart board-like diagrams to represent continuous probability spaces.

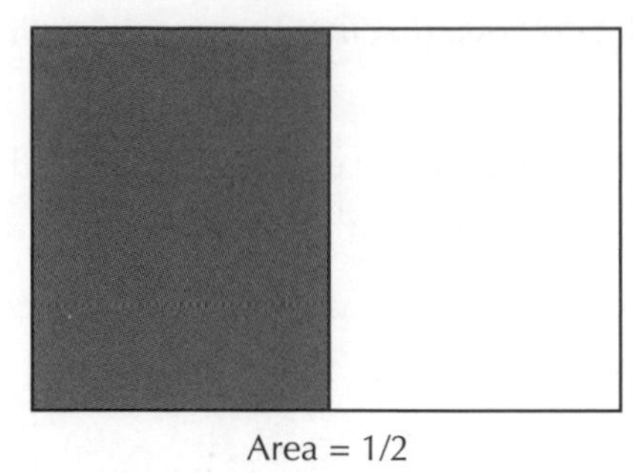

FIGURE 5–5

For another example of continuous probability space, consider this situation. Suppose the appliance repair person is going to visit sometime between 12 noon and 5 P.M., but you have no idea when. Then, for example, you can say that the probability is 1/2 that the person will arrive before 2:30, and the probability is 1/5 that the person will arrive in the first hour, 1/10 that the person will arrive in the first half-hour, and so on.

Unions of Three or More Events

There are times when we will be interested in the probability that any one of three events might occur. (Stated mathematically, we want to know the probability that the union of the three events will occur.) If the three events are mutually exclusive, then we can just add the three probabilities (see Theorem 4), but the situation is more complicated if the three events are not mutually exclusive. Consider the three regions A, B, and C on the dart board shown in Figure 5–6. Suppose that the area of the whole

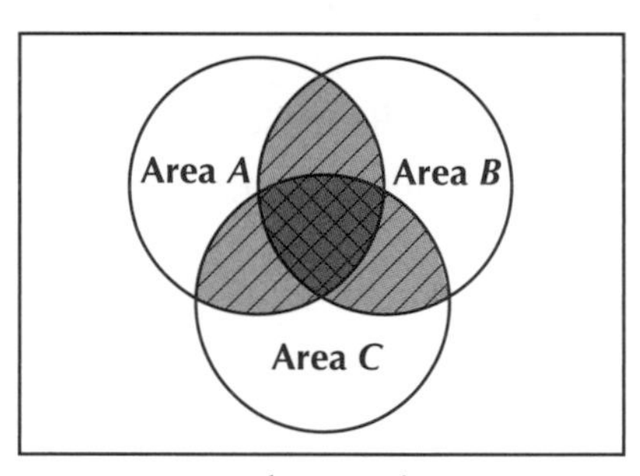

FIGURE 5–6

dart board is 1. Then $\Pr(A \cup B \cup C)$ is the entire area enclosed by the circles. If we add together $\Pr(A) + \Pr(B) + \Pr(C)$, we will double-count the striped area and triple-count the checked area. We can eliminate the double counting by subtracting, like this:

$$\Pr(A) + \Pr(B) + \Pr(C) - \Pr(A \cap B) - \Pr(B \cap C) - \Pr(A \cap C)$$

Now, however, we are not counting the checkered area at all, so we need to add that area back in:

$$\begin{aligned}\Pr(A \cup B \cup C) = {} & \Pr(A) + \Pr(B) + \Pr(C) \\ & - \Pr(A \cap B) - \Pr(B \cap C) - \Pr(A \cap C) \\ & + \Pr(A \cap B \cap C)\end{aligned}$$

This formula tells in general how to find the probability that the union of three events will occur. For example, suppose you are in a group of three people who are holding a Secret Santa draw. Each person is going to select randomly the name of one person in the group to give surprise presents to. What is the probability that one of the three people will select his or her own name?

First, we can easily calculate the probability that *you* will select your own name. Since there are three names you might choose, and one of them is your own, there is a 1/3 chance that you will draw your own name. Now we need to know the probability that at least one person in the group will draw his or her own name. If we let $E1$ be the event that you draw your own name and $E2$ and $E3$ be the events that person 2 and person 3 draw their own name, respectively, then we want to know the probability of $E1 \cup E2 \cup E3$. To use the formula, we need to know $\Pr(E1 \cap E2)$. $E1 \cap E2$ is the event that both person 1 and person 2 get their own names. The probability of that happening is 1/6. There are $3! = 6$ total possible ways of drawing the names, and there is only 1 way for both of them to get their own names. By the same reasoning, $\Pr(E1 \cap E3) = 1/6$, $\Pr(E2 \cap E3) = 1/6$, and $\Pr(E1 \cap E2 \cap E3) = 1/6$. Then, using our formula, we have:

$$\begin{aligned}\Pr(E1 \cup E2 \cup E3) &= \Pr(E1) + \Pr(E2) + \Pr(E3) \\ &\quad - \Pr(E1 \cap E2) - \Pr(E1 \cap E3) - \Pr(E2 \cap E3) \\ &\quad + \Pr(E1 \cap E2 \cap E3) \\ &= 1/3 + 1/3 + 1/3 - 1/6 - 1/6 - 1/6 + 1/6 \\ &= 2/3\end{aligned}$$

So there is a 2/3 probability that one of the three people will select his or her own name.

We could also have solved the problem by listing each of the six ways in which the names could have been drawn:

Name Drawn by 1	Name Drawn by 2	Name Drawn by 3
1	*2*	*3*
1	3	2
2	1	*3*
2	3	1
3	1	2
3	*2*	1

Asterisks mark those locations where a person has picked his or her own name, and we can see that four of the possible drawings have at least one person picking his or her own name.

If we want to know the probability of the union of more than three events, we end up with an awful formula.

PROBABILITY OF UNION OF *N* EVENTS

The basic pattern for the probability of the union of n events is the following:

1. Add the probabilities of all n events individually.
2. Subtract the probabilities of the intersections of all possible pairs of the events.
3. Add the probabilities of all possible intersections of the events taken three at a time.
4. Subtract the probabilities of all possible intersections of the events taken four at a time.

By now, you should have caught on to the pattern.

EXAMPLE Suppose that, during a backgammon game, your opponent has put one of your pieces on the bar, and your opponent has points 5 and 6 covered. In order to get back on the board, you need to roll a 1, 2, 3, or 4 on your next roll. What is the probability that you will do so?

SOLUTION

Let's say that $E1$ is the event of getting at least one 1, $E2$ is the event of getting at least one 2, and so on. Then we want to find the probability of $E1 \cup E2 \cup E3 \cup E4$. Using the method described above, we obtain:

$$\begin{aligned}\Pr(E1 \cup E2 \cup E3 \cup E4) = {} & \Pr(E1) + \Pr(E2) + \Pr(E3) + \Pr(E4) \\ & - \Pr(E1 \cap E2) - \Pr(E1 \cap E3) - \Pr(E1 \cap E4) \\ & - \Pr(E2 \cap E3) - \Pr(E2 \cap E4) - \Pr(E3 \cap E4) \\ & + \Pr(E1 \cap E2 \cap E3) + \Pr(E1 \cap E2 \cap E4) \\ & + \Pr(E1 \cap E3 \cap E4) + \Pr(E2 \cap E3 \cap E4) \\ & - \Pr(E1 \cap E2 \cap E3 \cap E4)\end{aligned}$$

We know that $\Pr(E1) = 11/36$, since it must be the same as $\Pr(E6)$, which we have already calculated. By the same reasoning, $\Pr(E2) = \Pr(E3) = \Pr(E4) = 11/36$.

The event $E1 \cap E2$ is the event of getting a 2 and a 1 on the dice. There are two ways for this to happen, so

$$\Pr(E1 \cap E2) = 2/36.$$

The event $E1 \cap E2 \cap E3$ is the event of getting a 1 and a 2 and a 3, which is obviously impossible, since we are only rolling two dice. So, using the formula, we get the following:

$$\begin{aligned}\Pr(E1 \cup E2 \cup E3 \cup E4) &= 11/36 + 11/36 + 11/36 + 11/36 \\ &\quad - 2/36 - 2/36 - 2/36 \\ &\quad - 2/36 - 2/36 - 2/36 \\ &= 32/36\end{aligned}$$

PROBABILITY SUMMARY

If there are s equally likely outcomes, and A is an event (that is, a group of outcomes), then

$$\Pr(A) = \frac{N(A)}{s}$$

where $N(A)$ is the number of outcomes in event A.

A^c (A complement) is the event that A does not happen:

$$\Pr(A^c) = 1 - \Pr(A)$$

A union B (written $A \cup B$) is the event that A or B occurs.

A intersect B (written $A \cap B$) is the event that both A and B occur.

If A and B are mutually exclusive events:

$$\Pr(A \text{ and } B) = 0$$

$$\Pr(A \text{ or } B) = \Pr(A) + \Pr(B)$$

If A and B are independent events:

$$\Pr(A \text{ and } B) = \Pr(A) \times \Pr(B)$$

In general:

$$\Pr(A \cup B) = \Pr(A) + \Pr(B) - \Pr(A \cap B)$$
$$\Pr(A \text{ or } B) = \Pr(A) + \Pr(B) - \Pr(A \text{ and } B)$$

EXERCISES

1. What is the probability of getting a prime number when you roll a die?
2. What is the probability of getting a 7 or an 11 when two dice are rolled?
3. Consider a roulette wheel that will stop on a number from 1 to 36, or 0 to 00. What is the probability that the wheel will stop at 7? Is this a discrete or a continuous probability space?
4. Consider a roulette wheel with a circumference of 38 inches. Assume that the wheel can stop at any point along the circumference. Is this a continuous or a discrete probability space? What is the probability that the wheel will land exactly on 7? What is the probability that the wheel will land between 7 and 10?
5. What is S^c? What is $\emptyset^c$?
6. Prove $\Pr(X) \leq 1$, for all events X.
7. Prove $\Pr(A^c) = 1 - \Pr(A)$.
8. Prove $\Pr(\emptyset) = 0$.
9. Prove $\Pr(A \cup B \cup C) = \Pr(A) + \Pr(B) + \Pr(C)$, if A, B, and C are all disjoint events.
10. Prove $\Pr(A \cup B) = \Pr(A) + \Pr(B) - \Pr(A \cap B)$.
11. The town of Wethersfield, Connecticut, has an area of about 14 square miles. What is the probability that a meteorite thrown randomly at the earth will hit Wethersfield? (The surface area of the earth is about 200,000,000 square miles.)
12. What is the probability that a meteorite thrown randomly at the earth will hit an ocean?
13. If you flip a coin five times, what is the probability that you will get heads on either the first, second, or third toss?
14. If you roll a die three times, what is the probability that you will get a 1 on at least one of the three tosses?

15. Seventeenth-century Italian gamblers thought that the chance of getting a 9 when they rolled three dice were equal to the chance of getting a 10. Calculate these two probabilities to see if they were right.

16. You're coach of a football team that has just scored a touchdown so you are now down by 8 points in the fourth quarter. You expect to score one more touchdown in the game. Should you try for 2-point conversion now, or should you kick a 1-point conversion now and then try a 2-point conversion after your next touchdown? Assume that you have a 100 percent chance of making the one-point conversion but only a 50 percent chance of making the two-point conversion.

17. Each person receives an eye-color gene from each parent. For simplicity, assume that the only two eye colors are blue and brown. A person with two brown-eye genes will have brown eyes, a person with two blue-eye genes will have blue eyes, and a person with one blue-eye gene and one brown-eye gene will have brown eyes. Assume that a child is equally likely to inherit either of the two genes from the mother, and either of the two genes from the father. What is the probability that the child will have blue eyes if the parents have the different gene types shown below?

Mother	Father
Br–Br	Br–Br
Br–Br	Br–Bl
Br–Br	Bl–Bl
Br–Bl	Br–Bl
Br–Bl	Bl–Bl
Bl–Bl	Bl–Bl

Is it possible for two parents with brown eyes to have a child with blue eyes? Is it possible for two parents with blue eyes to have child with brown eyes?

18. Suppose you know that all four grandparents of a child have one brown-eye gene and one blue-eye gene, but you don't know what gene types the two parents have. What is the probability that the child will have blue eyes?

19. Consider again the situation in the preceding problem. Suppose now that you are able to find out what color eyes the parents have, but you don't know what gene types they have. What is the probability that the child will have blue eyes if both parents have blue eyes? If both parents have brown eyes? If one parent has blue eyes and one parent has brown eyes?

20. If you were captain of a football team, how would you decide what to call at the opening coin toss?

21. If you toss a coin n times, what is the probability that you will get at least one head?

22. Assume that there are 21 one-hour slots for prime-time television shows during the week. Suppose that the networks select the time slots for their shows at random. If your two favorite shows are on different networks, what is the probability that they will conflict?

23. What is the probability that your two favorite shows will be on the same night?

24. What is the probability that your two favorite shows will both be on Monday?

25. Suppose your backgammon opponent has points 4, 5, and 6 covered, and you have two pieces on the bar. In order to get back on the board, you need to have both dice result in either 1, 2, or 3. What is the probability that both pieces will get back on the board?

26. Suppose you have a 40 percent chance of getting a job offer from your first choice firm, a 40 percent chance of getting a job offer from your second choice firm, and a 16 percent chance of getting a job offer from both firms. What is the probability that you will get a job offer from either firm?

27. Suppose 70 percent of the families in a certain town have children, 30 percent of the families have children under 6, and 60 percent of the families have children 6 or over. How many families have children both over 6 and under 6?

6

Calculating Probabilities by Counting Outcomes

In a situation when we have a set of outcomes that are all equally likely, the probability that an event A will occur is given by

$$\Pr(A) = \frac{N(A)}{s}$$

where $N(A)$ is the number of outcomes in A and s is the total number of possible outcomes. This means that if we can calculate these two numbers—$N(A)$ and s—then we're done. We can then directly calculate the probability. So an important part of probability involves figuring out how to count the outcomes corresponding to a given event. If there are not too many outcomes we can list them all; but for complicated situations the number of outcomes quickly becomes too large to list.

Multiplication Principle

Suppose we want to choose a car that has one of these four colors: red, blue, green, or white. We are interested in three different body types: 4-door, 2-door, or wagon. How many different types of cars do we need to consider? We can make a list:

red 4-door,	red 2-door,	red wagon
blue 4-door,	blue 2-door,	blue wagon
green 4-door,	green 2-door,	green wagon
white 4-door,	white 2-door,	white wagon

There are 12 possible types, since $12 = 4 \times 3$.

We can make a general statement of this principle. Suppose we are going to conduct two experiments. The first experiment can have any one of a possible outcomes, and the second experiment can have any one of b possible outcomes, and let's suppose that any outcome from the first experiment can be matched with any outcome from the second experiment. Then the total number of results of the two experiments is given by

$$a \times b$$

This rather obvious result is sometimes called the *multiplication principle*. Here are some examples:

- If you flip two coins, each flip has two possible outcomes, so the number of total outcomes is $2 \times 2 = 4$.
- Suppose you toss two dice. Since each die has six possible outcomes, the total number of possible outcomes from the two dice is $6 \times 6 = 36$.
- Suppose there are 5 candidates in the Republican primary for a particular office, and 6 candidates in the Democratic primary. Then the total number of possible general-election matchups is $5 \times 6 = 30$.

Sampling with Replacement

Now, suppose you have five sweaters in your drawer. Each morning you reach in and randomly select one sweater. In the evening you put the sweater back and mix the sweaters up again. How many different ways can you wear sweaters for the week?

We can use the same principle, only now there are more than two experiments. There are five possible sweaters for you to wear on Sunday. There are also five sweaters for you to wear on Monday, so there is a total of 25 possible different combinations of sweaters you can wear on Sunday and Monday. For each of these possibilities there are 5 more choices for Tuesday, so there are $25 \times 5 = 125$ possible sweater-wearing patterns for the first three days. In fact, for the first week there are

$$5 \times 5 \times 5 \times 5 \times 5 \times 5 \times 5 = 5^7 = 78{,}125$$

different ways of wearing the sweaters.

From this information we can calculate the probability that you will wear the same sweater every day. Since there are 78,125 possible outcomes, and only five outcomes in which you wear the same sweater every day, the probability of selecting the same sweater every day is

$$\frac{5}{78{,}125} = 0.000064$$

The sweater selection process described here is an example of what is called *sampling*. Sampling means choosing a few items from a larger group. The larger group is called the *population*. In this case the population consists of five sweaters. We are selecting a sample of one sweater on Sunday, one sweater on Monday, and so on, for a total of seven selections. This type of sampling is called *sampling with replacement*. It should be obvious why we use the words "with replacement," since we are replacing the sweater in the drawer each evening. (Later, we will discuss sampling without replacement.) In general, if you sample n times with replacement from a population of m objects, then there are m^n possible different ways to select the objects.

The key idea of sampling with replacement is that once an item has been selected it can still be selected again. Flipping a coin is an example of sampling with replacement. In this case the population is of size 2: head and tails. Just because you've selected heads once doesn't mean you can't select heads again the next time. So if you flip a coin n times there are 2^n possible results.

Rolling a die is another example of sampling with replacement. In this case the population is of size 6. If you roll a 5 on the die once, then nothing will prevent you from rolling a 5 the next time. Therefore, there are 6^n total possibilities if you roll a die n times.

Here are some more examples of sampling with replacement:

EXAMPLE Suppose you have to take a 20-question multiple choice exam in a subject you know absolutely nothing about. Each question has five choices. What is the probability that you will be able to get all of the answers right just by guessing?

SOLUTION

In this case we are sampling 20 times from a population of size 5, so the total number of possible ways of choosing the answers is $5^{20} = 9.5 \times 10^{13}$. There is only one possible outcome in which you have selected all of the right answers, so the probability of getting all of the answers right by pure guessing is $1/(9.5 \times 10^{13}) = 10^{-14}$ (approximately).

EXAMPLE Suppose you are trying to guess the license-plate number of a friend's car. (You haven't seen the car, but you do know that it doesn't have vanity license plates.)

SOLUTION

Assume that each license plate consists of three letters followed by three digits, such as DGM 235. First, calculate how many different possibilities there are for the three letters. That is the same as sampling 3 times with replacement from a population of 26 (since there are 26 letters in the alphabet), so there are $26^3 = 17{,}576$ ways of selecting the three letters. Since there are 10 possible digits there are $10^3 = 1{,}000$ ways of selecting the three digits. Each possible letter combination can be matched with each possible digit combination to give a valid license plate, so the total number of license plates is $17{,}576 \times 1{,}000 = 17{,}576{,}000$. Therefore, your chance of guessing the license plate correctly is $1/17{,}576{,}000 = 5.69 \times 10^{-8}$.

EXAMPLE Suppose you're with a group of 15 people and you are comparing birthdays. How many different possible patterns of birthdays are there in a group of 15 people?

SOLUTION

First, ignore people born on February 29, so then there are only 365 possibilities for the birthdays. Then you're sampling 15 times from a total population of 365, so the total number of possible birthday patterns is $365^{15} = 2.7 \times 10^{38}$.

Sampling Without Replacement

Now, consider a different situation. Suppose that you have seven T-shirts. Each morning you reach into the drawer and randomly select one T-shirt to wear that day. However, this time, instead of putting the T-shirt back in the drawer in the evening you put it in the bag of clothes to be washed. (The wash gets done only once per week.) How many ways can you select the seven shirts for the week?

On Sunday you have seven choices. However, on Monday there are only six clean T-shirts left, so you only have 6 choices. Therefore, there are $7 \times 6 = 42$ possible ways of choosing the T-shirts that you will wear during the first two days. On Tuesday there are only five shirts left, so there are $7 \times 6 \times 5 = 210$ ways of choosing the shirts for the first three days. Continuing the process for the rest of the week we can see that there are $7 \times 6 \times 5 \times 4 \times 3 \times 2 \times 1 = 5{,}040$ ways of selecting the T-shirts for the week.

In general, if you have n distinct objects, the number of ways of putting them in different orders is given by the product of all of the whole numbers from 1 up to n:

$$n! = n \times (n-1) \times (n-2) \times \cdots \times 4 \times 3 \times 2 \times 1$$

Because this quantity is used so often, it is given a special name: *n-factorial*. You can generate surprisingly large numbers with this function, which makes it easier to remember that this function is symbolized by an exclamation point. The Excel function FACT(*value*) will find the factorial of the specified value.

EXAMPLE How many ways are there of putting the letters A, B, and C in different orders?

SOLUTION

Calculate 3 factorial $= 3! = 3 \times 2 \times 1 = 6$. Here is a list of all of the ways:

ABC, ACB, BAC, BCA, CAB, CBA

EXAMPLE How many ways are there of putting the letters A, B, C, and D in different orders?

SOLUTION

Calculate 4 factorial $= 4! = 4 \times 3 \times 2 \times 1 = 24$. Here is a list of all of the ways:

ABCD	ABDC	ACBD	ACDB	ADBC	ADCB
BACD	BADC	BCAD	BCDA	BDAC	BDCA
CABD	CADB	CBAD	CBDA	CDAB	CDBA
DABC	DACB	DBAC	DBCA	DCAB	DCBA

EXAMPLE How many ways are there of putting the letters A and B in different orders?

SOLUTION

Calculate 2 factorial $= 2! = 2 \times 1 = 2$

EXAMPLE How many ways are there of putting the letter A in different orders?

SOLUTION

There is only one way: $1! = 1$.

EXAMPLE How many ways are there of putting no letters in different orders?

SOLUTION

This seems like a silly question, but it turns out to be useful to specify that there is only one way to put a set of objects in order if the set doesn't actually contain any objects. Therefore, 0 factorial is defined to be 1 (that is, $0! = 1$).

EXAMPLE How many different ways are there of shuffling a deck of 52 cards?

SOLUTION

There are 52 possibilities for the top card, 51 possibilities for the second card, and so on, so that altogether there are $52! = 8.07 \times 10^{67}$ ways of shuffling the deck.

EXAMPLE Suppose you have five different dinner menus to choose from for five days: hamburgers, hot dogs, pizza, macaroni, and tacos. In how many different orders can you arrange these five meals so that you don't repeat any meals during the five days?

SOLUTION

Since there are five meals, the number of different orderings is $5! = 120$.

EXAMPLE Suppose that your team is part of a 12-team league, and during the season you play each of the other teams once. How many different possible ways are there to arrange your schedule?

SOLUTION

Since there are 11 opponents, and the only question is what order you will play them in, there are $11! = 39{,}916{,}800$ possibilities for the schedule.

EXAMPLE Suppose you are having 20 people show up for a dinner party at your home. What is the probability that they will arrive at your house in alphabetical order?

SOLUTION

Since there are 20 guests there are $20! = 2.43 \times 10^{18}$ different possible orders in which they can arrive. Since there is only one way of putting them in alphabetical order, the probability that they will arrive in alphabetical order is therefore $1/(2.43 \times 10^{18}) = 4.1 \times 10^{-19}$.

EXAMPLE Suppose an indecisive baseball manager decides to try out every possible batting order before deciding on the order that is best for the team. How many games will it take to test every possible order?

SOLUTION

There are nine players, so there are $9! = 362{,}880$ different orders.

Permutations

Suppose now that you have ten T-shirts (and the T-shirts are still washed every week). How many different ways of selecting T-shirts are there during seven days? Note in this case you will not wear every shirt every week.

There are 10 choices for the shirt you wear on Sunday, then 9 choices for Monday, 8 choices for Tuesday, and so on down to 4 choices on Saturday. So the total number of choices is

$$10 \times 9 \times 8 \times 7 \times 6 \times 5 \times 4 = 604{,}800$$

We'd like a shorter way of writing that long expression, so we'll multiply by $\frac{3!}{3!}$

$$10 \times 9 \times 8 \times 7 \times 6 \times 5 \times 4 \times \frac{3!}{3!}$$
$$= \frac{10 \times 9 \times 8 \times 7 \times 6 \times 5 \times 4 \times 3 \times 2 \times 1}{3 \times 2 \times 1}$$

Both the top and bottom of this fraction can be rewritten with the factorial function:

$$= \frac{10!}{3!}$$

What we're doing is choosing a sample of size 7 without replacement from a population of size 10. It's obvious why we call this sampling without replacement, since this time we're *not* replacing the T-shirt in the drawer after it has been selected. Sampling without replacement means that once an item has been selected it cannot be selected again.

In general, suppose you're going to select j objects without replacement from a population of n objects. Then there are

$$\frac{n!}{(n-j)!}$$

ways of selecting the objects. Each way of selecting the objects is called a *permutation* of the objects, so the formula $n!/(n-j)!$ gives the number of permutations of n things taken j at a time. (The number of permutations counts each possible ordering of the selected objects separately. If you don't want to do that, see the next section.)

EXAMPLE Suppose you want to find out how many ways birthdays can be distributed among a group of 15 people so that none of them have the same birthday. This means that there are 365 possibilities for the first person's birthday, 364 choices for the second person's birthday, and so on.

SOLUTION

This is an example of selecting a sample of size 15 without replacement from a population of size 365. Therefore, there are

$$\begin{aligned}\frac{365!}{(365-15)!} &= \frac{365!}{350!} \\ &= 365 \times 364 \times 363 \times 362 \times \ldots \times 352 \times 351 \\ &= 2.03 \times 10^{38}\end{aligned}$$

ways of dividing birthdays among the 15 people so that no two have the same birthday.

EXAMPLE Suppose that you are attending an eight-horse race, and you are trying to guess the order of the top three finishers without knowing anything about the horses involved. What is the probability that you will guess right?

SOLUTION

Since three horses are to be chosen from the eight horses entered, this situation is equivalent to choosing a sample of size 3 without replacement from a population of size 8. Therefore, there are

$$\frac{8!}{(8-3)!} = \frac{8!}{5!} = 8 \times 7 \times 6 = 336$$

possible finishes. Your chance of randomly guessing the correct order is $1/336 = .003$.

Combinations

Suppose you are in a card game and you are going to be dealt a hand of 5 cards from a deck of 52 cards. There are 52 possibilities for the first card you will draw, then 51 possibilities for the second card, and so on. We can consider this an example of choosing a sample of size 5 without replacement from a population of 52 cards. (Once you've drawn a card, you can't draw that particular card again. If you draw the same card again, something suspicious is going on.) Therefore, you can find the number of ways of drawing the cards using the permutation formula:

$$\frac{52!}{(52-5)!} = \frac{52!}{47!}$$
$$= 52 \times 51 \times 50 \times 49 \times 48$$
$$= 311{,}875{,}200$$

However, suppose you drew these cards:

5C, 8D, 6H, AH, AD

For practical purposes that hand means exactly the same as it would if you drew the cards like this:

8D, 5C, 6H, AH, AD

The second hand contains exactly the same cards, the only difference being that they were picked in a different order. In many card games, the order in which you draw the cards doesn't matter; it's only *which* cards you draw that matters. However, in the way we have been counting the hands, we have counted these two possibilities as separate hands, because they were drawn in different orders. There are, of course, a lot of different orders in which this hand could be drawn. How many different orderings of these cards are there? We have already discussed that problem. Since there are 5 cards, there are $5! = 120$ different ways of arranging the cards in different orders. In fact, for every possible hand there are 120 different orderings. Our formula $52!/(52-5)!$ has told us the total number of different orderings of all the hands, but in this case we are only interested in the total number of *different* hands, regardless of the ordering. Therefore, our formula gives us 120 times too many hands, so we have to divide by 120. Therefore, the total number of distinct 5-card hands that can be drawn from a deck of 52 cards not counting different orderings is given by this formula:

$$= \frac{52!}{(52-5)!5!}$$
$$= 2{,}598{,}960$$

In general, suppose we are going to select j items without replacement from a population of n items and we're only interested in the total number of selections, without regard to their order. Then the number of possibilities is given by the formula

$$= \frac{n!}{(n-j)!j!}$$

The number of arrangements without regard to order is called the number of *combinations*. The formula $n!/[(n-j)!j!)]$ is said to represent the number of combinations of n things taken j at a time. We will use the formula $n!/[(n-j)!j!)]$ a lot in probability and statistics. We'd like a shorter way of writing this formula, so we will symbolize it like this:

$$\binom{n}{j} = \frac{n!}{(n-j)!j!}$$

This expression is also called the *binomial coefficient* because it is used in a mathematical formula called the binomial theorem.

The Excel function COMBIN(n,j) will calculate the number of combinations. Here is an example of the formula. Suppose we have 5 letter blocks and we are going to select 3 of them. In how many ways can we make the selection? If the blocks have the letters A, B, C, D, E, then we can list all of the possible permutations:

ABC	ACB	BAC	BCA	CAB	CBA
ABD	ADB	BAD	BDA	DAB	DBA
ABE	AEB	BAE	BEA	EAB	EBA
ACD	ADC	CAD	CDA	DAC	DCA
ACE	AEC	CAE	CEA	EAC	ECA
ADE	AED	DAE	DEA	EAD	EDA
BCD	BDC	CBD	CDB	DBC	DCB
BCE	BEC	CBE	CEB	EBC	ECB
BDE	BED	DBE	DEB	EBD	EDB
CDE	CED	DCE	DEC	ECD	EDC

There are 60 permutations in this list, which agrees with the formula

$$\frac{5!}{(5-3)!} = \frac{5!}{2!} = 5 \times 4 \times 3 = 60$$

However, if we look closely at the list we can see that all of the arrangements in a particular row contain exactly the same letters, only arranged in different orders. Each of the 10 rows has different letters in it, so there are 10 different combinations of the letters. That agrees with the formula for combinations:

$$\frac{5!}{(5-3)!3!} = \frac{5!}{3!2!}$$
$$= \frac{5 \times 4 \times 3 \times 2 \times 1}{3 \times 2 \times 1 \times 2 \times 1}$$
$$= 10$$

Here is the list of the 10 combinations:

ABC, ABD, ABE, ACD, ACE, ADE, BCD, BCE, BDE, CDE

Here are some special cases of the combinations formula.

There is one way of selecting zero objects: $\binom{n}{0} = 1$

There are n ways of selecting one object: $\binom{n}{1} = n$

The number of ways of selecting j objects is the same as the number of ways of not selecting the other $n - j$ objects, so

$$\binom{n}{j} = \binom{n}{n-j}$$
$$\binom{n}{n-1} = n$$
$$\binom{n}{n} = 1$$

If you need to calculate the value by hand, then you can take advantage of cancellation. For example, to find $\binom{12}{7}$, write it like this:

$$\frac{12!}{5!7!} = \frac{12 \times 11 \times 10 \times 9 \times 8 \times 7 \times 6 \times 5 \times 4 \times 3 \times 2 \times 1}{5 \times 4 \times 3 \times 2 \times 1 \times 7 \times 6 \times 5 \times 4 \times 3 \times 2 \times 1}$$

You can always directly cancel the larger of the factorials in the denominator (7! in this case) with several factors in the top. We are left with

$$\frac{12!}{5!7!} = \frac{12 \times 11 \times 10 \times 9 \times 8 \times 7 \times 6 \times 5 \times 4 \times 3 \times 2 \times 1}{5 \times 4 \times 3 \times 2 \times 1 \times 7 \times 6 \times 5 \times 4 \times 3 \times 2 \times 1}$$
$$= \frac{12 \times 11 \times 10 \times 9 \times 8}{5 \times 4 \times 3 \times 2 \times 1}$$

The 4×3 on bottom cancels out the 12 on top; the 5×2 on bottom cancels out the 10 on top. Therefore

$$\frac{12!}{5!7!} = 11 \times 9 \times 8 = 792$$

EXAMPLE How many 13-card hands can be dealt from a deck of 52 cards?

SOLUTION

Using the formula for the number of combinations of 52 objects taken 13 at time, we have

$$\frac{52!}{(52-13)!13!} = \frac{52!}{39!13!}$$
$$= 6.35 \times 10^{11}$$

Now we have all the tools we need to solve many probability problems. There's no sure-fire method that always works, because many problems contain subtle twists.

EXAMPLE Suppose you and your dream lover (whom you're desperately hoping to meet) are both in a group of 20 people, and five people are to be randomly selected to be on a committee. What is the probability that both you and your dream lover will be on the committee?

SOLUTION

The total number of ways of choosing the committee is

$$\binom{20}{5} = 15{,}504$$

Next, we need to calculate how many possibilities include both of you on the committee. If you've both been selected, then the three other committee members need to be selected from the remaining 18 people, and there are

$$\binom{18}{3} = 816$$

ways of doing this. Therefore, the probability that you will both be selected is 816/15,504 = .053.

EXAMPLE You have 4 cards labeled H1, H2, H3, and H4, and 3 other cards all labeled T. How many ways are there of arranging the cards?

SOLUTION

There are 7 slots where you can put H1; for each of these there are 6 slots left for H2, then 5 slots left for H3, 4 slots left for H4. Once these 4 slots have been chosen, the Ts will fill in the remaining slots. Therefore, there are $7 \times 6 \times 5 \times 4 = 840$ ways. Note that this result comes from the permutations formula: $7!/(7-4)! = 7!/3!$.

Here are the 20 arrangements that have H1 in the first slot and H2 in the second slot. (The complete table of all 840 arrangements would be 42 times as long since there are 42 different ways of arranging H1 and H2.)

H1	H2	H3	H4	T	T	T
H1	H2	H3	T	H4	T	T
H1	H2	H3	T	T	H4	T
H1	H2	H3	T	T	T	H4
H1	H2	H4	H3	T	T	T
H1	H2	T	H3	H4	T	T
H1	H2	T	H3	T	H4	T
H1	H2	T	H3	T	T	H4
H1	H2	H4	T	H3	T	T
H1	H2	T	H4	H3	T	T
H1	H2	T	T	H3	H4	T
H1	H2	T	T	H3	T	H4
H1	H2	H4	T	T	H3	T
H1	H2	T	H4	T	H3	T
H1	H2	T	T	H4	H3	T
H1	H2	T	T	T	H3	H4
H1	H2	H4	T	T	T	H3
H1	H2	T	H4	T	T	H3
H1	H2	T	T	H4	T	H3
H1	H2	T	T	T	H4	H3

However, what happens if the numbers on our cards get smudged so they are no longer visible? Then we have 4 cards with H and 3 cards with T. How many ways are there of arranging them? Since we can no longer distinguish H1, H2, H3, and H4, the number of arrangements will be much smaller. In fact, we will have to divide by $4! = 4 \times 3 \times 2 \times 1 = 24$ since there are 4! different ways of putting the 4 Hs in order.

$$\text{Number of ways of arranging 4 Hs and 3 Ts} =$$

$$\frac{7 \times 6 \times 5 \times 4}{4 \times 3 \times 2 \times 1} = \frac{\frac{7!}{3!}}{4!} = \frac{840}{24} = 35$$

Notice that this is the number of combinations of 7 things taken 4 at a time. Here is the list of possibilities:

1: H H H H T T T
2: H H H T H T T
3: H H H T T H T
4: H H H T T T H
5: H H T H H T T
6: H H T H T H T
7: H H T H T T H
8: H H T T H H T
9: H H T T H T H
10: H H T T T H H
11: H T H H H T T
12: H T H H T H T
13: H T H H T T H
14: H T H T H H T
15: H T H T H T H
16: H T H T T H H
17: H T T H H H T
18: H T T H H T H
19: H T T H T H H
20: H T T T H H H
21: T H H H H T T
22: T H H H T H T
23: T H H H T T H
24: T H H T H H T
25: T H H T H T H
26: T H H T T H H
27: T H T H H H T
28: T H T H H T H
29: T H T H T H H
30: T H T T H H H
31: T T H H H H T
32: T T H H H T H
33: T T H H T H H
34: T T H T H H H
35: T T T H H H H

EXAMPLE How many ways are there of tossing 4 heads out of 7 tosses?

SOLUTION

We suddenly experience deja vu—we just solved this problem in the preceding example. Use the combinations formula:

$$\text{Number of ways of tossing 4 heads out of 7 tosses} = \binom{7}{4} = \frac{7!}{4!3!} = 35$$

More generally, we can determine that the number of ways of tossing k heads out of n tosses is also given by the combinations formula:

$$\text{Number of ways of tossing } k \text{ heads out of } n \text{ tosses} = \binom{n}{k} = \frac{n!}{k!(n-k)!}$$

We have seen that there are 2^n total possible outcomes when you toss a coin n times. Of these, $\binom{n}{k}$ outcomes will have k heads, so the probability is therefore

$$\text{Probability of tossing } k \text{ heads out of } n \text{ tosses} = \frac{\binom{n}{k}}{2^n} = \binom{n}{k} \times .5^n = \frac{n!}{k!(n-k)!} \times .5^n$$

It turns out that this formula is a special case of a more general formula called the *binomial distribution*, which will be covered in more detail in Chapter 9.

EXAMPLE A deck of nine cards contains five low cards (A, 2, 3, 4, 5) and four high cards (9, J, Q, K). Five cards are to be drawn at random from this deck. What is the probability that three low cards will be drawn?

SOLUTION

Since five objects are to be selected from a group of nine objects, there are a total of $\binom{9}{5}$ ways of making the selection. Now we need to count how many of these ways include three low cards. Since the three low cards must be selected from the five low cards in the deck, there are $\binom{5}{3}$ ways of making this selection. A hand with three low cards must also include two high cards, and there are $\binom{4}{2}$ ways of selecting the high cards. Since any possible way of selecting the low cards can be matched with any possible way of selecting the high cards, we need to multiply to find the total number of possibilities:

$$(\text{Total possibilities with three low cards}) = \binom{5}{3} \times \binom{4}{2} = 10 \times 6 = 60$$

Here is a list of all of the possibilities with exactly three low cards and two high cards:

A,2,3,9,J	A,2,3,9,Q	A,2,3,9,K	A,2,3,J,Q	A,2,3,J,K	A,2,3,Q,K
A,2,4,9,J	A,2,4,9,Q	A,2,4,9,K	A,2,4,J,Q	A,2,4,J,K	A,2,4,Q,K
A,2,5,9,J	A,2,5,9,Q	A,2,5,9,K	A,2,5,J,Q	A,2,5,J,K	A,2,5,Q,K
A,3,4,9,J	A,3,4,9,Q	A,3,4,9,K	A,3,4,J,Q	A,3,4,J,K	A,3,4,Q,K
A,3,5,9,J	A,3,5,9,Q	A,3,5,9,K	A,3,5,J,Q	A,3,5,J,K	A,3,5,Q,K
A,4,5,9,J	A,4,5,9,Q	A,4,5,9,K	A,4,5,J,Q	A,4,5,J,K	A,4,5,Q,K
2,3,4,9,J	2,3,4,9,Q	2,3,4,9,K	2,3,4,J,Q	2,3,4,J,K	2,3,4,Q,K
2,3,5,9,J	2,3,5,9,Q	2,3,5,9,K	2,3,5,J,Q	2,3,5,J,K	2,3,5,Q,K
2,4,5,9,J	2,4,5,9,Q	2,4,5,9,K	2,4,5,J,Q	2,4,5,J,K	2,4,5,Q,K
3,4,5,9,J	3,4,5,9,Q	3,4,5,9,K	3,4,5,J,Q	3,4,5,J,K	3,4,5,Q,K

Note that each of the 6 columns corresponds to one of the 6 ways of selecting the two high cards; each of the 10 rows corresponds to one of the 10 ways of selecting the three low cards.

Therefore, the probability of selecting exactly three low cards is as follows:

$$\frac{\binom{5}{3} \times \binom{4}{2}}{\binom{9}{5}} = \frac{60}{126} = .476$$

EXAMPLE There are N people in the population: M Republicans and $N - M$ Democrats. If you select a random sample of n people from this population, what is the probability that k people in the sample will be Republicans and $n - k$ will be Democrats?

SOLUTION

Note that this is a generalization of the previous example. This is the type of question we need to answer to analyze polls. There will be $\binom{N}{n}$ possible ways of selecting the sample. There are $\binom{M}{k}$ ways of selecting k Republicans, and there are $\binom{N-M}{n-k}$ ways of selecting $n - k$ Democrats from among the $N - M$ Democrats in the population. Each possible way of selecting the Republicans can be matched with each of the possible ways of selecting the Democrats, so we need to multiply these two numbers to find the total number of ways of selecting the sample with k Republicans and $n - k$ Democrats. Therefore, the probability of this happening is given by this formula:

$$\frac{\binom{M}{k}\binom{N-M}{n-k}}{\binom{N}{n}}$$

We will see this formula again in Chapter 10.

EXAMPLE Suppose 18 people are going to be randomly divided up into two baseball teams. What is the probability that all 9 of the best players will be on the same team?

SOLUTION

There are $\binom{18}{9} = 48{,}620$ ways of choosing the team that will bat first.

Therefore, the chance of all the best players being on the team that bats first is 1/48,620. However, there is just as good a chance that all the best players will be on the team that bats second, so the chance they will all be on the same team is $2/48{,}620 = 4 \times 10^{-5}$.

EXAMPLE We can calculate the probabilities of the various hands that can occur in a poker game.

SOLUTION

We have already calculated that the total number of possible hands is $52!/[(52-5)!5!] = 2{,}598{,}960$. We can assume that each hand is equally likely, which is reasonable as long as the deck is well shuffled and the dealer is not cheating. Next, we need to figure out the number of possible ways of getting each type of hand.

- royal flush
  A royal flush must contain the cards 10, J, Q, K, A, and they must all be of the same suit. Since there are 4 suits to choose from there are 4 possible royal flushes.

- straight flush
  A straight flush must contain 5 cards in sequence, and they must all be of the same suit. We'll assume that we're playing under rules that allow the ace to be the low card in a straight as well as the high card. Then there are 9 possible straights other than the royal straight. (The 9 possible straights start with A, 2, 3, 4, 5, 6, 7, 8, or 9.) There are 4 choices for the suit, so there are $4 \times 9 = 36$ possible straight flushes.

- four of a kind
  There are 13 possible choices for the type of card that we will draw 4 of. The 5th card in the deck can be any one of the 48 remaining cards, so altogether there are $13 \times 48 = 624$ possible hands with 4 of a kind.

- full house
  A full house has 3 of one type of card and 2 of another. We have 13 choices for the type of card which will have 3. Those 3 cards must be chosen from the 4 cards of that type, so there are $\binom{4}{3} = 4$ ways of choosing them. Altogether, then, there are $13 \times 4 = 52$ ways of choosing the 3-of-a-kind cards. There are 12 possible choices left from which to choose the 2-of-a-kind type. Once we have chosen the type, there are $\binom{4}{2} = 6$ ways of selecting the cards of that rank, so there are $12 \times 6 = 72$ ways of choosing the 2-of-a-kind cards. That means that there are $52 \times 72 = 3{,}744$ hands that are full houses.

- flush
  A flush is a hand in which all 5 cards have the same suit. If the hand is to be all hearts (say), we need to choose 5 cards from the 13 hearts, so there are

$\binom{13}{5} = 1{,}287$ ways of doing this. Since a flush can be any one of the 4 suits, there are thus $4 \times 1{,}287 = 5{,}148$ total flushes. However, this total includes straight flushes, so we need to subtract 40 from this total to get 5,108 hands that are plain flushes.

- straight
  A straight consists of any 5 cards in sequence. As we saw earlier, there are 10 possible runs of 5 cards in sequence. For each of these possibilities there are $4^5 = 1{,}024$ ways of selecting the suits for the cards, so altogether there are 10,240 possible straights. However, once again we need to subtract the 40 straight flushes to get a total of 10,200 plain straights.

- three of a kind
  There are 13 ways to choose the rank of the card that we will select 3 of. Then there are 4 ways of selecting the cards of that rank. Then we have 48 choices for the 4th card in the hand, and then 44 choices for the 5th card (since the 5th card cannot match the 4th card). However, we have to divide by 2 because we don't care in which order we draw 4th and 5th cards. So altogether there are

  $$13 \times 4 \times 48 \times 44 \times \frac{1}{2} = 54{,}912 \text{ possible hands with 3 of a kind}$$

- two pair
  There are $\binom{13}{2} = 78$ ways of choosing the ranks of the cards that will make up the pairs. For each pair there are $\binom{4}{2} = 6$ ways of choosing the cards of that rank. Then, for each possible set of two pairs there are 44 ways of selecting the 5th card. Altogether, there are

  $$78 \times 6 \times 6 \times 44 = 123{,}552 \text{ hands with 2 pairs}$$

- pair
  There are 13 choices for the rank of the card that will make up the pair. Then there are $\binom{4}{2} = 6$ ways of choosing the cards of that rank. Now we need to figure out the number of ways of choosing the remaining 3 cards, which can be any nonmatching cards. There are 48 choices for the 3rd card, 44 choices for the 4th card, and 40 choices for the 5th card. We also have to divide by $3! = 6$, because we don't care about the order in which we draw the last 3 cards. Therefore, there are

  $$13 \times 6 \times 48 \times 44 \times 40 \times \frac{1}{6} = 1{,}098{,}240 \text{ hands with 1 pair in them}$$

- nothing
  The number of hands with 5 cards all of different ranks is $52 \times 48 \times 44 \times 40 \times 36$ divided by 5! different orderings, which equals 1,317,888. Then subtract the 15,348 hands that are either straights or flushes, leaving 1,302,540 hands that are worth nothing. We can make a table of these results to get the probability of each hand occurring:

TABLE 6–1: PROBABILITIES OF POKER HANDS

Type of Hand	Number of Possible Hands	Probability
Royal flush	4	1.54×10^{-6}
Straight flush	36	1.39×10^{-5}
Four of a kind	624	2.40×10^{-4}
Full house	3,744	.00144
Flush	5,108	.00197
Straight	10,200	.00392
Three of a kind	54,912	.02113
Two pairs	123,552	.04754
Pair	1,098,240	.42257
Nothing	1,302,540	.50118

(Notice that the ranking of the hands has been scientifically designed so that the hands with lower probability have higher rankings.)

In the interests of science, we decided to check these results by dealing out 400 five-card hands. (Sometimes science is a lot of work.) Here are our results: 222 nothings (.555), 157 pairs (.393), 17 two-pairs (.043), and 4 three-of-a-kinds (.010).

EXAMPLE In Chapter 5, we found that there is a 2/3 probability that somebody will pick his or her own name if three people take part in a Secret Santa drawing. That isn't very good, since it's no fun to be your own Secret Santa. Let's add more people, so now there are n people participating in the drawing. First, what is the probability that everybody will draw his or her own name? There are $n!$ ways of making the drawing, and there is only one way where everybody draws his or her own name. So the probability of everybody drawing his or her own name is $1/n!$.

Second, what is the probability that you will draw your own name? Since you are picking from n names, the probability of that happening is $1/n$. Third, what is the probability that Andrew or Ann, the two people first in alphabetical order, will both draw their own names? If Andrew and Ann have both drawn their own names, then there are $(n - 2)!$ ways for the other $n - 2$ people to draw their names, so the probability is $(n - 2)!/n!$ that Andrew and Ann will draw their own names.

Fourth, what is the probability that the first j people in alphabetical order will all draw their own names? There is only one way for the names to be drawn by those j people, and then there are $(n - j)!$ ways for the other $n - j$ people to draw their names. Therefore, the probability that the first j people will draw their own names is $(n-j)!/n!$.

Finally, what is the probability that at least one of the n people will draw his or her own name? If we let Ei be the event that person i draws his or her own name, then we need to find the union of $E1, E2, E3 \ldots$, and so on, up to En. We have to use the horrible formula described in Chapter 5. First, we have to add up the probability of each event individually. $\Pr(Ei) = 1$, as we saw, and there are n of these events, so we get $n(1/n)$ as the first part of the sum. Next, we need the probabilities of all of the intersections of the events taken 2 at a time. We already saw that $\Pr(E1 \cap E2) = (n-2)!/n!$. By the same reasoning we used for Ann and Andrew, we can show that $\Pr(Ei \cap Ek) = (n-2)!/n!$ for any two people in the group (that is, for any values of i and k). How many possible intersections are there between two sets in a group of n? That is $\binom{n}{2}$.

Going on, we can see that there are $\binom{n}{j}$ ways of choosing groups of j from the set. The probability of the intersections of these events taken j at a time is $(n-j)!/n!$, as we saw. Therefore, we can use this formula:

$$\Pr(E1 \cup E2 \cup E3 \cup \cdots \cup En)$$
$$= \binom{n}{1}\frac{(n-1)!}{n!} - \binom{n}{2}\frac{(n-2)!}{n!} + \binom{n}{3}\frac{(n-3)!}{n!} - \binom{n}{4}\frac{(n-4)!}{n!} + \cdots$$

If we simplify this expression, it turns out to be equal to

$$\frac{1}{1} - \frac{1}{2!} + \frac{1}{3!} - \frac{1}{4!} + \frac{1}{5!} - \frac{1}{6!} + \cdots + \frac{(-1)^{n-1}}{n!}$$

We can make a table of these results (see Table 6–2).

TABLE 6–2

Number of People in Secret Santa Drawing	Probability at Least One Person Gets Own Name
1	1.00000
2	.50000
3	.66667
4	.62500
5	.63333
6	.63194
7	.63214
8	.63211
9	.63212
10	.63212

Note that the probability that at least one person will draw his own name does not go to zero, as you might expect, as n becomes large. Instead, it approaches a constant value of about .63212. This is an interesting number for mathematicians, since it is about $1 - 1/2.71828$. The number 2.71828 ... is a very special number that is given the name e. We will run into it again later.

EXAMPLE Suppose that a group of 16 people is going to be divided into roommates with two people in each room. The group consists of 8 pairs of best friends. If the selection is made entirely at random, what is the probability that all of the best friends will be roommates?

SOLUTION

First, we need to calculate the number of different ways a group of n people can be divided into roommates. If there are two people, then there is obviously only one way. If there are four people, then there are three ways. If the people are labeled A, B, C, and D, then we can list all of the ways:

AB, CD
AC, BD
AD, BC

Now, suppose that we have n people. Let's make up a function called Roommate, such that Roommate(n) equals the number of ways of dividing a group of n people into roommates. (Note that n obviously must be an even number, because there is no way to divide an odd number of people into groups of 2.) Then, we have seen that Roommate(2) = 1, and Roommate(4) = 3. Now, let's consider Roommate(n). There are $(n - 1)$ choices for the first person's roommate. After the first person's roommate has been chosen, $(n - 2)$ people are left; they can be divided into roommates in Roommate($n - 2$) ways. Therefore:

$$\text{Roommate}(n) = (n - 1) \times \text{Roommate}(n - 2)$$

From this rule we can figure out that Roommate(6) = 5 × Roommate(4) = 15; Roommate(8) = 7 × Roommate(6) = 105; and, in general:

$$\text{Roommate}(n) = 1 \times 3 \times 5 \times 7 \times 9 \times 11 \times \ldots \times (n - 1).$$

In particular, we are interested in Roommate(16), which is equal to 2,027,025. Therefore, there are 2,027,025 total ways for the 16 people to be divided into roommates, and there is only one way in which all of the pairs of best friends are together, so the probability of the best friends being together if the selection is made totally at random is $1/2{,}027{,}025 = 4.93 \times 10^{-7}$.

We can write our formula in a shorter form:

$$\begin{aligned}
&\text{Roommate}(n)\\
&= 1 \times 3 \times 5 \times 7 \times 9 \times \cdots \times (n-1)\\
&= \frac{1 \times 2 \times 3 \times 4 \times 5 \times 6 \cdots \times n}{2 \times 4 \times 6 \times \cdots \times n}\\
&= \frac{n!}{(2\times 1)\times(2\times 2)\times(2\times 3)\times(2\times 4)\times\cdots\times(2\times n/2)}\\
&= \frac{n!}{2^{n/2}(1\times 2\times 3\times 4\times 5\times\cdots\times n/2)}\\
&= \frac{n!}{2^{n/2}(n/2)!}
\end{aligned}$$

Multinomial Formula

This formula is an example of a more general formula called the *multinomial formula*. Suppose we have n objects that are to be put into boxes, with each box having k objects. (Therefore, we will need n/k boxes.) Then there will be

$$\frac{n!}{(k!)^{n/k}}$$

ways to split the objects up into the boxes.

(This formula assumes that it makes a difference what box each group is put into. If we care only about which objects are grouped together, then we need to divide by $(n/k)!$ For example, in the roommate problem we were concerned only about the number of ways of dividing the people into pairs, without also worrying about the $(n/2)!$ different ways that the pairs could be put in rooms.)

In general, the multinomial formula says that if you are going to put n objects in m groups, with k_1 objects in the first group, k_2 objects in the second group, and so on, up to k_m objects in group m, then there are

$$\frac{n!}{k_1!k_2!k_3!\cdots k_m!}$$

ways of doing this. (Note that $k_1 + k_2 + k_3 + \ldots + k_m = n$.) If $m = 2$, then the multinomial formula becomes the same as the combinations formula.

EXAMPLE Consider an artist who is painting a still life that will consist of three apples, five oranges, two bananas, and one plum, arranged in a row. In how many different ways can these fruits be arranged?

SOLUTION

Clearly, if each fruit is distinguishable, then there are 11! ways of arranging them. However, if all of the fruits of each type have the same size and shape, then we need to divide by 3!5!2! since there are 3! ways of arranging the apples, 5! ways of arranging the oranges, and 2! ways of arranging the bananas. So there are $11!/(3!5!2!) = 27{,}720$ arrangements.

MULTIPLICATION PRINCIPLE

If the first experiment has a outcomes, and the second has b outcomes, and any outcome of the first experiment can be matched with any outcome of the second experiment, then there are ab possible results of the two experiments.

SAMPLING WITH REPLACEMENT

If you sample n times with replacement from m objects there are m^n different ways of selecting the objects.

FACTORIAL

There are $n! = n \times (n-1) \times (n-2) \times \ldots \times 3 \times 2 \times 1$ different ways to put n objects in order.

PERMUTATIONS

Select j objects without replacement from a group of n objects. There are $\frac{n!}{(n-j)!}$ ways of selecting the objects (assuming that each different ordering of the selected objects is counted separately).

COMBINATIONS

Select j objects without replacement from a group of n objects. There are

$$\binom{n}{j} = \frac{n!}{(n-j)!j!}$$

ways of selecting the objects (assuming that each different ordering of the selected objects is not counted separately).

EXERCISES

1. If 24 pieces of sausage are randomly put onto a pizza that is sliced into 8 pieces (with none of the sausages getting cut), what is the probability that your slice will have 3 pieces of sausage? 4 pieces?

2. A hostess has so many guests coming that she needs to use more than one set of plates. She has 22 guests coming, and she has 10 plates in one set and 12 plates in the other set. All of the people will sit at a round table. How many different ways can the plates be set at the table?

3. How many ways can all 32 chess pieces be arranged in a row? (Each color has 8 pawns, 2 rooks, 2 bishops, 2 knights, 1 king, and 1 queen.)

4. Suppose you suddenly find that you have to cook for yourself for one week. You have the following kinds of frozen dinners: 4 beef dinners, 2 chicken dinners, and 1 turkey dinner. How many ways can you arrange your menus for the week?

5. A choreographer is planning a dance routine consisting of four star dancers and an eight-member chorus line. The indifferent choreographer regards the four stars as indistinguishable, and the chorus line members as indistinguishable. The dance will end with all 12 dancers in a row. How many different ways can that row be arranged?

6. How many ways can the 4 aces be located in a deck of 52 cards?

7. Suppose you are getting dressed in a dark room. In your drawer you have four red socks, three blue socks, and two brown socks. If you randomly select two socks, what is the probability that you will get two socks that match?

8. If you have five pennies and four dimes in your pocket, and you reach in and pick two coins, what is the probability that you will get 20 cents?

9. Messrs. Smith, Jones, and Brown go to a garage sale. At the sale there are nine different bicycle horns. If Smith, Jones, and Brown each buy one horn, how many different possible purchases are there?

10. At a meeting, name tags are being made for four people named John, two people named Julie, and two people named Jane. How many different ways can the name tags be distributed?

11. At a party five people have blue jackets, four people have brown jackets, and two people have red jackets. If, at the end of the party, they each randomly select a jacket of the correct color, in how many different ways can the jackets be mixed up?

12. If you have 10 large chairs and 5 small chairs to be arranged at a round table, how many different ways are there to arrange them?

13. If you are picking four cards from a standard deck, what is the probability that you will pick an ace, 2, 3, and 4?

14. At a picnic the three indistinguishable Smith children and the two indistinguishable Jones children are sitting on a bench. In how many distinguishable ways can they be arranged on the bench?

15. If you are going to buy something from a vending machine that costs 55 cents, and you have five nickels and three dimes in your pocket, in how many different ways can you put change in the machine?

16. If you have 20 blue balls and 30 red balls in a box, and you randomly pull out 20 balls, what is the probability that they will all be blue?

17. What is $\binom{n}{0}$? Explain intuitively.

18. What is $\binom{n}{1}$? Explain intuitively.

19. What is $\binom{n}{n}$? Explain intuitively.

20. What is $\binom{n}{n-1}$? Explain intuitively.

21. Compare $\binom{n}{j}$ and $\binom{n}{n-j}$. Explain intuitively.

22. How many different possible five-letter words are there?

23. A combination for a combination lock consists of three numbers from 1 to 30. What is the probability that you could randomly guess the combination?

24. Suppose a monkey spends a long time at a typewriter. What is the probability that it will type the *Encyclopaedia Britannica*? What is the probability that it will type *E–Z Statistics*?

25. Have It Your Way Burgers, Inc., likes to give its customers a lot of choices, and it likes to keep all possible combinations on hand. Customers have their choice between these options: cheese/no cheese, onion/no onion, pickle/no pickle, well done/medium, ketchup/no ketchup, and there are six possible sizes: tiny, small, junior, medium, hefty, and jumbo. If any option can be selected with any other option, how many hamburgers will the hamburger place have to keep on hand in order to have every single possibility available?

26. At a Chinese restaurant, you have a choice between five items in column 1, six items in column 2, and four items in column 3. How many possible choices do you have?

27. If you roll three dice, what is the probability of getting at least two numbers the same?

28. Suppose you roll a die six times. What is the probability that no number will occur twice during the six rolls?

29. Suppose you roll a die n times. What is the probability that no number occurs twice during the n rolls?

30. If you roll three dice, what is the probability that they will all turn up the same?

31. If you roll five dice, what is the probability that they will all turn up the same?

32. If you roll a die n times, what is the probability that at least one 1 will turn up?

33. If you draw cards from a well-shuffled deck, what is the probability that you will draw all four kings before drawing the ace of spades?

34. What is the probability that you will draw all four kings before you draw a single ace?

35. Suppose you are playing a card game where you need to get either four in a row of the same suit, or four of a kind. You currently have 5D, 5C, 5H, 6H, and 7H, and you must discard one of these cards. Which card should you discard?

36. Consider an 8-team league, where each team plays the other teams in the league once during the season. How many different ways can the schedule be arranged?

37. If there are n people in a room, what is the probability that a least one was born on February 29? (Assume that leap year occurs every four years.) How large must n be to make this probability larger than 1/2?

38. Consider the example where a committee is to be selected from a group of 20 people, only this time the size of the committee is different from 5. If there are n people on the committee, make a table that shows the probability that both you and your dream lover will be on the committee for values of n from $n = 2$ to $n = 20$.

39. Suppose n people are voting in an election with two candidates. Assume that each voter is equally likely to vote for either of the two candidates. What is the probability that each candidate will get exactly the same number of votes?

40. Suppose a group of 16 people is being divided into roommates. Half of the people are day people (that is, they like to get up early) and half of the people are night people (that is, they like to stay up late and sleep late). If the selection is made completely at random, what is the probability that every day person will be matched with a night person?

41. In the situation described in the previous exercise, what is the probability that no roommate pairs will contain both a day person and a night person?

42. Explain intuitively why the probability that one person will draw his or her own name does not go to zero as the number of people in the Secret Santa drawing becomes very large.

☆ 43. Derive the multinomial formula. Start by asking yourself: How many ways are there to select k_1 items in the first group? Then, how many ways are there to choose the k_2 items for the next group from the remaining $n - k_1$ items, and keep going like that.

7

Conditional Probability

With random events we're often totally in the dark about what will happen, as we have seen. However, it sometimes happens that we can get some information that sheds light on the issue by telling us whether a particular random event is more or less likely to occur.

For example, suppose we want to know the probability that an 8 will turn up when we roll two dice. Ordinarily, we know that the probability this will happen is 5/36. However, suppose we roll one of the dice first. Then we'll have a better idea about how likely we are to get an 8. For example, suppose we get a 5 on the first die. Then, to get a total of 8, we need to roll a 3 on the second die, and we know that the probability of that happening is 1/6. Therefore, once we are given the fact that the first die was 5, our chances of rolling an 8 have improved from 5/36 to 1/6 (see Figure 7–1).

(1,1)	(1,2)	(1,3)	(1,4)	(1,5)	(1,6)
(2,1)	(2,2)	(2,3)	(2,4)	(2,5)	(2,6)
(3,1)	(3,2)	(3,3)	(3,4)	(3,5)	(3,6)
(4,1)	(4,2)	(4,3)	(4,4)	(4,5)	(4,6)
(5,1)	(5,2)	(5,3)	(5,4)	(5,5)	(5,6)
(6,1)	(6,2)	(6,3)	(6,4)	(6,5)	(6,6)

FIGURE 7–1

On the other hand, suppose that the first die came up 1. Then we know that there is no way to roll an 8, no matter what happens with the second die. Therefore, the probability that we will roll an 8 given that we rolled a 1 on the first die is 0.

EXAMPLE Suppose we're interested in the probability of getting a royal flush.

SOLUTION

Ordinarily this probability is only $4/2{,}598{,}960 = 1.54 + 10^{-6}$ (see Chapter 6). However, suppose that we already have drawn the ace of hearts and the king of hearts, and we are

going to draw three more cards. Then there are $\binom{50}{3} = 19{,}600$ ways of drawing the remaining three cards, so our probability of getting a royal flush has improved to $1/19{,}600 = 5.1 + 10^{-5}$.

Calculating Conditional Probabilities

All of these situations are examples of *conditional probability*. A conditional probability answers the question, "What is the probability that one particular event will occur, if we already know that another specific event has occurred?" In particular, suppose we know that event B has occurred and we want to know the probability that event A will occur. The conditional probability that event A will occur given that event B has occurred is written like this:

$$\Pr(A \mid B)$$

The vertical line means "given that."

Now we have to figure out how to calculate conditional probabilities. In ordinary circumstances the probability that event A will occur is $N(A)/s$, where s is the total number of equally likely outcomes and $N(A)$ is the number of events in A. However, we know that not all of these outcomes are possible. We know that event B has occurred, so only those outcomes in event B need to be considered. So the number of possibilities is $N(B)$. The next question is: How many of these remaining possibilities also have event A occurring? Ordinarily there are $N(A)$ ways for event A to occur, but not all of these are possible now. The outcomes that are in A but not in B cannot occur. Therefore, the number of possible outcomes in which event A can occur is equal to the number of outcomes that are in both event A and event B. But we have already given a name to that event:

$$A \text{ and } B = A \cap B$$

The event in which both A and B occur is called A intersect B. Therefore, the probability that event A will occur given that event B occurs is given by

$$\Pr(A \mid B) = \frac{N(A \cap B)}{N(B)}$$

We can rewrite this formula by dividing both the top and bottom by s:

$$\Pr(A \mid B) = \frac{N(A \cap B)/s}{N(B)/s}$$
$$= \frac{\Pr(A \cap B)}{\Pr(B)}$$

In words, the probability that event A will occur, given that event B occurs, is equal to the probability that both A and B will occur, divided by the probability the B will occur. (Note that this definition does not work when $\Pr(B) = 0$, because then we would be dividing by 0, which is no help at all.)

Let A = event of getting an 8 on a pair of dice. Let B = event of getting a 5 on the first roll; $\Pr(B) = 1/6$. Then $A \cap B$ = event of getting a 5 on first roll and 8 total. The event $A \cap B$ can occur only if we roll 5, 3, so $\Pr(A \cap B) = 1/36$. Thus,

$$\Pr(A \mid B) = \frac{1/36}{1/6} = 6/36 = 1/6$$

Here are three examples where one event is a subset of the other:

EXAMPLE Let A be the event of getting 5 heads in a row.
Let B be the event of getting 2 heads in the first two rolls.

$$\Pr(B) = 1/4$$
$$\Pr(A \cap B) = 1/32$$
$$\Pr(A \mid B) = \frac{1/32}{1/4} = 4/32 = 1/8$$

EXAMPLE Let A be the event of getting a royal flush.
Let B be the event of getting AH, KH on the first two cards.

$$\Pr(B) = \frac{\binom{50}{3}}{\binom{52}{5}}$$
$$= \frac{19{,}600}{2{,}598{,}960}$$
$$= 7.54 \times 10^{-3}$$

The intersection $A \cap B$ is the event of getting both AH, KH, and a royal flush, so it means getting the hand AH, KH, QH, JH, 10H.

$$\Pr(A \cap B) = 1/2{,}598{,}960$$

Therefore,

$$\Pr(A \mid B) = \frac{1/2{,}598{,}960}{19{,}600/2{,}598{,}960} = 1/19{,}600$$

EXAMPLE What is the probability that you will draw a king, given that you have drawn a face card?

$$\Pr(\text{king} \mid \text{face card}) = \frac{\Pr(\text{king} \cap \text{face card})}{\Pr(\text{face card})} = \frac{4/52}{12/52} = \frac{4}{12} = \frac{1}{3}$$

(See Figure 7–2.)

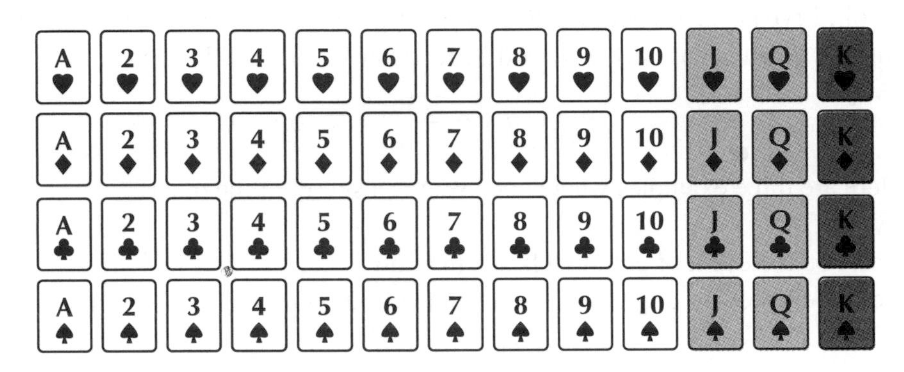

FIGURE 7–2

In general, if A is a subset of B, then

$$\Pr(A \mid B) = \frac{\Pr(A)}{\Pr(B)}$$

Next, consider an example with two mutually exclusive events.

EXAMPLE What is the probability that you will draw a king, given that you have drawn a non-face card?

SOLUTION

Since there is no way to draw a card that is both a king and a non-face card, this probability is zero.

In general, if A and B are mutually exclusive events, then $\Pr(A \mid B) = 0$.

Independent Events

We have seen how conditional probability allows you to use your knowledge that one event has occurred to revise your estimate of the probability that another event occurs. However, suppose that the two events don't have any effect on each other. In that case, knowing that one of them occurs does not provide any information about whether the other event has occurred. The two events are said to be *independent*.

For example, suppose you know that a family just had a baby girl. What is the probability that their next baby will be a girl? In this case, knowing about the last baby does not give you any information about the next baby.

Or, suppose that you roll a 3 on the first roll of a die. What is the probability that you will roll a 5 on the next roll? Knowing that the first roll came up a 3 does not help you know what will happen on the next roll. In this case, if A is the event of getting a 3 on the first roll and B is the event of getting a 5 on the second roll, then $\Pr(A) = 1/6$, $\Pr(B) = 1/6$, and $\Pr(A \mid B) = 1/6$, since the fact that B has occurred does not affect the probability that A will occur.

The formal definition of independence follows:

Events A and B are independent if $\Pr(A \mid B) = \Pr(A)$.

EXAMPLE What is the probability that you will draw a king, given that you have drawn a red card?

SOLUTION

$$\Pr(\text{king} \mid \text{red card}) = \frac{\Pr(\text{king} \cap \text{red card})}{\Pr(\text{red card})} = \frac{2/52}{26/52} = \frac{2}{26} = \frac{1}{13}$$

Note that 1/13 is also the ordinary probability of drawing a king. Since $\Pr(\text{king}) = \Pr(\text{king} \mid \text{red card})$, the two events are independent. In other words, if a foolish cheater rigged a device allowing him to see the color of the card, without seeing the type of card, it would be no help in determining whether it was a king or not. (Note that two events can be independent even if they are not physically distinct. In this case, one physical event—drawing a card—determines both whether or not you have drawn a king, and whether or not you have drawn a red card. However, the two events are independent because of the nature of the deck of cards: the fraction of red cards that are kings is the same as the fraction of nonred cards that are kings.)

There is an interesting result that we can get from the formula for conditional probability. Note that, in general,

$$\Pr(A \mid B) = \frac{\Pr(A \cap B)}{\Pr(B)}$$

And, if A and B are independent, then

$$\Pr(A \mid B) = \Pr(A)$$

Therefore, when A and B are independent we can write:

$$\Pr(A) = \frac{\Pr(A \cap B)}{\Pr(B)}$$

Therefore:

$$\Pr(A \cap B) = \Pr(A) \times \Pr(B), \text{ or } \Pr(A \text{ and } B) = \Pr(A) \times \Pr(B)$$

This result gives us a nice, simple rule to find the probability of $A \cap B$ if A and B are independent.

EXAMPLE There is a 1 percent chance that the primary navigation system on a spacecraft will fail. In case this happens, the backup navigation system can be used. There is a 1 percent chance that the backup system will fail. What is the probability that they both will fail?

SOLUTION

For this problem, we need to know a crucial piece of information: are the two systems independent? If they are, then we know that we can find the probability that they both fail by multiplying. If A is the event the primary system fails, and B is the event the backup fails, then $\Pr(A \text{ and } B) = \Pr(A) \times \Pr(B) = 0.01 \times 0.01 = 0.0001$. This calculation is reassuring to the crew, because it indicates that the chance of both systems failing is very small. However, this result does not work if the systems are not independent. For example, if a power failure could occur that would knock out both systems, then events A and B are not independent, and the probability of both events occurring is greater than 0.01×0.01. In order to figure out the exact value, it would be necessary to have some information about how the two events are related.

There is an important lesson in this result: If your business maintains a backup data system in case the primary system fails, then the backup system should be at a different location than the primary system. Then it is much less likely that a single event would ruin both systems. As long as the two systems are independent, and the probability of either one of them failing is small, then the probability that they both fail is very small.

EXAMPLE Suppose you draw one card that is a king. What is the probability that the next card you draw will be a king?

SOLUTION

Since there are 3 kings left among the 51 remaining cards, this probability is 3/51. (Note that this is different from the probability of drawing the first king, which was

4/52.) Therefore, the two events (drawing a king on the first card and drawing a king on the second card) are not independent. However, it is important to note that in this case our sample was selected without replacement (that is, we did not put the first card back in the deck after we drew it).

EXAMPLE Suppose you will select two cards from a deck with replacement. You draw the first card and see that it is a king. Then you replace that card in the deck, shuffle the cards again, and draw a second card. What is the probability that the second card will be a king?

SOLUTION

That probability is 4/52, the same as it was when you drew the first card. Therefore, the event of drawing a king on the second card *is* independent of the event of drawing a king on the first card, if the sample is selected with replacement.

EXAMPLE Suppose you toss seven dice. What is the probability that no sixes will appear?

SOLUTION

The probability of not getting a six on one die is 5/6. The probability of not getting any sixes on two dice is $(5/6)^2$, since the two dice are independent. By the same reasoning, the probability of no sixes on seven dice is $(5/6)^7 = .279$.

In general, if a process is repeated in n independent trials with the probability of "success" each time equal to p, then the probability that you will succeed all n times is given by p^n.

Bayes's Rule

Suppose 2 percent of the population has a particular disease. There is a test for the disease, but it is not perfectly accurate, as 3.2 percent of the population tests positive for the disease. There is a 75 percent chance that a person with the disease will test positive. We need to calculate the probability that a person who does test positive really does have the disease.

Let $D+$ be the event of having the disease; $D-$ be the event of not having the disease; $T+$ be the event of testing positive; and $T-$ be the event of testing negative. Then we know:

$$\Pr(D+) = .020$$
$$\Pr(T+) = .032$$
$$\Pr(T+ \mid D+) = .750$$

We need to find $\Pr(D+ \mid T+)$. We can set up a table as follows:

	D+	*D–*	Total
T+			.032
T–			.968
Total	**.020**	**.980**	**1.000**

We can fill in the totals for the rows and columns with the information we're given, but we don't yet know the remaining numbers in the table. However, we can find

$$\begin{aligned}\Pr\left[(D+)\cap(T+)\right] &= \Pr(D+ \text{ and } T+)\\ &= \Pr(T+ \mid D+)\times\Pr(D+)\\ &= .75\times .02\\ &= .0150\end{aligned}$$

In other words, the fraction of the population that has the disease and will test positive is .015. Once we know this number, we can fill in its place in the table:

	D+	*D–*	Total
T+	.015		.032
T–			.968
Total	**.020**	**.980**	**1.000**

Since we know the totals for each row and column, we can now fill in the remaining numbers in the table:

	D+	*D–*	Total
T+	.015	.017	.032
T–	.005	.963	.968
Total	**.020**	**.980**	**1.000**

From the table we can see that $\Pr(D+ \mid T+) = .015/.032 = .469$. Therefore, fewer than half the people who test positive actually have the disease.

It is important to remember that the way you state a conditional probability is important: the probability that you will test positive, given that you have the disease [$\Pr(T+ \mid D+)$], is not the same as the probability that you will have the disease, given that you tested positive [$\Pr(D+ \mid T+)$].

We can also solve this problem with a rule known as Bayes's rule, which states that for any two events A and B, the following hold true:

$$\Pr(B \mid A) = \frac{\Pr(A \mid B)\Pr(B)}{\Pr(A \mid B)\Pr(B) + \Pr\left(A \mid B^c\right)\Pr\left(B^c\right)}$$

Here B^c means B complement; that is, the event that B does not occur.

To use Bayes's rule for our example, we need to find

$$\Pr(T+ \mid D-) = \frac{\Pr(T+ \text{ and } D-)}{\Pr(D-)} = \frac{.017}{.980} = .01735.$$

Now we can calculate $\Pr(D+ \mid T+)$:

$$\Pr(D+ \mid T+) = \frac{\Pr(T+ \mid D+)\Pr(D+)}{\Pr(T+ \mid D+)\Pr(D+) + \Pr(T+ \mid D-)\Pr(D-)}$$

$$\Pr(D+ \mid T+) = \frac{.750 \times .020}{.750 \times .020 + .01735 \times .98} = .469$$

CONDITIONAL PROBABILITY SUMMARY

General definition:

$$\Pr(A \mid B) = \frac{\Pr(A \text{ and } B)}{\Pr(B)}$$

$\Pr(A \mid B)$ is read "the probability that event A will occur, given that event B has occurred," or, more concisely, "the probability of A given B."

If A and B are mutually exclusive events:

$$\Pr(A \mid B) = 0$$

If A is a subset of B:

$$\Pr(A \mid B) = \frac{\Pr(A)}{\Pr(B)}$$
$$\Pr(A \text{ and } B) = \Pr(A)$$

If A and B are independent events:

$$\Pr(A \mid B) = \Pr(A)$$
$$\Pr(A \text{ and } B) = \Pr(A) \times \Pr(B)$$

Bayes's rule:

$$\Pr(B \mid A) = \frac{\Pr(A \mid B)\Pr(B)}{\Pr(A \mid B)\Pr(B) + \Pr(A \mid B^c)\Pr(B^c)}$$

EXERCISES

1. Suppose you have two nickels in your pocket. You know that one is fair and one is two-headed. If you take one out, toss it, and get a head, what is the probability that it was the fair coin?

2. In blackjack, each player is given two cards to start with, and then tries to get a numerical total of 21 in the following way: 2's through 10's are worth their face value, face cards are worth 10, and an ace can be worth either 1 or 11, depending on the player's preference. The player can take more cards, trying to get as close to 21 as possible without going over (in which case the game is lost). Suppose you are dealt a 4 and a 9. If the dealer is dealing from a single deck of 52 cards, and the 4 and the 9 are the only cards that have been dealt from that deck that you know the value of, should you take another card? In other words, what is the probability that if you take another card you won't go over 21?

3. In some versions of blackjack, one of the two cards dealt each player is turned up. Suppose you can see that your fellow players have been dealt two aces, a 5, and a king. You again have a 4 and a 9. Should you draw another card?

4. What is the probability that you will get a royal flush in your next four cards if your first card is the ace of hearts? What is the probability if the next two cards you draw are the king of hearts and queen of hearts? What is the probability if the fourth card you draw is the jack of hearts?

5. What is the probability the you will draw a straight if you have already drawn a 2 and a 3?

6. Suppose there is a probability p that your team will beat its opponent in any particular World Series game. What is the probability that your team will win the World Series? (The teams in the World Series keep playing until one of the teams has won four games.)

7. If you flip a fair coin twice, and know that at least one head come up, what is the probability of there being two heads?

8. Suppose you remove all the diamonds from a 52-card deck, put the ace of diamonds back in, shuffle the remaining 40 cards, and pick a card at random.

Given that the card you pick is an ace, what is the probability that it is the ace of diamonds?

9. Suppose you roll two dice and pick a card at random from a 52-card deck. Suppose further that the number rolled on the dice is the same number on the card. What is the probability that the number is 6?

10. What is the probability that you win a lottery with a three-digit number, given that you have two digits of the winning number?

11. What is the probability of getting four aces in a poker hand given that two of the cards in your hand are the ace and king of spades?

12. What is the probability of drawing a jack from a 52-card deck, given that you've drawn a face card?

13. Suppose you roll three dice. Given that one of the dice shows a 5, what is the probability of rolling a 14?

14. Consider the experiment of tossing two coins. Let A be the event of getting a head on the first toss.

    (a) Can you name an event that is disjoint from but not independent of A?
    (b) Can you name an event that is independent of but not disjoint from A?
    (c) Can you name an event that is both disjoint from and independent of A?
    (d) Can you name an event that is neither disjoint from nor independent of A?

15. Suppose that an election is being conducted with two candidates, Smith and Jones. Of the people in the city, 2/3 support Jones, but 5/9 of the people from the country support Smith. Half of the people live in the country and half live in the city. If you randomly start talking with a voter who turns out to be a Jones supporter, what is the probability that that voter lives in the country?

☆ 16. Derive Bayes's rule.

17. Suppose that 5 percent of the people with blood type O are left-handed, 10 percent of those with other blood types are left-handed, and that 40 percent of the people have blood type O. If you randomly select a left-handed person, what is the probability that that person will have blood type O?

18. Suppose that 70 percent of the people with brown eyes have brown hair, 20 percent of the people with green eyes have brown hair, and 5 percent of the people with blue eyes have brown hair. Also, suppose 75 percent of the people have brown eyes, 20 percent have blue eyes, and 5 percent have green eyes. What is the probability that a randomly selected person with brown hair will also have green eyes?

8

Discrete Random Variables

It often happens in probability that the events we're interested in involve counting something or measuring something. For example, we have been interested in the number of heads that appears when we flip a coin or the sum that appears on a pair of dice. In these cases it is easier to talk about *random variables* rather than probability spaces and events. If X is the number of heads that appear when you flip a coin three times, then X is a random variable. If Y is the number that appears when you toss one die, then Y is also a random variable. If W represents the number of times that the word "tennis" is used on the 11 o'clock news, then W is a random variable (which can take on values anywhere from zero during the dead of winter to 50 during Wimbledon). We'll use capital letters to stand for random variables, to avoid confusion with the ordinary variables used in algebra.

Suppose we roll a die and then write down the number Y that appears. This process is called *observing* (or *measuring*) the value of Y. If we roll the die ten times, then we have ten observations of the random variable Y.

As we have seen, a random event is something for which we don't know for sure whether or not it will happen, but we can often calculate the probability that it will happen. By analogy, a random variable is a variable such that we're not sure what it will equal, but for which we can often calculate the probability that it will equal a particular value. (Formally, a random variable is a variable that takes on a specified value when a particular random event occurs, so it is a function from sets to numbers.)

Although we usually cannot tell exactly what the value of a random variable will be, we often can determine what its values will not be. For example, the tennis variable W cannot be $3\frac{1}{3}$ or π or any other weird number. The variable W must have a whole-number value. The number on the die Y must also be a whole number, but it can have only one of six possible values: 1, 2, 3, 4, 5, or 6. Random variables that can only take on isolated values are called *discrete random variables*. (We'll later talk about *continuous random variables*.)

Discrete random variables don't have to take just whole-number values, though. Suppose we roll two dice, and let T be the average of the two numbers that appear. Then T has the possible values $1,\ 1\frac{1}{2},\ 2,\ 2\frac{1}{2},\ 3,\ 3\frac{1}{2},\ 4,\ 4\frac{1}{2},\ 5,\ 5\frac{1}{2}$, and 6.

Probability Functions

Now let's figure out what we need to know about random variables. With ordinary variables, about all we need to know is their values. However, with random variables it is much more complicated. First, we need to know which values are possible and which are impossible. For example, if a random variable X can never take the value of 3/2, we can write

$$\text{Probability that } X = 3/2 \text{ is } 0.$$

We can write that in a shorter fashion:

$$\Pr(X = 3/2) = 0$$

Once we've made a list of all the possible values, the next thing we would like to know is this: How likely are these different values? In the case of tossing one die, the situation is very simple: there are only six possible values and they are all equally likely. We can make a list of these values and their probabilities:

$$\begin{aligned} \Pr(Y = 1) &= 1/6 \\ \Pr(Y = 2) &= 1/6 \\ \Pr(Y = 3) &= 1/6 \\ \Pr(Y = 4) &= 1/6 \\ \Pr(Y = 5) &= 1/6 \\ \Pr(Y = 6) &= 1/6 \end{aligned}$$

We can also make a list of the probabilities for the random variable X defined as the number of heads in three tosses:

$$\begin{aligned} \Pr(X = 0) &= 1/8 \\ \Pr(X = 1) &= 3/8 \\ \Pr(X = 2) &= 3/8 \\ \Pr(X = 3) &= 1/8 \end{aligned}$$

In both of these cases we understand the process that is generating the random variable, so it is easy to calculate the probability of each possible value. In other circumstances, such as the tennis example, we cannot calculate the probabilities because we do not understand the process well enough. (However, later we will discuss ways to estimate the probabilities of these occurrences.)

To make things a little more convenient, we will define a *probability function* or *probability density function* for a random variable. The value of the probability function for a particular number is just the probability that the random variable will equal that number. We'll use a small letter f to stand for the probability function. So we can make this definition:

$$f(a) = \Pr(X = a)$$

(The probability function is also sometimes called the *probability mass function*.) When we start talking about more than one random variable at a time, we will write the probability function as $f_x(a)$ to make it clear that f is the probability function for the random variable X.

Here is the probability function for the toss of one die:

$$f(1) = 1/6$$
$$f(2) = 1/6$$
$$f(3) = 1/6$$
$$f(4) = 1/6$$
$$f(5) = 1/6$$
$$f(6) = 1/6$$

(This function is graphed in Figure 8–1.)

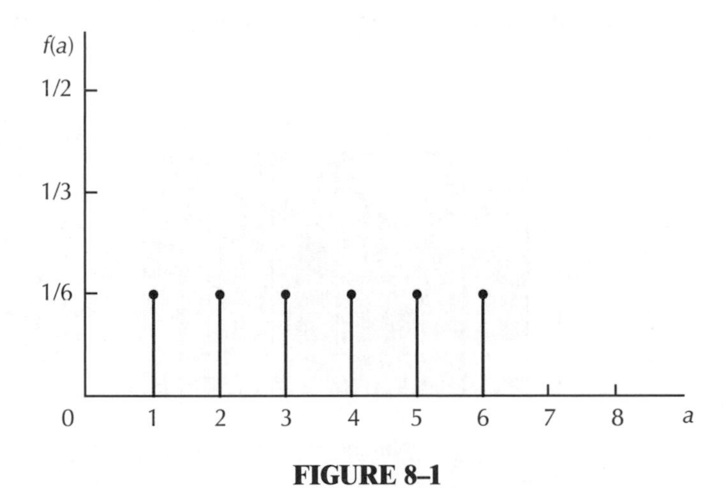

FIGURE 8–1

Here is the probability function for flipping a coin three times:

$$f(0) = 1/8$$
$$f(1) = 3/8$$
$$f(2) = 3/8$$
$$f(3) = 1/8$$

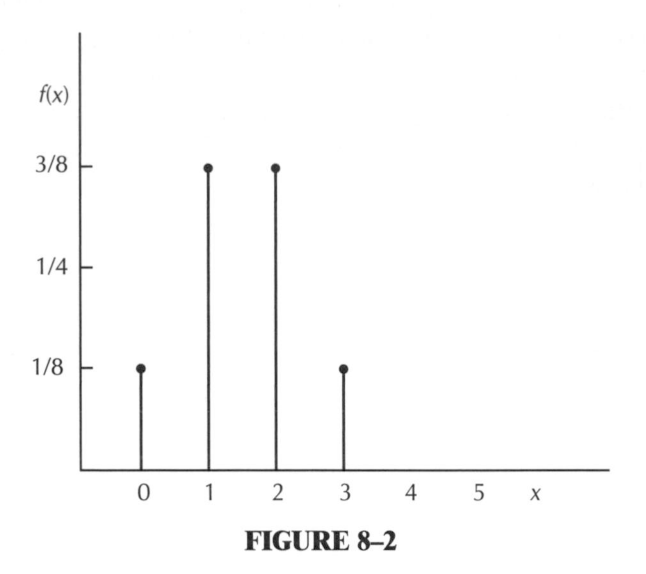

FIGURE 8–2

Figure 8–2 shows a graph of this function. (In both of these cases $f(a) = 0$ for all values of a except the ones that are listed.)

There is a close connection between the probability function of a random variable and the frequency histogram for the numbers in a sample. For example, suppose we roll a die 6,000 times and then make a frequency histogram showing the number of times that each possible result appears (Figure 8–3). The frequency histogram has approximately the same shape as the probability function.

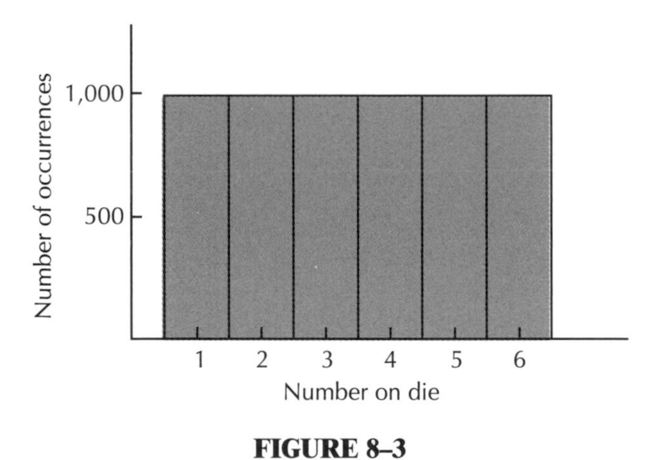

FIGURE 8–3

We can quickly establish two obvious properties that a probability function must satisfy

$$f(a) \leq 1 \quad \text{for all possible values of } a$$
$$f(a) \geq 0 \quad \text{for all possible values of } a$$

These two statements just point out that there is no such thing as a probability that is greater than 1 or less than 0.

Suppose we would like to know the probability that a random variable will have either one of two values. For example, what is the probability that X will equal 2 or 3? We can write it like this:

$$\Pr[(X=2) \quad \text{or} \quad (X=3)]$$

We can rewrite that probability as the probability of a union of two events:

$$\Pr[(X=2) \cup (X=3)]$$

The two events $(X = 2)$ and $(X = 3)$ are clearly mutually exclusive. (Just try to make X equal both 2 and 3 at the same time!) Therefore, we can rewrite the probability:

$$\begin{aligned}\Pr[(X=2) \text{ or } (X=3)] &= \Pr(X=2)+\Pr(X=3)\\ &= f(2)+f(3)\end{aligned}$$

In general, when we want to know the probability that X will equal either a or b, we can take the sum of $f(a) + f(b)$.

Now, suppose we make a list of all of the possible values and add up their probabilities. If the probabilities add up to less than 1 or more than 1, then we know that something is wrong. The probability must be exactly 1 that X will equal one of its possible values. Therefore, in order for a function f to be a valid probability function for a discrete random variable, the sum of its possible values must be 1: If the possible values are $a_1, a_2, a_3, \ldots, a_n$, then

$$f(a_1)+f(a_2)+f(a_3)+\ldots+f(a_n)=1$$

For example, when X represents the number of heads that appear on 3 coin tosses, we can add all of the probabilities:

$$1/8 + 3/8 + 3/8 + 1/8 = 8/8 = 1$$

This shows that the sum does indeed equal 1.

It's nice to know what the probability will be that X will take on any particular value, but sometimes we don't need to know all of that. Sometimes we'd like to know what the probability is that X will be less than or equal to a particular value a_k. For example, if you're playing blackjack and you've already drawn a 7 and an 8, then you're mainly interested in the probability that the value of the next card you draw will be less than or equal to 6.

To find the probability that the random variable Y in the case of rolling a die will be less than or equal to 3, we need to add $\Pr(Y = 1) + \Pr(Y = 2) + \Pr(Y = 3) = 1/2$. In general, if $a_1, a_2, a_3, \ldots, a_k$, are all of the possible values of X that are less than or equal to a_k, then

$$\Pr(X \le a_k) = f(a_1)+f(a_2)+f(a_3)+\ldots+f(a_k)$$

We will give a special name to the function that tells the probability that X will be less than or equal to a particular value. We'll call that function the *cumulative distribution function*, and we'll represent it by a capital F:

$$F(a) = \Pr(X \leq a)$$

We can derive some properties that a cumulative distribution function must satisfy:

1. $F(a)$ must be between 0 and 1 all of the time, since $F(a)$ is itself the probability of an event (the event that $X \leq a$).
2. If we keep letting a become smaller and smaller, eventually we will find $F(a) = 0$. Sooner or later a will become less than the smallest possible value of X, and there is 0 probability that X can be less than a number if that number is smaller than the smallest possible value of X. Likewise, if we make a big enough, eventually we must have $F(a) = 1$. Formally, we can say that $\lim_{a\to\infty} F(a) = 1$ and $\lim_{a\to-\infty} F(a) = 0$ (In this paragraph we have assumed that X has only a finite number of possible values, but the same property holds even if X has an infinite number of possible values.)
3. If $a \leq b$, then $F(a) \leq F(b)$. That means that, as you go from left to right along the number line, $F(a)$ must always be either getting bigger or staying the same. (Formally, F is a *monotone increasing function*.)
4. A graph of a cumulative distribution function looks like an irregular staircase. Figure 8–4 shows the graph of a typical cumulative distribution function. (Formally, this means that $F(a)$ is *piecewise constant*. It will stay flat for a while until it comes to one of the possible values, and then it will increase in a jump.)

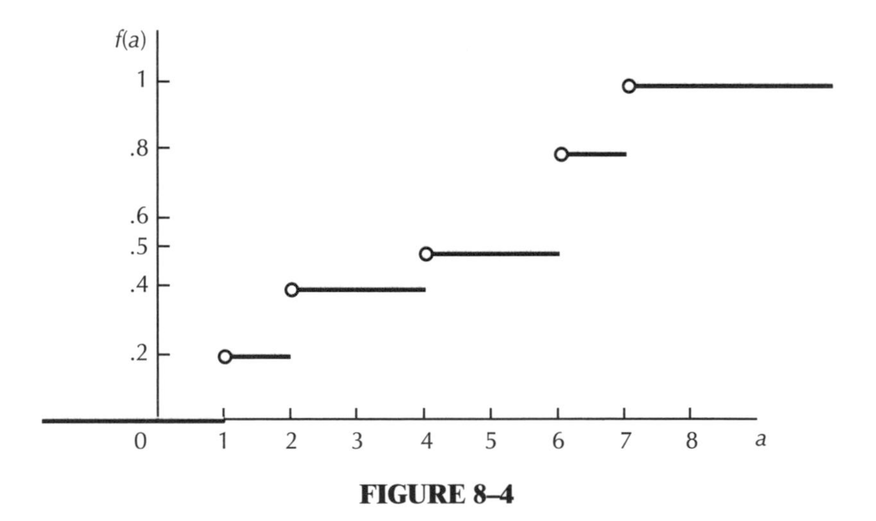

FIGURE 8–4

We can calculate the cumulative distribution function for the number of heads X on three coin tosses:

$$F(a) = \begin{pmatrix} 0 & \text{if } a < 0 \\ 1/8 & \text{if } 0 \leq a < 1 \\ 1/2 & \text{if } 1 \leq a < 2 \\ 7/8 & \text{if } 2 \leq a < 3 \\ 1 & \text{if } 3 \leq a \end{pmatrix}$$

As we can see, this function satisfies the properties that a cumulative distribution function must satisfy.

Expectation

Once we know the density function or the cumulative distribution function for a particular random variable, we know just about everything we might possibly need to know about it. However, many times we would like to summarize the information that we have about the random variable. We showed that we can summarize a group of numbers by a single number (the average). We will develop a similar concept for a random variable.

Consider the variable X (the number of heads in 3 coin tosses.) Suppose we measure X eight million times. How many times are we likely to get no head, 1 head, 2 heads, or 3 heads? You can guess that we will likely observe the value $X = 0$ close to one million times, the value $X = 1$ close to three million times, the value $X = 2$ three million times, and the value $X = 3$ one million times. In reality, we would probably not get exactly these numbers, but let us assume for a moment that we performed this experiment eight million times and did get exactly these results. (In general, if you measure a random variable N times, then the number of times that you can expect to get the value a will be equal to $f(a) \times N$.)

Now, we have eight million numbers written down (one number for each of the eight million repetitions of our experiment.) Since we don't want to carry around all eight million numbers, we'd like to summarize all of these values by taking their average:

$$\begin{array}{ll} (0 + 0 + 0 + 0 + \ldots & \text{(one million zeros)} \\ + 1 + 1 + 1 + 1 + \ldots & \text{(three million ones)} \\ + 2 + 2 + 2 + 2 + \ldots & \text{(three million twos)} \\ + 3 + 3 + 3 + 3 + \ldots)/8{,}000{,}000 & \text{(one million threes)} \end{array}$$

We can rewrite that in a shorter fashion:

$$\begin{array}{l} (1{,}000{,}000 \times 0 \\ +3{,}000{,}000 \times 1 \\ +3{,}000{,}000 \times 2 \\ +1{,}000{,}000 \times 3)/8{,}000{,}000 \end{array}$$

It's even shorter to write it like this:

$$1/8 \times 0 + 3/8 \times 1 + 3/8 \times 2 + 1/8 \times 3 = 1\frac{1}{2}$$

So the average of all the values is $1\frac{1}{2}$. We can also write that expression in terms of the probability function *f*:

$$\text{average} = f(0) \times 0 + f(1) \times 1 + f(2) \times 2 + f(3) \times 3$$

We will call this quantity the *expected value* of X, or the *expectation of* X. The expectation of a random variable tells us the average of all the values we would expect to get if we measured the random variable many times. We'll use the capital letter E to stand for expectation, and write it like this:

$$E(X) = \text{expectation of } X$$

We can write a general formula for calculating an expectation:

$$\begin{aligned} E(X) &= f(a_1)a_1 + f(a_2)a_2 + f(a_3)a_3 + \cdots + f(a_n)a_n \\ &= \sum_{i=1}^{n} f(a_i)a_i \end{aligned}$$

Here $a_1, a_2, \ldots, a_n$ are all of the possible values of the random variable X.

The expectation of X is also called the *mean* of X, or the mean of the distribution of X, and is usually symbolized by the Greek letter μ (mu). Note that $E(X) = \mu$ is not itself a random variable. It is a regular constant number.

The expectation of a random variable does not itself have to be one of the possible values of the random variable. For example, we found that the expectation value for the number of heads in three tosses is $1\frac{1}{2}$, but we strongly caution you against betting that the number of heads will ever come out to be $1\frac{1}{2}$ (unless you make the bet with us).

The expectation of the number (Y) that shows up on a die is easy to calculate:

i	$Pr(Y = i)$	$Pr(Y = i)$	$i \times Pr(Y = i)$	$i \times Pr(Y = i)$
1	1/6	.1667	1/6	.1667
2	1/6	.1667	2/6	.3333
3	1/6	.1667	3/6	.5000
4	1/6	.1667	4/6	.6667
5	1/6	.1667	5/6	.8333
6	1/6	.1667	6/6	1.0000
Total	**6/6**	**1**	**21/6**	**3.5**

The expected value is 3.5.

There are two important properties of expectations that we can establish. If c is a constant number (in other words, not a random variable), then

$$E(cX) = cE(X)$$

For example, suppose you are fortunate enough to play a game where you roll a die and will be paid $100 times the number that appears. How much will you expect to win?

Let W represent your winnings, and let X represent the number that appears on the die. Then $W = 100X$, so $E(W) = E(100X) = 100E(X) = 100 \times 3.5 = 350$. Therefore, your expected winnings are $350. We must again stress this vitally important point: your winnings on any particular play of the game are unpredictable, but if you were to play this game many times you would expect your average winnings to be $350.

Suppose we have two random variables X and Y, and we form a new random variable V which is equal to $V = X + Y$. In general, putting two random variables together creates a lot of complications, and we will not talk about these problems much until Chapter 14. However, one property we can establish is that

$$E(X + Y) = E(X) + E(Y)$$

There will also be times when we would like to calculate the expectation of a particular function of a random variable. For example, suppose $Z = X^2$, and we want to calculate $E(Z) = E(X^2)$. We can use this formula:

$$E\left(X^2\right) = \sum_{i=1}^{n} x_i^2 f\left(x_i\right)$$

For example, when X is the number of heads that appear on three coins, we can calculate $E(X^2)$:

$$\begin{aligned} E\left(X^2\right) &= 0^2 \times 1/8 + 1^2 \times 3/8 + 2^2 \times 3/8 + 3^2 \times 1/8 \\ &= 0 \times 1/8 + 1 \times 3/8 + 4 \times 3/8 + 9 \times 1/8 \\ &= 24/8 = 3 \end{aligned}$$

Note that it would have been easier if we could say that $E(X^2) = [E(X)]^2$, but as we can see that formula does not work. In general, though, if $g(x)$ is any function,

$$E\left[g(X)\right] = \sum_{i=1}^{n} g(x_i) f\left(x_i\right)$$

Variance

Let Y be the number that appears on one die, and let A_2 be the average number that appears when two dice are tossed. Figure 8–5 shows the probabilities for these two random variables. In each case the sum of the probabilities is one. However, note that the probabilities are spread out more in the case of one die than they are with two dice.

In other words, if you were forced to make a guess about the value of these two random variables, you would rather guess the value of A_2 because it involves less uncertainty.

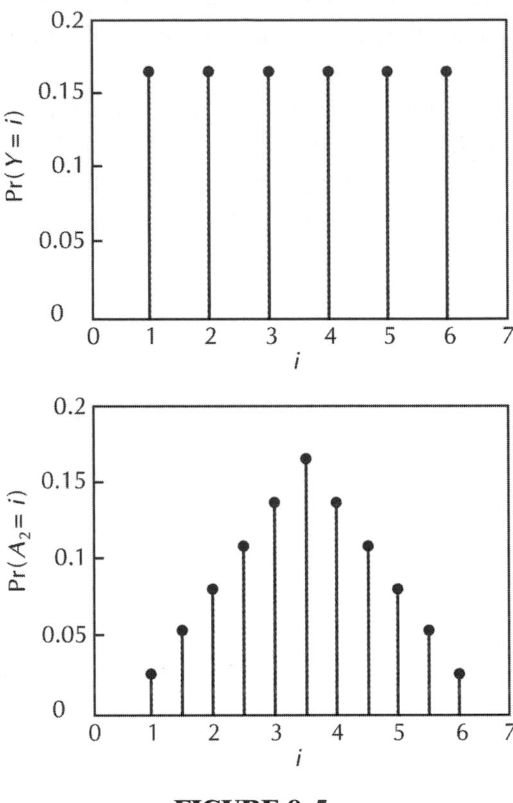

FIGURE 8–5

We need a way of measuring the degree of uncertainty for a random variable. Another way of saying it is that we need a way to measure the degree to which the probabilities of that random variable are spread out. As you recall from Chapter 4, we used the concept of the variance to measure the degree to which a list of numbers was spread out:

$$\text{Var}(x) = \frac{(x_1 - \bar{x})^2 + (x_2 - \bar{x})^2 + \cdots + (x_n - \bar{x})^2}{n}$$

Here $x_1, x_2, \ldots x_n$ are the n numbers in the list. We will use the same concept here by looking at how far apart the possible values of the random variable are. You might initially be tempted to define the variance like this:

$$\text{Var}(X) = \frac{(a_1 - E(X))^2 + (a_2 - E(X))^2 + \cdots + (a_n - E(X))^2}{n}$$

Here $a_1, a_2, \ldots, a_n$ are the n possible values of the random variable X. However, there is a problem with doing it this way: not all possible values are equally likely. We need to attach greater weight to those values that have greater probability. Therefore, we define the variance of a discrete random variable as follows:

$$\text{Var}(X) = \Pr(X = a_1)[a_1 - E(X)]^2 + \Pr(X = a_2)[a_2 - E(X)]^2 + \cdots + \Pr(X = a_n)[a_n - E(X)]^2$$

We can also write this definition using the probability function f:

$$\text{Var}(X) = f(a_1)[a_1 - E(X)]^2 + f(a_2)[a_2 - E(X)]^2 + \cdots + f(a_n)[a_n - E(X)]^2 = \sum_{i=1}^{n} f(a_i)[a_i - E(X)]^2$$

The variance is also symbolized by σ^2, and the square root of the variance is called the standard deviation (symbolized by σ).

Here are the calculations for the variance of Y (which is the number that appears on one die). (We have previously found that $E(Y) = 3.5$.)

i	Pr(*Y* = *i*)	i - E(*Y*)	[*i* - E(*Y*)]^2	[*i* - *EY*)]^2 × Pr(*Y* = *i*)
1	1/6	−2.5	6.25	1.0417
2	1/6	−1.5	2.25	0.3750
3	1/6	−0.5	0.25	0.0417
4	1/6	0.5	0.25	0.0417
5	1/6	1.5	2.25	0.3750
6	1/6	2.5	6.25	1.0417
Total	**6/6**	**0**	**17.5**	**2.9167**

Here are the steps:

1. Calculate $i - E(Y) = i - 3.5$ for each possible value.
2. Square these possible values.
3. Multiply these squares by the probability (see the last column).
4. Add together the final column; the result is Var(Y).

In this case, Var(Y) = 2.91667.

There is a slightly shorter way for finding Var(Y). Create a column with the values of i^2; multiply each of these squares by the probability; then find the sum of these values. The result is the expected value of the square of Y, written $E(Y^2)$:

i	$\Pr(Y = i)$	$\Pr(Y = i)$	i^2	$i^2 \times \Pr(Y = i)$	$i^2 \times \Pr(Y = i)$
1	1/6	0.1667	1	1/6	0.1667
2	1/6	0.1667	4	4/6	0.6667
3	1/6	0.1667	9	9/6	1.5000
4	1/6	0.1667	16	16/6	2.6667
5	1/6	0.1667	25	25/6	4.1667
6	1/6	0.1667	36	36/6	6.0000
Total	**6/6**	**1**		**91/6**	**15.1667**

Therefore, $E(Y^2) = 15.1667$. Once this is known, find the variance from this formula:

$$\begin{aligned} \text{Var}(Y) &= E(Y^2) - [E(Y)]^2 \\ &= 15.16667 - 3.5^2 \\ &= 2.91667 \end{aligned}$$

This is the same answer we found from the other method.

Before looking for $\text{Var}(A_2)$, we'll look for the variance of T_2, where T_2 is the sum of the numbers that appear on two dice.

i	No. of Outcomes	$\Pr(T_2 = i)$	$i^2 \times \Pr(T_2 = i)$	$i^2 \times \Pr(T_2 = i)$
2	1	1/36	4 × 1/36	0.1111
3	2	2/36	9 × 2/36	0.5000
4	3	3/36	16 × 3/36	1.3333
5	4	4/36	25 × 4/36	2.7778
6	5	5/36	36 × 5/36	5.0000
7	6	6/36	49 × 6/36	8.1667
8	5	5/36	64 × 5/36	8.8889
9	4	4/36	81 × 4/36	9.0000
10	3	3/36	100 × 3/36	8.3333
11	2	2/36	121 × 2/36	6.7222
12	1	1/36	144 × 1/36	4.0000
Total	**36**	**36/36**	**1974/36**	$E(T_2^2) = 54.8333$

We find that $\text{Var}(T_2)$ is $54.83333 - 7^2 = 5.83333$, which happens to be $2.91667 + 2.91667$. Since $\text{Var}(T_2) = \text{Var}(X_1 + X_2) = \text{Var}(X_1) + \text{Var}(X_2)$ we're tempted to state a general rule about the variance of a sum. This rule does work, but only under the right condition.

If X and Y are any two *independent* random variables, then $\text{Var}(X + Y) = \text{Var}(X) + \text{Var}(Y)$.

Two random variables are independent if knowledge of the value for one of them provides no help in predicting the value of the other one. (Note that this definition is similar to the definition of independent events.) Since the number on one die is not affected by the number on the other die, these two random variables qualify as being independent.

Now let's find the variance of A_2, the average of the numbers that appear on two dice. Using the same method as before, the result turns out to be 1.4583 (see Exercise 21). However, we can find the result more quickly by using this rule:

$$\text{Var}(cX) = c^2\text{Var}(X)$$

Here c is a constant. Since $A_2 = \frac{1}{2}T_2$, it follows that $\text{Var}(A_2) = \text{Var}\left(\frac{1}{2}T_2\right) = \left(\frac{1}{2}\right)^2 \text{Var}(T_2) = \frac{1}{4} \times 5.83333 = 1.4583$. Note how the variance of the average number on two dice is less than the variance of the number that appears on one die. This vitally important fact will be investigated further in Chapter 15.

Next, we can calculate the variance of X, the number of heads in three coin tosses, since we have already found that $E(X^2) = 3$ and $E(X) = 1\frac{1}{2}$. Therefore,

$$\text{Var}(X) = E(X^2) - [E(X)]^2$$
$$\text{Var}(X) = 3 - \left(1\frac{1}{2}\right)^2$$
$$= 3/4.$$

Bernoulli Trials

Now we'll look at another simple random variable that illustrates all of these concepts. Let us consider a mad scientist who repeats a particular experiment each day. There is a probability p that the experiment will succeed on any particular trial; let's suppose the $p = 1/5$. (An experiment like this that can have only two possible results, success or failure, is called a *Bernoulli trial*). Let Z be the random variable that is equal to the number of successes on a particular day. Then Z has only two possible values: 0 and 1. We can easily calculate the complete probability function:

$$f(0) = \Pr(Z = 0) = 1 - p$$
$$f(1) = \Pr(Z = 1) = p$$
$$f(a) = 0 \text{ for all other values of } a$$

The expectation can be found from

$$E(Z) = 0 \times (1 - p) + 1 \times p = p$$

We can calculate the variance two ways. First, we can use the definition

$$\begin{aligned}\text{Var}(Z) &= f(0)(0-p)^2 + f(1)(1-p)^2 \\ &= (1-p)p^2 + p(1-p)^2 \\ &= p^2 - p^3 + p - 2p^2 + p^3 \\ &= p - p^2 \\ &= p(1-p) \\ &= .16(\text{when } p = 1/5)\end{aligned}$$

We can also use the short-cut formula to calculate the variance:

$$\begin{aligned}\text{Var}(Z) &= E(Z^2) - [E(Z)]^2 \\ &= 0^2(1-p) + 1^2 p - p^2 \\ &= p(1-p)\end{aligned}$$

Note that if $p = 0$, we know for sure that $Z = 0$; and if $p = 1$, we know for sure that $Z = 1$. In either case, Var(Z) is 0, which we know must be the case. The variance is largest if $p = 1/2$; in that case there is the greatest uncertainty as to whether the experiment will be a success or not.

Choosing One Item from a List of Items

EXAMPLE Suppose you have n people whose heights are $h_1, h_2, \ldots, h_n$. You will randomly select one person and let X be the height of the person selected. What is the expectation and variance of X?

SOLUTION

The probability that X will equal h_1 is $1/n$, since all n people are equally likely to be chosen. Therefore,

$$\begin{aligned}E(X) &= \left(\frac{1}{n}\right)h_1 + \left(\frac{1}{n}\right)h_2 + \cdots + \left(\frac{1}{n}\right)h_n \\ &= \frac{h_1 + h_2 + \cdots + h_n}{n}\end{aligned}$$

We recognize that expression as being the mean (or average) height in the population (call $\overline{h}$.)

$$\begin{aligned}E(X^2) &= \left(\frac{1}{n}\right)h_1^2 + \left(\frac{1}{n}\right)h_2^2 + \cdots + \left(\frac{1}{n}\right)h_n^2 \\ &= \frac{h_1^2 + h_2^2 + \cdots + h_n^2}{n}\end{aligned}$$

This expression is the average value of h_i^2 $\left(\text{call it } \overline{h^2}\right)$. We can find the variance:

$$\text{Var}(X) = E(X^2) - [E(X)]^2 = \overline{h^2} - \bar{h}^2$$

The expression $\overline{h^2} - \bar{h}^2$ gives us the variance of the heights of the people in this population. In general, once we know the average and variance of a list of numbers, then we also know the expected value and variance of a random variable found by choosing one number from that list:

List of Numbers	Random Variable
$x_1, x_2, \ldots x_n$	X equals one number chosen at random from the list
Average $= \bar{x}$	Expected value $= E(X) = \bar{x}$
Variance $= \overline{x^2} - \bar{x}^2$	Variance $= \overline{x^2} - \bar{x}^2$

DISCRETE RANDOM VARIABLE SUMMARY

The probability function $f(a)$ for a random variable X gives the probability that X equals a given value a.

$$f(a) = \Pr(X = a)$$

The cumulative distribution function $F(a)$ gives the probability that X is less than or equal to a given value:

$$F(a) = \Pr(X \le a)$$

If $a_1, a_2, \ldots a_n$ are the possible values, then

$$f(a_1) + f(a_2) + \cdots + f(a_n) = \sum_{i=1}^{n} f(a_i) = 1$$

The expected value is found from

$$E(X) = a_1 f(a_1) + a_2 f(a_2) + \cdots + a_n f(a_n)$$
$$= \sum_{i=1}^{n} a_i f(a_i)$$
$$E(g(X)) = g(a_1) f(a_1) + g(a_2) f(a_2) + \cdots + g(a_n) f(a_n)$$
$$= \sum_{i=1}^{n} g(a_i) f(a_i)$$

Properties of the expectation:

$$E(cX) = cE(X) \quad \text{if } c \text{ is a constant}$$

$$E(X + Y) = E(X) + E(Y)$$

The variance is found from

$$\text{Var}(X) = \sum_{i=1}^{n} f(a_i)[a_i - E(X)]^2 = E(X^2) - [E(X)]^2$$

Properties of the variance:

$$\text{Var}(cX) = c^2 \text{Var}(x) \quad \text{if } c \text{ is a constant}$$

$$\text{Var}(X + Y) = \text{Var}(X) + \text{Var}(Y) \text{ if } X \text{ and } Y \text{ are independent}$$

NOTE TO CHAPTER 8

It is possible for a discrete random variable to have an infinite number of possible values, just so long as all of the probabilities still add up to 1. For example, suppose the probability function for a random variable X looks like this: $f(2) = 1/2, f(4) = 1/4, f(8) = 1/8, f(16) = 1/16$, and so on. Then all the probabilities add up to 1. However, if we try to calculate the expectation of X, the result turns out to be infinity. In cases such as this when the expectation formula leads to a value of infinity, it is said that the expectation does not exist.

EXERCISES

1. List five quantities not yet mentioned in this book that can be represented by a discrete random variable (such as the number of slices cut from a pizza), and give their possible values.

2. In a given city in a given year, let X be the number of days that it rained and Y be the time (measured in seconds) that it rained. Which of these are discrete random variables? (Check the possible values.)

3. Let X be the number rolled by three fair dice. Determine $f_x(k)$, the probability function of X.

4. Let X be a discrete random variable. If $\Pr(X < 7) = 1/3$, and $\Pr(X > 7) = 1/5$, then what is $fx(7)$?

5. Suppose X is a discrete random variable. If $\Pr(X < 5) = 5/6$, and $\Pr(X \leq 5) = 11/12$, then what is $f_x(5)$?

6. Suppose X is a discrete random variable, and suppose that at every point the value of the probability function for X is either 0 or 1/5. How many possible values are there for X?

7. Flip a fair coin a few times, and let X be the number of heads you get before your first tail. Determine the probability function for X.

8. Let X be a random variable with the following probability function:

$$\begin{aligned} f(-1/2) &= 1/2 \\ f(1/2) &= 1/6 \\ f(2) &= 1/3 \\ f(k) &= 0 \text{ otherwise} \end{aligned}$$

   Make a graph of $Fx(a)$.

9. Let X be a discrete random variable such that $F(a)$ is a constant for $a \geq 10$. What can you say about $f(k)$ if $k \geq 10$?

10. Show that $E(X + Y) = E(X) + E(Y)$ for any two random variables.

11. Show that $\text{Var}(c) = 0$ if c is a constant number.

12. Calculate $\text{Var}(Y)$ when Y represents the number that appears when three dice are thrown.

13. Verify that $\text{Var}(X + Y + Z) = \text{Var}(X) + \text{Var}(Y) + \text{Var}(Z)$ if X, Y, and Z are the results of tossing three different dice.

14. Suppose you toss a die n times. Let X represent the total of all of the numbers that appear. What is $E(X)$? What is $\text{Var}(X)$?

15. You and a friend play the following game: You flip a fair coin. If it comes up heads, you pay \$1. If it comes up tails, your friend rolls a die. If the result is an even number, you are paid \$2; if the result is an odd number, then you pay your friend \$3. What is your expected payoff—that is, the average amount you will win or lose?

16. You and a friend play the following game: You pay your friend \$3 each turn and then flip a fair coin. If it's tails, your friend pays you \$ (2^n), where n is the number of times you've flipped the coin, and the game ends. If it's heads, you have the choice of stopping or continuing. If you have m dollars to start with, and you play the game either until you win or until you have no money left, what will you win on the average?

17. Let X be a random variable that equals 1 on Sunday, 2 on Monday, and so on, up to 7 on Saturday. Calculate $E(X)$ and $\text{Var}(X)$.

18. If X is a discrete random variable with the cumulative distribution:

$$F(a) = \begin{cases} 0 & a < -2 \\ .4 & -2 \leq a < 4.5 \\ .85 & 4.5 \leq a < 9 \\ 1 & 9 \leq a \end{cases}$$

what is the probability function for X?

19. Let X be a discrete random variable representing one half of the number of seconds that have elapsed since this book was purchased. What is the set of possible values for X?

20. Let X be a discrete random variable. If $f(0) = .5, f(1) = .2, f(2) = .1, f(3) = c$, and $f(x) = 0$ everywhere else, then what is the value of c?

21. Let X be a discrete random variable representing the average of the numbers rolled on two dice. What is the probability function for X? Use the probability function to find $E(X)$ and $\text{Var}(X)$.

22. Let X be a discrete random variable representing the number of the television channel that you last watched. What are the possible values for X?

23. Let X be the numerical value of the card you draw from a 52-card deck. Suppose that kings, jacks, and queens are worth 10. Find $E(X)$.

24. What is the variance of a Bernoulli-trial variable when the probability of success is .6 and the probability of failure is .4?

25. Let X and Y be two random variables representing the numbers that come up on two dice. Let $Z = X + Y$. Calculate the probability function of Z and then calculate $E(Z)$. Show that $E(Z) = E(X) + E(Y)$.

26. Show that $E(cX) = cE(X)$ for any random variable X and any constant c.

27. Show that $\text{Var}(cX) = c^2\text{Var}(X)$ for any random variable X and any constant c.

28. Show that $E(X + Y + Z) = E(X) + E(Y) + E(Z)$, for any three random variables X, Y, and Z.

29. Derive the probability function for the random variable whose cumulative distribution function is illustrated in Figure 8–4.

30. (a) Let T_2 be a random variable equal to the sum of the numbers that appear when you toss two dice, and let T_{50} be the sum of the numbers that appear when you toss 50 dice. Calculate $E(T_2)$, $\text{Var}(T_2)$, $E(T_{50})$, and $\text{Var}(T_{50})$.

    (b) Let A_2 be a random variable equal to the average of the numbers that appear when you toss two dice, and let A_{50} be the average of the numbers that appear when you toss 50 dice. Calculate $E(A_2)$, $\text{Var}(A_2)$, $E(A_{50})$, and $\text{Var}(A_{50})$.

☆ 31. Prove that any cumulative distribution function F is nondecreasing.

☆ 32. Prove that any cumulative distribution function F is piecewise constant.

☆ 33. Show that $E(X^2) \geq [E(X)]^2$ for any random variable X.

☆ 34. Let X be a discrete random variable with the following probability function:

$$f(k) = c2^{-k} \quad \text{if } k \text{ is a positive integer}$$
$$f(k) = 0 \quad \text{if } k \text{ is not a positive integer}$$

What is the value of c?

9

The Binomial Distribution

Suppose you are taking a 20-question multiple-choice exam. Each question has four possible answers, so the probability is .25 that you will be able to answer a question correctly by guessing. What is the probability that you can get at least 10 questions right purely by guessing?

The answer can be found by using a type of discrete random variable distribution known as the *binomial distribution*. These are the conditions where the binomial distribution applies:

1. You are conducting an "experiment" that has a probability p of being a success. (The word *experiment* here is used in a very general sense; it can be applied to any process where there are two outcomes that can be called a "success" and a "failure.")
2. This experiment will be repeated n times, and each trial will be independent of all the other trials.
3. The probability of success remains constant for all trials (equal to p).
4. Let X be a random variable representing the number of successes out of the n trials. Then X is said to have a binomial distribution with parameters n and p.

The Excel function BINOMDIST(*successes, trials, probability,* FALSE) will calculate the probability of getting the specified nunber of successes out of this number of trials with this probability of success. Later in this chapter we'll derive the formula used to calculate these probabilities. The function BINOMDIST(*successes, trials, probability,* TRUE) will calculate the cumulative probability—that is, the probability that the number of successes will be less than or equal the specified number of successes. (Note that the final argument is TRUE if you want the cumulative probability, and FALSE if you want the probability for a specific value.)

Here is the table of values:

k	Pr($X = k$)	Cumulative Probability Pr($X \leq k$)
0	.00317	.00317
1	.02114	.02431
2	.06695	.09126
3	.13390	.22516
4	.18969	.41484
5	.20233	.61717
6	.16861	.78578
7	.11241	.89819
8	.06089	.95907
9	.02706	.98614
10	.00992	.99606
11	.00301	.99906
12	.00075	.99982
13	.00015	.99997
14	.00003	1.00000
15	3.426E–06	1.00000
16	3.569E–07	1.00000
17	2.799E–08	1.00000
18	1.555E–09	1.00000
19	5.457E–11	1.00000
20	9.095E–13	1.00000

The table gives the probabilities rounded to 5 decimal places, except that exponential notation is used for very small values. For example, 3.426E–06 means $3.426 \times 10^{-6} = 0.000003426$. By looking at the cumulative column, we can see that there is a probability of .98614 that you will get 9 or fewer questions right. Therefore, the probability of getting at least 10 right is $1 - .98614 = .01386$. It would seem in this case that studying is a better strategy than random guessing. (Of course, you can look at the bright side—there is only about a 9 percent probability that you will get fewer than 3 answers correct.)

EXAMPLE If you toss a coin n times, the number of heads will have a binomial distribution with $p = .5$ (since there is a 50 percent chance of tossing heads on any particular trial, and all of the tosses are independent).

Here are some other examples of applications of the binomial distribution.

EXAMPLE Suppose you have three red sweaters and two blue sweaters in a drawer. Every day you randomly pull out one sweater (and you put it back at the end of the day). If X is the number of red sweaters you select during a week, then X has a binomial distribution with parameters $n = 7$ and $p = .6$. We can calculate the probabilities:

k	$\Pr(X = k)$
0	.001
1	.017
2	.077
3	.193
4	.290
5	.261
6	.130
7	.027

EXAMPLE An airline has discovered that 5 percent of the people who make reservations do not show up for their flights. Therefore, it knows that it will likely end up with empty seats if the number of reservations it takes is exactly the same as the number of available seats. It decides to live dangerously by taking more reservations then there are seats, hoping that enough people fail to show up so that an overflow does not occur. If the planes seat 350 people, and the airline takes 362 reservations, what is the probability of an overflow?

SOLUTION

Call it a success when a person shows up for the flight. Use the binomial distribution with $n = 362$ and $p = .95$:

i	$\Pr(X = i)$	Cumulative Probability
351	0.02227	0.02227
352	0.01322	0.03549
353	0.00712	0.04260
354	0.00344	0.04604
355	0.00147	0.04751
356	0.00055	0.04806
357	0.00018	0.04824
358	0.00005	0.04829
359	0.00001	0.04830
360	0.00000	0.04830
361	0.00000	0.04830
362	0.00000	0.04830

Note that the chance of 360 or more people showing up is very small. However, we can see that there is a 4.8 percent chance that 351 or more people will show up, which means that an overflow will occur. If we need to reduce the probability of an overflow, then we need to take fewer reservations; if we are willing to live with a larger probability of overflow, then we can take more reservations.

Expectation, Variance, and Probability Function of the Binomial Distribution

If your experiment has a 70 percent chance of success, and you conduct 100 independent trials, you would guess that you would have about 70 successes out of those trials. Your intuition here correctly leads you to the expectation of the binomial distribution:

$E(X) = np$, where X has a binomial distribution with n trials and a probability of success equal to p.

To prove that, consider the Bernoulli trial random variable Z from the previous chapter. $Z = 1$ if the experiment is a success, and 0 if it fails. We found $E(Z) = p$, where p is the probability of success. Now repeat this experiment n times and let Z_i be the Bernoulli random variable that will be 1 if trial i results in success, and 0 if trial i results in failure. Then, let

$$X = Z_1 + Z_2 + Z_3 + \cdots + Z_n$$

All of the Zs will be either 1 or 0. The number of 1's will be the number of successes out of the n trials, so therefore X will have a binomial distribution with parameters n and p. Now we can find the expected value of X:

$$E(X) = E(Z_1 + Z_2 + Z_3 + \cdots + Z_n)$$

We found in the previous chapter that the expectation of a sum of random variables is the sum of their expectations:

$$E(X) = E(Z_1) + E(Z_2) + E(Z_3) + \cdots + E(Z_n)$$

Each Z_i has expectation equal to p:

$$E(X) = p + p + p + \cdots + p = np$$

since there are n different ps added together.

Using a similar method, we can find the variance of a binomial random variable:

$$\text{Var}(X) = \text{Var}(Z_1 + Z_2 + Z_3 + \cdots + Z_n)$$

We can break the variance of the sum into the sum of the variances (however, note that this step works only because the Z's are independent of each other).

$$\text{Var}(X) = \text{Var}(Z_1) + \text{Var}(Z_2) + \text{Var}(Z_3) + \cdots + \text{Var}(Z_n)$$

Each Z_i has variance of $p(1 - p)$ (as found in the previous chapter, page 100):

$$\text{Var}(X) = p(1-p) + p(1-p) + p(1-p) + \cdots + p(1-p)$$

$$= np(1-p)$$

Now we'll work on finding the probability that the binomial random variable X will take on a specific value i, where $0 \leq i \leq n$.

There are two situations where the answer is relatively easy. First, we'll find the probability of n successes. Since there are n independent trials, we multiply the probability of success each time, giving us p^n. Second, consider the probability of 0 successes (that is, n failures). We have to multiply the probability of failure $(1-p)$ together n times, so we have $(1-p)^n$. Here are our results so far:

$$\Pr(X=0)=(1-p)^n$$

$$\Pr(X=n)=p^n$$

Now we'll find the probability of one success among the n trials [that is, $\Pr(X = 1)$]. Before we answer that question, we'll answer a slightly simpler question: What is the probability that the first trial will be a success, and the remaining $n-1$ trials are all failures? We need to multiply p (the probability of success on the first trial) by $(1-p)^{n-1}$ (the probability that the remaining $n-1$ trials all fail):

$$\Pr(\text{first trial succeeds and all others fail}) = p(1-p)^{n-1}$$

Now, what is the probability that the second trial succeeds and all others fail? That is the same as in the previous case:

$$\Pr(\text{second trial succeeds and all others fail}) = p(1-p)^{n-1}$$

In general, we can say:

$$\Pr(\text{any one specified trial succeeds and all others fail}) = p(1-p)^{n-1}$$

Now to return to our original question: What is the probability that there will be exactly one success among the n trials? Notice that when we phrase the question this way, we don't care whether the success occurs on the first trial or the last trial or anywhere in between—we only care that there is exactly one success. If there is one success, then it must occur in one of those spots:

$$\begin{aligned}
\Pr(\text{1 trial succeeds}) \quad &= \quad \Pr[(\text{1st trial succeeds, all others fail}) \\
&\text{or} \quad (\text{2nd trial succeeds, all others fail}) \\
&\text{or} \ldots \text{or} \quad (n\text{th trial succeeds, all others fail})]
\end{aligned}$$

Since these are mutually exclusive events, we can add up the probabilities:

$$\begin{aligned}
\Pr(\text{1 trial succeeds}) \quad &= \quad \Pr(\text{1st trial succeeds, all others fail}) \\
&\quad\quad \Pr(\text{2nd trial succeeds, all others fail}) + \ldots \\
&\quad\quad \Pr(n\text{th trial succeeds, all others fail}) \\
&= \quad p(1-p)^{n-1} + p(1-p)^{n-1} + \ldots + p(1-p)^{n-1} \\
&= \quad np(1-p)^{n-1}
\end{aligned}$$

Therefore, $\Pr(X=1) = np(1-p)^{n-1}$.

Now consider the general case of i successes and $n - i$ failures. First, ask "What is the probability that the first i trials succeed and the remaining $n - i$ trials fail?" Again we can multiply the probabilities, with the result $p^i(1-p)^{n-i}$. The same formula can be used to give the probability that any one specified group of i trials succeed and the others fail. This means the probability of i successes can be found by multiplying together two factors:

- the probability that any specified i trials succeed; we have found this probability is $p^i(1-p)^{n-i}$.
- the number of different ways of arranging i successes among the n trials

The second question should sound a bit familiar. We need to choose i trials from the group of n trials; the number of different ways of doing this is given by $\binom{n}{i}$, the formula for combinations. Therefore, we have completed the formula for i successes among the n trials:

$$\Pr(X = i) = \binom{n}{i} p^i (1-p)^{n-i}$$

The above formula gives the probability function for a random variable with the binomial distribution with parameters n and p. Fortunately, it is consistent with our previous work.

If $i = 0$:

$$\Pr(X = 0) = \binom{n}{0} p^0 (1-p)^n = (1-p)^n$$

If $i = n$:

$$\Pr(X = n) = \binom{n}{n} p^n (1-p)^{n-n} = p^n$$

If $i = 1$:

$$\Pr(X = 1) = \binom{n}{1} p^1 (1-p)^{n-1} = np(1-p)^{n-1}$$

If $p = \frac{1}{2}$:

$$\begin{aligned} \Pr(X = i) &= \binom{n}{i}\left(\frac{1}{2}\right)^i\left(1-\frac{1}{2}\right)^{n-i} \\ &= \binom{n}{i}\left(\frac{1}{2}\right)^{i+n-i} \\ &= \binom{n}{i}\left(\frac{1}{2}\right)^n = \frac{\binom{n}{i}}{2^n} \end{aligned}$$

Recall that there are 2^n outcomes if you toss a coin n times. Of these, the combinations formula $\binom{n}{i}$ gives the number of outcomes with i heads, so the probability of i heads is $\dfrac{\binom{n}{i}}{2^n}$.

Here is an example of the binomial distribution with $n = 10$ and $p = .4$:

i	$\Pr(X = i)$
0	$1 \times .4^0 \times .6^{10} = .00605$
1	$10 \times .4^1 \times .6^9 = .04031$
2	$45 \times .4^2 \times .6^8 = .12093$
3	$120 \times .4^3 \times .6^7 = .21499$
4	$210 \times .4^4 \times .6^6 = .25082$
5	$252 \times .4^5 \times .6^5 = .20066$
6	$210 \times .4^6 \times .6^4 = .11148$
7	$120 \times .4^7 \times .6^3 = .04247$
8	$45 \times .4^8 \times .6^2 = .01062$
9	$10 \times .4^9 \times .6^1 = .00157$
10	$1 \times .4^{10} \times .6^0 = .00010$

Figure 9–1 shows a graph of the probability function. We will show later that as n becomes large, the binomial probability function can be approximately represented by a bell-shaped curve known as the *normal* curve.

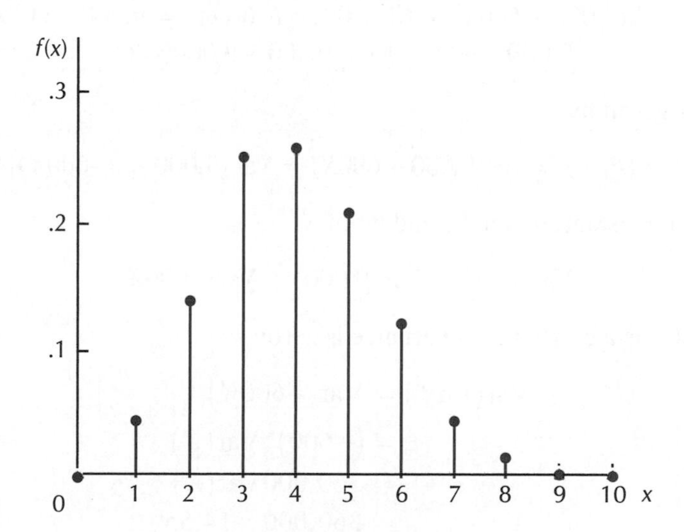

Binomial density function with $n = 10$, $p = .4$

FIGURE 9–1

The probabilities of the various outcomes must add up to 1 if the binomial distribution is a legitimate probability distribution. To prove that they do, we can use a useful theorem called the binomial theorem, which states that

$$(x+y)^n = \sum_{i=0}^{n} \binom{n}{i} x^i y^{n-i}$$

Using this, with $x = p$ and $y = 1 - p$, gives

$$\begin{aligned} \sum_{i=0}^{n} \Pr(x=i) &= \sum_{i=0}^{n} \binom{n}{i} p^i (1-p)^{n-i} \\ &= (p+1-p)^n \\ &= 1^n \\ &= 1 \end{aligned}$$

EXAMPLE An insurance company with 500 customers estimates there is a probability of .03 that each customer will file a claim, in which case they are paid \$600. Each customer pays a premium of \$18. Assuming that each customer is independent, then the number of claims X will be given by a binomial distribution with $n = 500$ and $p = .03$. The expected value of the number of claims is $E(X) = np = 500 \times .03 = 15$; the variance is $\mathrm{Var}(X) = np(1-p) = 500 \times .03 \times .97 = 14.55$.
We can find a formula for the profit:

$$Prof = 18n - 600X = 9{,}000 - 600X$$

The expected value is given by

$$E(Prof) = E(9{,}000 - 600X) = E(9{,}000) - E(600X) = 9{,}000 - 600E(X) = 9{,}000 - 600 \times 15 = 9{,}000 - 9{,}000 = 0$$

The variance is given by

$$\mathrm{Var}(Prof) = \mathrm{Var}(9{,}000 - 600X) = \mathrm{Var}[9{,}000 + (-600X)]$$

Since 9,000 is a constant, it is independent of X:

$$\mathrm{Var}(Prof) = \mathrm{Var}(9{,}000) + \mathrm{Var}(-600X)$$

Also, since 9,000 is a constant, its variance is zero:

$$\begin{aligned} \mathrm{Var}(Prof) &= \mathrm{Var}(-600X) \\ &= (-600)^2\,\mathrm{Var}(X) \\ &= 360{,}000\,\mathrm{Var}(X) \\ &= 360{,}000 \times 14.55 \\ &= 5{,}238{,}000 \end{aligned}$$

Here is a table of probabilities for the number of claims:

i	$\Pr(X = i)$	Cumulative
0	2.431E–07	2.431E–07
1	3.760E–06	4.003E–06
2	.00003	.00003
3	.00015	.00018
4	.00057	.00075
5	.00176	.00251
6	.00448	.00699
7	.00978	.01677
8	.01864	.03541
9	.03152	.06693
10	.04786	.11479
11	.06593	.18072
12	.08310	.26382
13	.09648	.36030
14	.10379	.46409
15	.10401	.56810
16	.09751	.66560
17	.08586	.75146
18	.07125	.82272
19	.05590	.87862
20	.04158	.92020
21	.02940	.94960
22	.01979	.96939
23	.01272	.98212
24	.00782	.98994
25	.00461	.99454

We will not extend the table past this point because the chance of there being more than 25 claims becomes vanishingly small. This is very fortunate for the insurance company, since it could be ruined if too many people filed claims. Insurance companies rely on a careful study of probability to make sure that the risk they are taking is reasonably small.

However, the binomial distribution can only be used if the trials are independent. In this case, each customer counts as a trial, so we have to assume that the probability that one customer files a claim is independent of the probabilities of all of the other customers filing claims. This assumption is valid for many types of insurance, but it clearly is not valid if there is a single disastrous event that would cause many people to file a claim at the same time. As a result, it is much harder to get insurance against big events such as earthquakes than it is to get insurance against events that only affect one customer at a time.

In reality, calculating the probabilities associated with insurance is far more complicated, because there are many variables that are considered. An *actuary* is a professional whose job is to study this area.

BINOMIAL DISTRIBUTION SUMMARY

Suppose an experiment is repeated n times, where p is the probability of success on each trial. Assume each trial is independent of the others. Let X be the number of successes on the n trials. Then X has a binomial distribution.

$$\Pr(X = i) = \binom{n}{i} p^i (1 - p)^{n-i}$$
$$E(X) = np$$
$$\operatorname{Var}(X) = np(1 - p)$$

EXERCISES

1. Suppose that 1,000 meteorites hit the earth each year. What is the probability that the town of Wethersfield, Connecticut, will be struck by two meteorites in 11 years? (See Chapter 5, Exercise 11.)

2. What is the probability of getting three primes in five rolls of a die?

3. In tossing a fair coin, what is the probability of getting at least four heads in five tosses?

4. Pennsylvania has a daily lottery. A three-digit number is chosen every night. What is the probability of getting a number less than 100 more than five times in one week?

5. You're hunting Moby Dick. Each day you sent out one small boat with harpooners from your ship. (You never catch Moby Dick, just because Moby Dick *is* Moby Dick.) The probability is 2/3 that the small boat will be sunk on any particular day. You plan to hunt Moby Dick for four days. What is the probability that you will lose three or more small boats?

6. Assume that you've been given a 100-question true/false exam on a subject that you know nothing about. If you guess randomly, what is the probability of getting at least 75 answers correct?

7. How many times must you toss a fair coin for the probability to be greater than 1/2 that you will get two heads?

8. Assume that 10 percent of the population is left-handed. If three people are chosen at random, what is the probability that at least one will be left-handed?

9. What is the probability that two of the next three presidents of the United States will have been born on a Sunday?

10. Assume that 2/5 of the population have O+ blood type. If you randomly choose six people, what is the probability that four of them are O+?

11. Suppose X_1 has a binomial distribution with parameters n_1 and p, and X_2 has a binomial distribution with parameters n_2 and p. Show that $X_1 + X_2$ has a binomial distribution with parameters $(n_1 + n_2)$ and p.

12. Assume that 45 percent of the Smiths in the world are women. If you randomly run into three Smith siblings, what is the probability that at least two are sisters?

13. Suppose you are running an insurance company. You have N customers. There is a probability $p = .05$ that any particular customer will file a claim in a year, in which case you have to pay $C = \$1{,}000$. You collect a premium of $50 per year from each customer.

    (a) What is the expected value for your profits each year?

    (b) Suppose you have $N = 20$ customers. What is the probability that your profits will be $2,000? What is the probability they will be $1,000? 0? –$1,000? –$2,000? –$3,000?

    (c) Repeat the above calculations for $N = 50$ and $N = 100$.

14. Consider the total profits of the insurance company in the preceding question over a ten-year period. What is the expected value of total profit? Calculate the probability that the total profit will equal these values: $2,000, $1,000, 0, –$1,000, –$2,000?, –$3,000, for three different values of N: $N = 20$, $N = 50$, $N = 100$.

15. Suppose you are running an airline company. You know that there is a probability $p = .07$ that a customer with a reservation will not show up for a particular flight. The plane can hold 200 people. You would like the plane to be full, but you know that if you take only 200 reservations there will probably be some empty seats. So you decide to overbook—that is, take more reservations than you have room for, and then hope that fewer than 201 people will show up. Suppose you accept R reservations. Let X be the number of people who actually show up at the plane. What should R be so that the chance of an overflow crowd is less than 5 percent?

☆ 16. Calculate the mean and the variance of a binomial random variable using the binomial probability function.

Other Discrete Distributions

There are several other important special types of discrete random-variable distribution.

The Poisson Distribution

Let X be the number of telephone calls that an office receives in a given hour. What kind of distribution would X have?

One way of thinking of it would be to list all of the people that would be likely to call (clients, relatives of employees, and so forth). Suppose that there are n of them. They wouldn't all be likely to call at once. Some would call during that hour, but most wouldn't. The easiest way to model the situation would be to say that all of the n people have the same probability p of calling during that hour.

We've just described a binomial distribution. Unfortunately, it's a huge binomial distribution. The formula

$$f(k) = \binom{n}{k} p^k (1-p)^{n-k}$$

becomes very hard to calculate for large n, even for computers.

We do know that the expected value of X should be around np. Let's try to use this number in our calculations. Typically n will be huge and p very small, but np is just right, at least in terms of calculation. We'll let $\lambda = np$ (where λ is the Greek letter lambda).

Then

$$f(k) = \frac{n!}{k!(n-k)!}p^k(1-p)^{n-k}$$
$$= \frac{n(n-1)(n-2)\cdots(n-(k-1))}{k!}p^k(1-p)^{n-k}$$
$$= (np)((n-1)p)((n-2)p)\cdots((n-(k-1))p)\frac{(1-p)^{n-k}}{k!}$$
$$= np(np-p)(np-2p)\cdots(np-(k-1)p)\frac{(1-p)^{n-k}}{k!}$$

If p is very small, this will be close to

$$\underbrace{(np)(np)\cdots(np)}_{k\text{ times}}\frac{(1-p)^{n-k}}{k!} = \frac{\lambda^k(1-p)^{n-k}}{k!}$$

We've simplified most of the calculation, but we still have to deal with the $(1-p)^{n-k}$ part. Using some calculus (something called l'Hôpital's rule) you can show that as n gets larger and larger, $(1-p)^{n-k} = \left(1-\frac{\lambda}{n}\right)^{n-k}$ approaches $e^{-\lambda}$. So, our formula becomes

$$f(k) = \frac{e^{-\lambda}\lambda^k}{k!}$$

We derived this formula as an approximation, but it seems to model situations like the office and its phone calls rather well. Well enough to have its own name: it's called the probability function for the *Poisson distribution*. X is called a *Poisson random variable* with parameter λ.

If the average number of calls that the office gets per hour is 5, we can let $\lambda = 5$ to get the following probabilities:

k	**Pr($X = k$) (Probability of Getting Exactly k Phone Calls)**
0	.006
1	.033
2	.084
3	.140
4	.175
5	.175
6	.146
7	.104
8	.065
9	.036
10	.018
11	.008
12	.003

Figure 10–1 shows a graph of a Poisson density function.

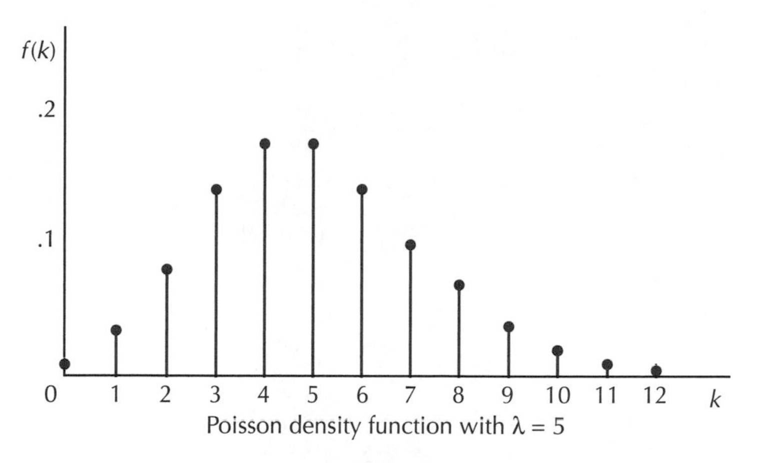

FIGURE 10–1

Note that, in theory, there is an infinite number of possible values for X, but the probability that X will equal k becomes very small as k becomes large.

For another example, suppose we have 500 students each of whom has a probability of .00002 of cutting a little finger on his or her test paper during finals. The calculation of the probability of i successes using the binomial density function becomes unmanageable. If we let $\lambda = np$, then the binomial density function can be approximated by the Poisson distribution:

$$\Pr(X = i) = e^{-\lambda}\frac{\lambda^i}{i!}$$

We need to show that this is a legitimate probability distribution with all of the probabilities adding up to 1 (see Exercise 20).

In the example given above, $\lambda = np = 500 \times (.00002) = .01$, so the probability of two people getting paper cuts from their finals is

$$e^{-.01}(.01)^2(1/2) = 4.95 \times 10^{-5}$$

Other examples where a Poisson distribution is applicable include:

- the number of novas in our galaxy in a given decade
- the number of movies to gross over 25 million dollars in a year
- the number of Ph.D. students who don't finish their dissertations on time
- the number of people who have bought this book who bought it in New York City

The expectation of a random variable with a Poisson distribution can be found in this way:

$$\begin{aligned} E(X) &= \sum_{i=0}^{\infty} ie^{-\lambda}\frac{\lambda^i}{i!} \\ &= \sum_{i=1}^{\infty} e^{-\lambda} i\frac{\lambda^i}{i!} \\ &= \sum_{i=1}^{\infty} e^{-\lambda}\frac{\lambda^i}{(i-1)!} \end{aligned}$$

Let $j = i - 1$. Then

$$\begin{aligned} E(X) &= \sum_{j=0}^{\infty} e^{-\lambda}\frac{\lambda^{j+1}}{j!} \\ &= \lambda\sum_{j=0}^{\infty} e^{-\lambda}\frac{\lambda^j}{j!} \\ &= \lambda \end{aligned}$$

The last step is true because the probabilities sum to 1. This agrees with our intuition, if we interpret λ as np. The variance can be found as follows:

$$\begin{aligned} E(X^2) &= \sum_{i=0}^{\infty} i^2 e^{-\lambda}\frac{\lambda^i}{i!} \\ &= \sum_{i=0}^{\infty} (i^2 - i)e^{-\lambda}\frac{\lambda^i}{i!} + \sum_{i=0}^{\infty} ie^{-\lambda}\frac{\lambda^i}{i!} \end{aligned}$$

The second sum equals λ, as we have just shown above. Now

$$\begin{aligned} \sum_{i=0}^{\infty} (i^2 - i)e^{-\lambda}\frac{\lambda^i}{i!} &= \sum_{i=2}^{\infty} i(i-1)e^{-\lambda}\frac{\lambda^i}{i!} \\ &= \sum_{i=2}^{\infty} e^{-\lambda}\frac{\lambda^i}{(i-2)!} \end{aligned}$$

Let $j = i - 2$. Then

$$\begin{aligned} \sum_{i=2}^{\infty} e^{-\lambda}\frac{\lambda^i}{(i-2)!} &= \sum_{j=0}^{\infty} e^{-\lambda}\frac{\lambda^{j+2}}{j!} \\ &= \lambda^2 \end{aligned}$$

Again, this is true because the probabilities sum to 1. Thus:

$$\begin{aligned} E(X^2) &= \lambda^2 + \lambda \\ \text{Var}(X) &= (\lambda^2 + \lambda) - \lambda^2 = \lambda \end{aligned}$$

Therefore, the variance is λ. The Poisson distribution has the very peculiar property that its expectation is equal to its variance.

The Geometric and Negative Binomial Distributions

Suppose that we still have the same unfair coin we had when we discussed the binomial distribution. In that case, we fixed the number of tosses and counted the number of heads that resulted. This time we're going to fix the number of heads that we want and then we're going to count the number of tosses until we get that number of heads. More precisely, we want to know what the probability is that it will take n tosses to get i heads (assuming again that the probability of getting a head on one toss is p).

First, note that the last toss must be a head. Otherwise, we would have had i heads already, so we wouldn't have had to make that last toss. Then there must be $i - 1$ heads in the first $n - 1$ tosses, in any combination. The probability of this happening can be found from the binomial distribution formula

$$\binom{n-1}{i-1} p^{i-1}(1-p)^{n-i}$$

The probability of getting i heads in the n tosses is exactly p times this last expression (since that is the probability that toss number n will be a head). Therefore, this probability is

$$\binom{n-1}{i-1} p^{i}(1-p)^{n-1}$$

If $i = 1$, this expression reduces to

$$p(1-p)^{n-1}$$

In general, a random variable X representing the number of independent trials necessary to obtain i successes, where each trial has a probability $p(p > 0)$ of success, is said to have a *negative binomial distribution*. The corresponding probability is

$$\Pr(X = n) = \binom{n-1}{i-1} p^{i}(1-p)^{n-1}$$

If $i = 1$, the distribution is called a *geometric distribution*. The proof that the sum of the probabilities for a geometric distribution is 1 is very simple. Since

$$\sum_{i=0}^{\infty} x^i = \frac{1}{1-x} \qquad \text{for } |x| < 1,$$

$$\begin{aligned}
\sum_{n=1}^{\infty} p(1-p)^{n-1} &= p\sum_{n=1}^{\infty}(1-p)^{n-1} \\
&= p\left(\frac{1}{1-(1-p)}\right) \\
&= \frac{p}{p} \\
&= 1
\end{aligned}$$

The expectation of a negative binomial random variable is i/p. For example, if the probability of success is 1/3, then you can expect to make 30 attempts before attaining 10 successes. The variance of a negative binomial random variable is $i(1-p)/p^2$. For the geometric distribution, the mean is $1/p$ and the variance is $(1-p)/p^2$.

The Hypergeometric Distribution

Suppose you are given a box containing 10 pieces of candy, all of which look alike on the outside. Suppose further that you know that 8 of the pieces are marshmallow-filled (which you love) and 2 contain almonds (which you despise). If you take 5 pieces out of the box, what is the probability that you will get exactly 3 marshmallow-filled pieces?

This is a case of the probability being the number of "successes" divided by the number of possible outcomes. First, we need to know the total number of ways of picking the 5 pieces from the box of 10. We can use the combinations formula for this:

$$\binom{10}{5} = \frac{10!}{5!5!} = 252$$

(See Chapter 6.)

Now we need to calculate how many of these possibilities have exactly 3 marshmallow-filled pieces. Since there are 8 possible marshmallow-filled candies to choose from, there are $\binom{8}{3}$ ways of picking the 3 marshmallow candies. We need to multiply this by the number of possible ways of picking the two almond-flavored candies from the two almond candies in the box, and this will be just $\binom{2}{2}$ (which is, of course, equal to 1).

Therefore, the probability of picking exactly 3 marshmallow candies is

$$\frac{\binom{8}{3}\binom{2}{2}}{\binom{10}{5}} = \frac{56}{252} = \frac{2}{9}$$

Let's generalize. The 10 pieces of candy become N objects. The 8 marshmallow candies become the M objects in the desired state, with the almond candies becoming the $N-M$ objects in the undesired state. (Note that $M < N$). The 5 pieces that you take become n trials (selections without replacement from the N objects.) The 3 pieces of marshmallow candy become the i desired objects selected. (Note that $i < M$, the total number of desired objects, and that $i < n$, the total number of objects selected.) Letting X be the random variable standing for the number of desired objects selected, we have

$$\Pr(X = i) = \frac{\binom{M}{i}\binom{N-M}{n-i}}{\binom{N}{n}}$$

(This holds for $0 \le i \le n$, and $i \le M$; $\Pr(X = i) = 0$ otherwise). The random variable X is said to have the *hypergeometric distribution* with parameters $n, N,$ and M.

Examples of hypergeometric distributions include:

- the number of defective merchandise items in a random sample of a large shipment
- the number of persons you will meet in your lifetime with the name Fred
- the number of pennies drawn out of a jar filled with M pennies and $(N - M)$ nickels. If only one draw is made, then $n = 1$, and the probability of getting a penny is M/N, as is to be expected.

Another important application comes when you are conducting an opinion poll. The people who are asked questions during the poll are analogous to the candies that are chosen from the box, and the entire population of people is analogous to the whole box of candies. When we conduct an opinion poll we need to know how likely it is that the proportion of people with a particular opinion in the sample is the same as the proportion of people with that opinion in the population. We will discuss this question in Chapter 19.

We can figure out the expectation of the hypergeometric distribution by intuition. Suppose there are 1,000 marbles in a vat (600 of them red, 400 blue). We will randomly select 10 of these; how many red marbles would you expect? Since 60% of the marbles in the population are red, we would expect 60 percent of the marbles in the sample to be red, so

$$E(X) = 10\left(\frac{600}{1{,}000}\right) = 6$$

(Again the usual warning applies: the expected value does not have to equal the actual value in any one particular selection—it is the average value if you select these marbles many times from this same situation).

In general, let $p = M/N$ (that is, the proportion of desired objects in the population). The expectation of the hypergeometric random variable X comes from this formula:

$$E(X) = n\left(\frac{M}{N}\right) = np$$

The variance is given by this formula:

$$n\left(\frac{M}{N}\right)\left(1 - \frac{M}{N}\right)\left(\frac{N-n}{N-1}\right) = np(1-p)\left(\frac{N-n}{N-1}\right)$$

This formula should look a bit familiar. Suppose we draw n candies from the box, only this time we put each candy back after we draw it. We'll call it a success if we draw the kind of candy we want. The probability of success is therefore M/N. If X is the number of successes in the n draws, we know that X has a binomial distribution with parameters n and $p = M/N$, and we know that its variance is $np(1 - p)$. That is the same as the variance of the hypergeometric distribution when N is very large, since $\frac{N-n}{N-1}$ approaches 1 as N becomes large. This fact illustrates the difference between the two distributions. With the binomial distribution, each draw is independent of the others, because you always put the candy you draw back in the box. With the hypergeometric distribution, you're not putting the candy back, so each draw is not independent of the others. The probability of success of each draw depends on how many of each type of candy are left in the box, which depends in turn on what candies you removed in the preceding draws. However, if the number of candies in the box is very, very large, then removing a few candies is not going to change the probabilities for future draws very much. In that case it doesn't make too much difference whether you draw the candies with replacement (and use the binomial distribution) or draw the candies without replacement (and use the hypergeometric distribution).

EXAMPLE Draw 10 cards from a 52-card deck. Let X equal the number of hearts that you draw. Then X will have a hypergeometric distribution with $N = 52, M = 13$, and $n = 10$. Figure 10–2 shows the probabilities.

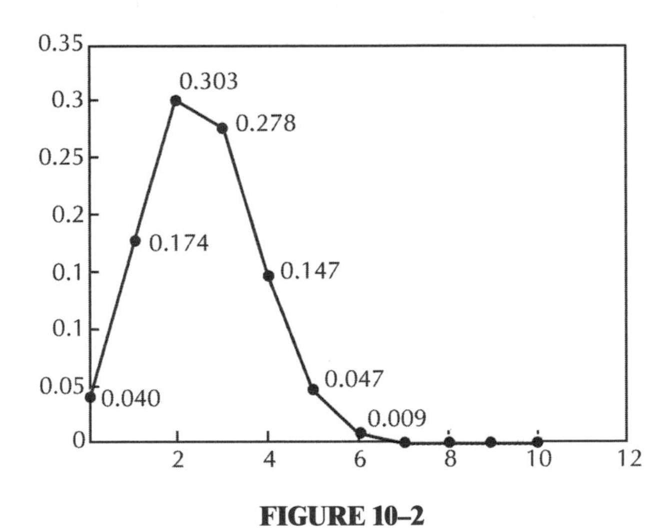

FIGURE 10–2

The expected value is $E(X) = nM/N = 10 \times 13/52 = 2.5$. Note how the probabilities on the graph reach their peak near the expected value.

DISCRETE RANDOM VARIABLE DISTRIBUTION SUMMARY

Poisson Distribution

$$\Pr(X = k) = \frac{e^{-\lambda}\lambda^k}{k!}$$
$$E(X) = \lambda$$
$$\text{Var}(X) = \lambda$$

Negative Binomial Distribution

X is the number of trials needed to obtain i successes.
(p is the probability of success on each independent trial.)

$$\Pr(X = n) = \binom{n-1}{i-1} p^i (1-p)^{n-i}$$
$$E(X) = i/p$$
$$\text{Var}(X) = i(1-p)/p^2$$

Geometric Distribution

This is the same as the negative binomial distribution with $i = 1$.

Hypergeometric Distribution

A population contains M desirable objects and $N - M$ undesirable objects. Select a random sample without replacement of size n, and let X be the number of desirable objects in the sample. Then

$$\Pr(X = i) = \frac{\binom{M}{i}\binom{N-M}{n-i}}{\binom{N}{n}}$$
$$E(X) = n\left(\frac{M}{N}\right)$$
$$\text{Var}(X) = n\left(\frac{M}{N}\right)\left(1 - \frac{M}{N}\right)\left(\frac{N-n}{N-1}\right)$$

EXERCISES

1. Calculate the mean and variance of a hypergeometric random variable with parameters $N = 1{,}000$, $M = 300$, and $n = 25$.

2. Calculate the mean and variance of a negative binomial random variable with parameters $p = 1/3$ and $i = 25$.

3. Calculate the mean and variance of a negative binomial random variable with parameters $p = 1/4$ and $i = 16$.

4. Suppose that you pull 15 balls out of a jar containing 30 white balls and 15 black balls. How many white balls would you expect to pull out on the average?

5. Let X be a random variable representing the number of times that you have to roll two dice until you get ten 11's. Calculate the mean and variance of X.

6. Let X be a random variable representing the number of times the word "platypus" is said on a given day. Assume X has a Poisson distribution with parameter $\lambda = 1/2$. What is $\Pr(X > 1)$?

7. What is the maximum value for $\Pr(X = n)$ if X is a Poisson random variable with parameter $\lambda > O$?

8. If X is a Poisson random variable with parameter $\lambda = 10$, what is $\Pr(1 \leq X \leq 3)$?

9. If X is a geometric random variable with parameter $p = 1/3$, what is $\Pr(X \leq 4)$?

10. If X is the same as in the preceding problem, what is the smallest n such that $\Pr(X \leq n) \geq 1/2$?

11. Fastburgers, Inc. starts to give away free soft drinks in the following fashion. With each purchase it gives you a card containing an X or an O (under ink that has to be rubbed off, so that the card's contents aren't visible), and it gives a small soft drink for five cards containing Xs. If the probability of getting an X on a given card is 1/3, what is the probability of getting your fifth X on your tenth card?

12. Your cruel building superintendent refuses to turn on the heat in your building until it has snowed three times this winter. If the probability that it will snow on a given winter day is 1/5, what is the probability that your building superintendent won't turn the heat on until the 13th day of winter?

13. Given that 20 books in a shipment of 200 books (for a bookstore) contains misprints, and you buy three of them, what is the probability that one of your books will contain a misprint?

14. You're dressing in the dark because of a power failure. You have two black socks and six red socks in a drawer. You pull out three. What is the probability that two are black?

15. Let X be a Poisson random variable with parameter $\lambda = 3$, representing the number of people who use a given dictionary in a given library on a given day. If $F(a)$ is the cumulative distribution function, what is $F(4)$?

16. Suppose that the probability is 1/3 that it will rain on a given day in your neighborhood. What is the probability that it will be four days before it rains next?

17. Suppose that you want to collect eight toy boats (all the same kind) that are included in one out of every three Sweet-Tooth cereal boxes. What is the probability that you'll find the last one in your 20th cereal box?

18. A dictionary has 300 pages. What is the probability that if you look up five words at random, two of them will be on pages with page numbers ending in a zero? (Assume that the two words you look up are on different pages.)

19. In the Gobbler, a new video game, an alien monster roams the screen intending to eat your character. You're armed with a laser pistol. If you have a 1 out of 12 chance of hitting the monster if you fire randomly, what is the probability that it will take you 20 shots to hit the monster?

☆ 20. Show that all of the probabilities for the Poisson distribution add up to 1.

☆ 21. Derive the formula for the mean of a hypergeometric random variable.

☆ 22. Derive the formulas for the mean and variance of a negative binomial random variable.

11

Continuous Random Variables

Let us suppose that we randomly select a name from the phone book and then measure the height of the person selected. If H is the height in feet of the person, we can regard H as a random variable. However, it is different from the other random variables that we have done up to now. Suppose we try to list all of the possible values for H. There are clearly some values that are not possible. For example, H can never be less than 1/4 or greater than 9. However, we'll find that we can't list all of the possible values. The height might be 5 feet, or it might be 5.1 feet, or 5.00001 feet, or 5.000000001 feet. In fact, assuming that we can measure the height with perfect accuracy (oh, well, this is theory—not the real world) there is an infinite number of possible values for the height. A discrete random variable cannot be used in a case like this, where the result can be any number in a particular range. Instead, we need to use a *continuous random variable*.

Examples of continuous random variables include

- the height above the floor at the point where a dart hits a dart board
- the length of time until a light bulb burns out
- the length of time until a radioactive atom decays
- the length of the life of a person

Discrete random variables are easier to understand intuitively. However, continuous random variables are usually easier to handle mathematically. If a discrete distribution has many possible values that are close together, then it can usually be approximated by a continuous distribution.

Continuous Cumulative Distribution Functions

Now we have to figure out how to describe the behavior of continuous random variables. There are many similarities between discrete random variables and continuous random variables, but there are some important differences. We can estimate the probability that the person we select will have height less than 6 feet. Or we could calculate the probability that the person will have a height greater than 50 feet (which is, of course, zero). Therefore, we can define a cumulative distribution function for a continuous random variable, just the same as we did for a discrete random variable. We'll use a capital letter, such as F, to stand for a cumulative distribution function, so we can make the definition

$$F(a) = \Pr(X \leq a)$$

where X is the random variable we are discussing.

A continuous cumulative distribution function satisfies the same requirements that we found for a discrete cumulative distribution function:

1. $F(a)$ is always between 0 and 1.
2. As a becomes very large, $F(a)$ approaches 1.
3. As a becomes very small (approaches minus infinity), $F(a)$ approaches 0.
4. $F(a)$ is never decreasing.

Here are two important practical properties:

If we want to find the probability that a continuous random variable X will be greater than a particular value a, we can use the formula

$$\Pr(X > a) = 1 - \Pr(X < a) = 1 - F(a)$$

(Note: It is *not* necessarily true that $\Pr(X > a) = 1 - \Pr(X < a)$ if X is a discrete random variable.)

If we want to find the probability that X will be between two particular values b and c, we can use the formula

$$\Pr(b < X < c) = F(c) - F(b)$$

Figure 11–1 shows a cumulative distribution function for the heights of a group of people.

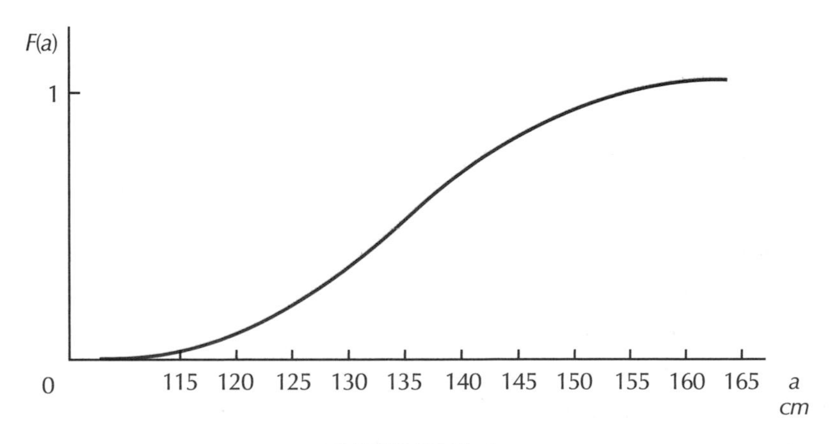

FIGURE 11–1

One example of a continuous random variable is a uniform random variable. That is a random variable that is equally likely to take on any value within a particular interval. For example, let's consider the random variable Y that has an equal chance of taking any value between 0 and 3. Then the probability that Y will be less than 1 is 1/3, the probability that Y will be between 1 and $1\frac{1}{2}$ is 1/6, and so on. If we make a graph of the cumulative distribution function for Y, it looks like the function shown in Figure 11–2.

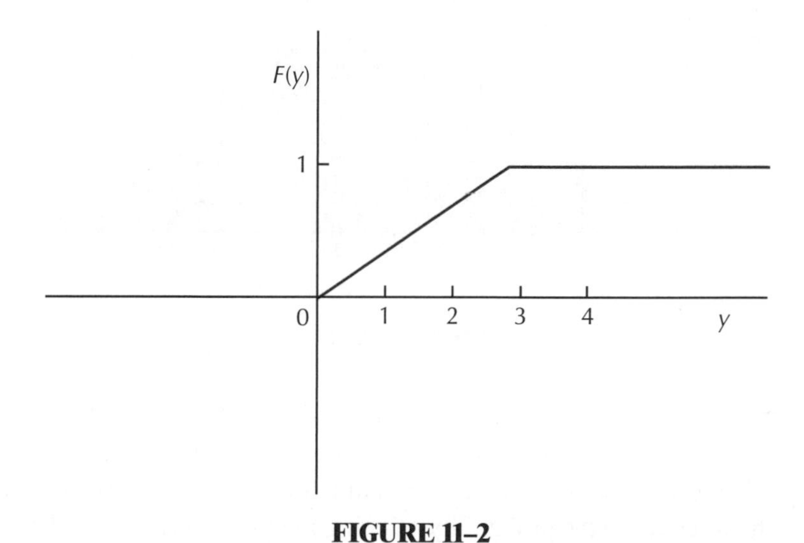

FIGURE 11–2

Continuous Probability Density Functions

Now, let's find out the probability that Y will be exactly equal to 2. Any number from 0 to 3 has an equal chance of being selected, so if we let N be the number of numbers from 0 to 3, then $\Pr(Y = 2) = 1/N$. However, there is an infinite number of numbers between 0 and 3 (for example, 0.01, 0.011, 0.0111, 0.01111, and so on). This means that

$$\Pr(Y = 2) = \frac{1}{\infty} = 0$$

This property holds in general for continuous random variables: The probability that *any* continuous random variable will take on *any* specific precise value is zero!

Therefore, we can't define a density function for a continuous random variable in the same way that we defined the probability function for a discrete random variable. We need to develop a new approach. To start with, remember that there is a connection between the density function for a discrete random variable and the frequency histogram for a sample. So we'll start by drawing a frequency histogram for the weights of people in a particular sample (see Figure 11–3). Note that the height of each bar is not the number of people whose weight equals a particular amount.

Instead, it is the number of people whose weight is between two specified values. For example, the height of the bar between 140 and 145 pounds is the number of people in the sample whose weights are between 140 and 145. We'll call the width of each bar Δx. (In this case, $\Delta x = 5$.)

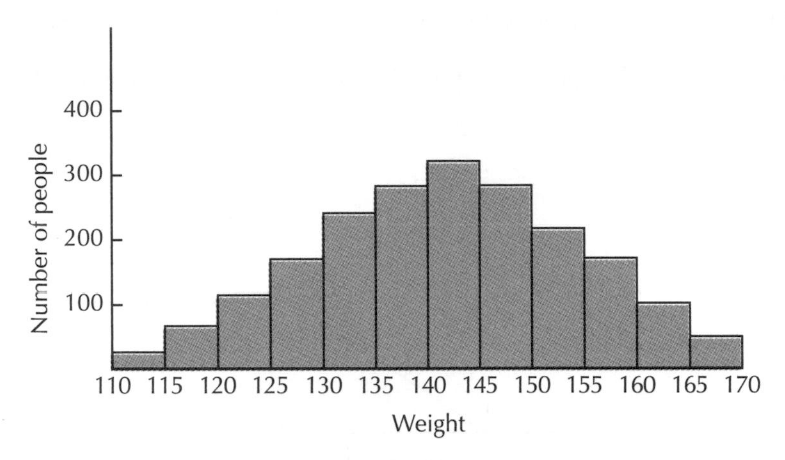

FIGURE 11–3

We can, by analogy, draw an approximate density function such that the height of the function in any given interval is equal to the probability that the random variable will have a value within that interval. It turns out to be more convenient to make the height of each bar equal to the probability of being in that interval divided by Δx, the width of the bar (see Figure 11–4).

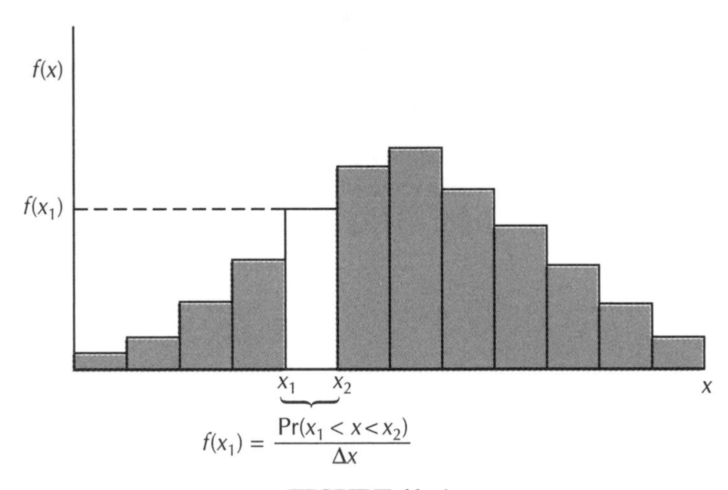

FIGURE 11–4

Then:

$$\text{(height of the bar between } a \text{ and } a + \Delta x = \frac{\Pr(a < X < a + \Delta x)}{\Delta x}$$

And therefore,

$$\Pr(a < X < a + \Delta x) = (\text{height of bar}) \times \Delta x$$

We'll let $f(a)$ stand for the height of the bar from a to $a + \Delta x$, so

$$\Pr(a < X < a + \Delta x) = f(a)\Delta x$$

Suppose we need to know the probability that X will be between two values a and b. We need to add up the heights of all the bars from a to b and then multiply by Δx. However, since $f(x)$ is the height of each bar and Δx is the width, $f(x)\Delta x$ is the area of the bar. Therefore, the probability that X will be between a and b is just equal to the area of all of the bars between a and b (see Figure 11–5).

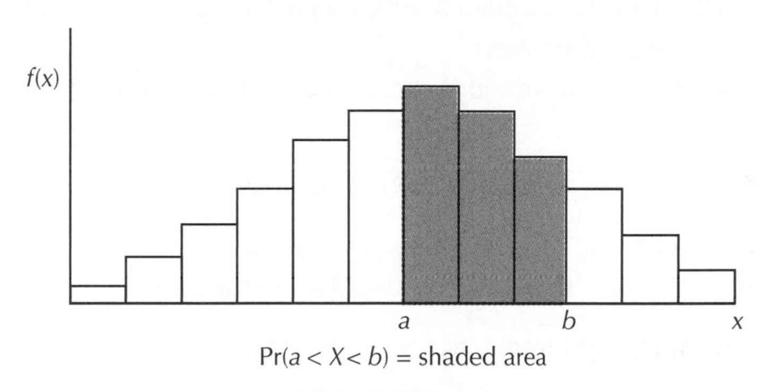

FIGURE 11–5

$$\Pr(a < X < b) = (\text{area of all of the rectangles between } a \text{ and } b)$$

This is the basic defining feature of the density function for a continuous random variable: The area under the function between two values is the probability that the random variable will be between those two values. However, the bar diagram is only an approximate representation of the density function for a continuous random variable. We can get a better approximation of the true nature of the continuous random variable by making the bars narrower and narrower. When the bars become very, very narrow, the density function looks like a smooth curve (see Figure 11–6).

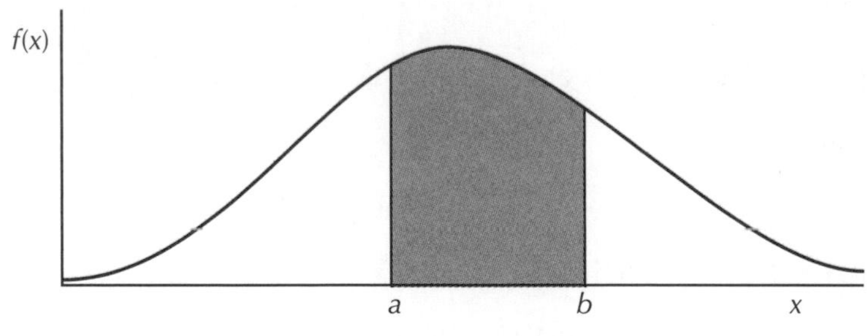

Pr($a < X < b$) = shaded area

FIGURE 11–6

We will make this definition: The function $f(x)$ is a density function for the random variable X if it satisfies the property that the area under the curve $y = f(x)$, to the left of the line $x = b$, to the right of the line $x = a$, and above the x axis, is equal to $\Pr(a < X < b)$. (Remember that capital letters represent random variables and small letters represent ordinary variables.)

It would help to be able to write this area expression in a shorter fashion, so we will just write

$$\text{area under } f(x) \text{ from } a \text{ to } b$$

to mean "the area under the curve $f(x)$ between a and b."

We know that if $F(x)$ is the cumulative distribution function, then

$$\text{area under } f(x) \text{ from } a \text{ to } b = \Pr(a < X < b) = F(b) - F(a)$$

Now, suppose we look at the interval from minus infinity to plus infinity. We know that $F(+\infty) - F(-\infty) = \Pr(-\infty < X < \infty) = 1$, since the value of X must be somewhere between $-\infty$ and $+\infty$. (It doesn't have any other choice.) This means that

$$\text{area under } f(x) \text{ from } -\infty \text{ to } +\infty = 1$$

In other words, the total area under the function $f(x)$ must be equal to 1. If $f(x)$ doesn't have this property, then it can't be a legitimate probability density function. We can show that this condition is met for the density function of the uniform variable Y (see Figure 11–7).

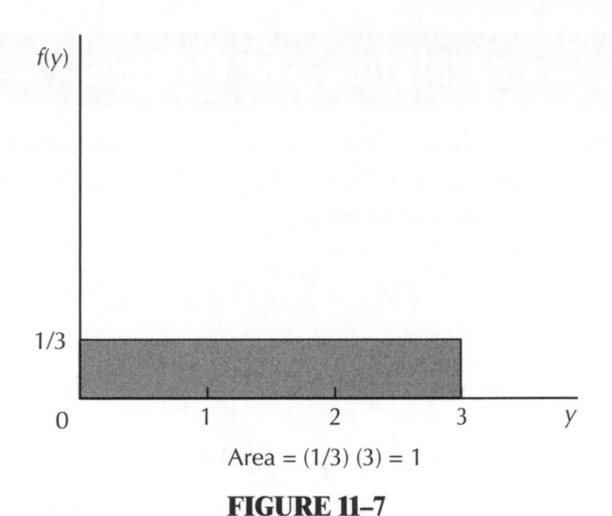

FIGURE 11–7

We would like an even shorter way to write the area under the function $f(x)$ between a and b. We'll symbolize area with a squiggle, like this:

$$\text{area under } f(x) \text{ from } a \text{ to } b = \int_a^b f(x)\,dx$$

In this notation, the function whose area is to be found [in this case $f(x)$] is placed in the middle, surrounded by the squiggle on the left and the dx on the right. The left-hand limit of the area is written at the bottom of the squiggle, and the right-hand limit is written at the top of the squiggle.

Therefore, by definition, these four quantities are all equal:

$$\text{area under } f(x) \text{ from } a \text{ to } b$$

$$\int_a^b f(x)\,dx$$

$$F(b) - F(a)$$

$$\Pr(a < X < b)$$

Of course, if you know calculus, you will realize that $\int$ is the symbol for an *integral* and you will know how to calculate explicit values for the area if you are given a specific form for the function $f(x)$. However, many of the density functions that we use in statistics (such as the normal density function) cannot be integrated by any easy method, so in that respect the people who know calculus don't have much of an unfair advantage over the people who don't. They have to use a computer or look up the values in the tables just like everybody else.

Expectation and Variance

We would like to be able to calculate the expectation and variance for a continuous random variable, just as we did for a discrete random variable. For discrete random variables, we defined the expectation like this:

$$E(Y) = \sum_{i=1}^{n} y_i f(y_i)$$

So we will make an analogous definition for continuous random variables:

$$E(X) = \int_{-\infty}^{\infty} x f(x) dx$$

This expression says to set up the function $x \times f(x)$, then find the area under that function from $x = -\infty$ to $x = +\infty$. For example, let's say that X is a uniform random variable that can have any value from 0 to 5. Then $f(x) = 1/5$ if $0 < x < 5$ and $f(x) = 0$ everywhere else. So we need to find the area under the function $xf(x) = x/5$ from $x = 0$ to $x = 5$. If we make a graph of that function, we can see that it is just a triangle (see Figure 11–8).

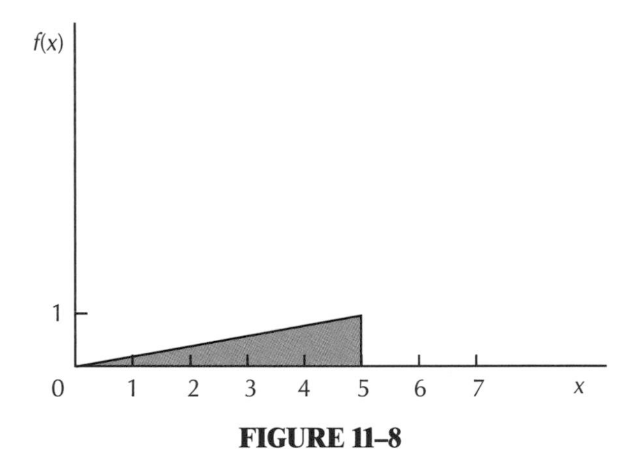

FIGURE 11–8

The area of the triangle is $(1/2)(5)(1) = 2\frac{1}{2}$, so $E(X) = 2\frac{1}{2}$. Of course, you could have figured that out on your own. In general, the expectation of any uniform random variable will be the point halfway between the two boundaries.

You'll note that the function $xf(x)$ is negative whenever x is negative. So how do you calculate the area? What you do is just subtract the total area enclosed by the curve below the horizontal axis from its total area above the axis. For example, suppose T is a uniform continuous random variable with possible values between –1 and 5. Then the value of $f(t)$ is 1/6 if t is between –1 and 5, and 0 otherwise. The function $t\,f(t)$ is shown in Figure 11–9. The total positive area is 25/12 and the total negative area is 1/12. If you subtract, you can see that $E(T) = 24/12 = 2$.

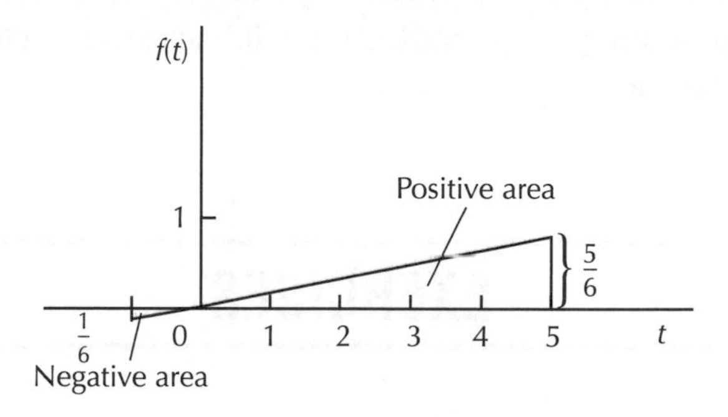

FIGURE 11–9

If you know calculus, then you're used to calculating these types of integrals. If not, and you trust us, then you can just take our word when we tell you what the expectation of a continuous random variable is.

The variance of a continuous random variable is defined in exactly the same way as is the variance of a discrete random variable:

$$\text{Var}(X) = E\left\{[X - E(X)]^2\right\} = E(X^2) - [E(X)]^2$$

For example, here is how to find the variance of a uniform random variable that can take on values between 0 and a:

$$\begin{aligned}
E(X) &= a/2 \\
\text{density function} &= f(x) = 1/a \\
E(X^2) &= \int_0^a x^2 (1/a)\,dx \\
&= \left(\frac{1}{a}\right)\left(\frac{1}{3}\right)x^3\Big|_0^a = \frac{a^2}{3} \\
\text{Var}(X) = E(X^2) - [E(X)]^2 &= \frac{a^2}{3} - \frac{a^2}{4} = \frac{a^2}{12}
\end{aligned}$$

NOTE TO CHAPTER 11

It is also possible for a random variable to have a mixed distribution—that is, a distribution that is part discrete and part continuous. For example, suppose a scale can represent only weights up to 250 pounds. If the true weight is greater than 250, then the scale will display the value 250. If X represents the weight as measured by this scale, then there is a certain probability p that X will exactly equal 250. The value of X will be less than 250 with probability $1 - p$, in which case its value can be characterized by a continuous density function.

EXERCISES

1. Given a person chosen at random, which of the following data constitute continuous random variables: height, social security number, weight, temperature?

2. Under what circumstances can the cumulative distribution function of a continuous random variable have a maximum? a minimum?

3. $f(x) = 1/2$ when $-1 \leq x \leq 1$ and 0 otherwise. Draw a graph of $F(a)$.

4. Suppose that you are throwing a dart at a circular dart board of radius 30 centimeters. Let R be a random variable representing the distance from the center to the point where the dart strikes. What is the cumulative distribution function for R?

☆ 5. Derive the density function for the random variable in the preceding problem.

☆ 6. Let $f(x) = cx^n$ for $0 < x < 1$, and 0 otherwise. (Assume $n > 0$.) What must be the value of c if f is to be a probability density function?

☆ 7. Let $f(x) = 1/x$ if $x \geq -1$, and 0 otherwise. What is $F(a)$ for $a = 0$, $a = 1/2$, and $a = 10$?

☆ 8. If $F(a) = 1 - e^{-a}$ if $a \geq 0$, and 0 otherwise, and $f(0) = 0$, what is $f(x)$?

9. Why can't the function

$$g(a) = \begin{cases} 0 & \text{if } a \le 0 \\ a & \text{if } 0 \le a \le 2 \\ 4 - a & \text{if } 2 \le a \le 3 \\ 1 & \text{if } a \ge 3 \end{cases}$$

be a cumulative distribution function?

☆ 10. If $f(x) = (1/\pi)[1/(1 + x^2)]$, what is $\Pr(-1 \le X \le 1)$?

☆ 11. Let $f(x) = |\sin x|$ if $-\pi/3 \le X \le \pi/3$, and 0 otherwise. What is $F(a)$?

☆ 12. If $f(x) = (3/4)(1 - x^2)$ if $-1 \le x \le 1$, and 0 otherwise, what is $F(x)$?

☆ 13. Calculate the variance of the random variable Y that is a uniform random variable between 0 and 3.

☆ 14. Calculate the variance of the random variable that has a uniform distribution between two fixed numbers a and b.

☆ 15. If $f(x)$ is the density function for X, and if $f(c - x) = f(c + x)$ for all values of x, then X is said to be *symmetric* about the point $x = c$. Show that $E(X) = c$.

☆ 16. If $F(x)$ is the cumulative distribution function of X, and if X is symmetric about c, then show that $F(c - x) = 1 - F(c + x)$.

☆ 17. Let X be a continuous random variable with density function $f(x) = 2e^{-x}$ if $x \ge a$. What is a?

☆ 18. Let X be a continuous random variable with density function $f(x) = x^{-2}$ if $x > 1$. What is $\Pr(5 < X < 6)$?

☆ 19. Let X be a continuous random variable with density function $f(x) = 0$ if $x < 0$, $f(x) = x$ if $0 < x < 1$, and $f(x) = x^{-3}$ if $x \ge 1$. What is the cumulative distribution function?

☆ 20. Let X be a continuous random variable with cumulative distribution function $F(a) = 0$ if $a \le -1$, $F(a) = 1/2(a + 1)(a + 2)$ if $-1 < a \le 0$, and $F(a) = 1$ if $a > 0$. What is the density function?

☆ 21. Let X be a continuous random variable with cumulative distribution function $F(a) = 0$ if $a \le 0$, $F(a) = a$ if $0 < a \le 1$, and $F(a) = 1$ if $1 \le a$. Where is $f(x)$ nonzero?

22. Suppose that a certain train always pulls into a certain station between 1 P.M. and 1:05 P.M. Let X be a random variable representing the number of minutes after 1 P.M. that the train arrives. Suppose that the density function of X is equal to c times the number of seconds after noon that the train would arrive if $X = x$. What is the value of c? What is the cumulative distribution function for X?

12

The Normal Distribution

The Normal Density Function

Suppose you make a graph of the probabilities of the number of heads you will expect to see if you repeatedly flip a coin 15 times (see Figure 12–1). Or suppose you select 1,000 people off the street and make a frequency diagram of their heights (see Figure 12–2).

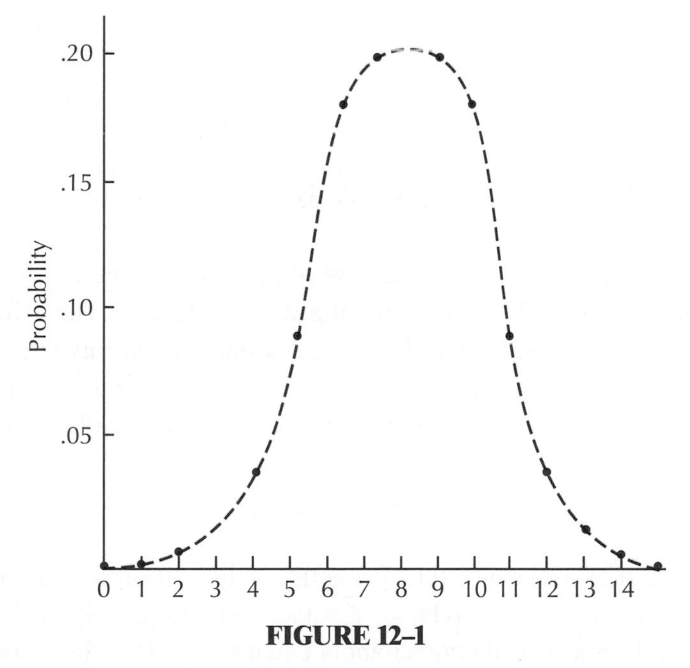

FIGURE 12–1

These graphs look similar. Their bell-shaped curve is the most important density function in probability and statistics. A random variable whose density function looks like this is called a *normal* random variable.

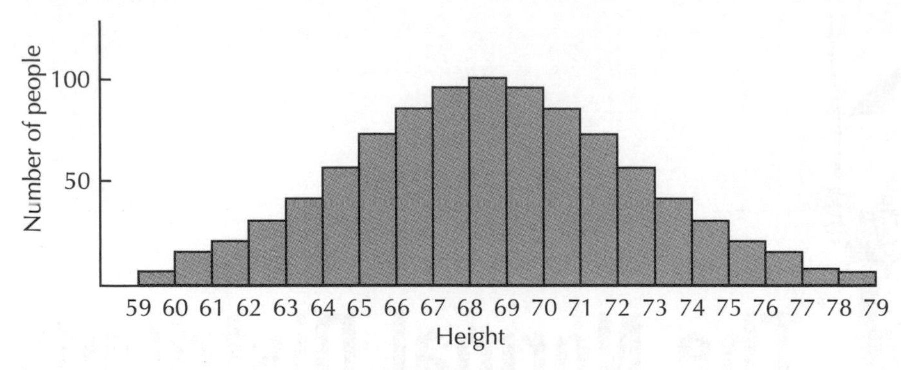

FIGURE 12–2

Let's see if we can make up a mathematical function that has this kind of shape. The function $f(x) = e^{-(1/2)x^2}$ looks right (see Figure 12–3).

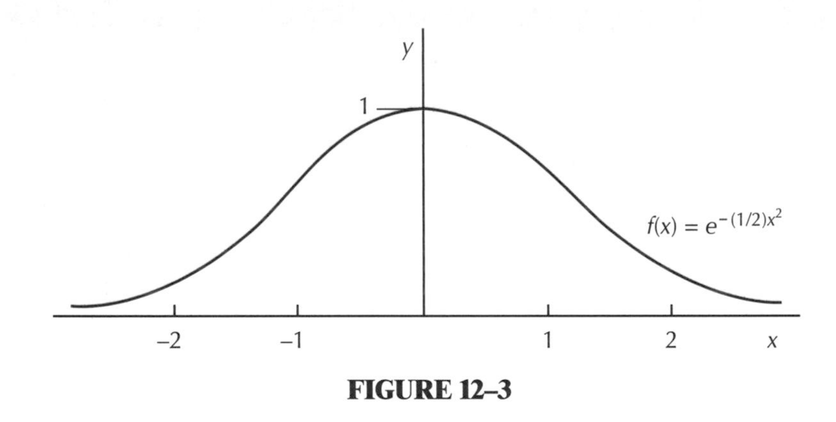

FIGURE 12–3

We could use any positive number as the base of this function, but it is most convenient to use the special number symbolized by the letter e. The value of e is about 2.71828. This function is very close to the normal probability curve, but we need to make a couple of adjustments. First, we want to be able to adjust the peak and shape of the distribution, so we need to put in two parameters (called μ and σ):

$$f(x) = e^{-1/2\left(\frac{x-\mu}{\sigma}\right)^2}$$

This function can be a true probability function only if the area under it is 1. The area under the function turns out to be $\sqrt{2\pi}\sigma$ (see Exercise 27). Pi (π) is a Greek letter used to stand for a special number about equal to 3.14159. So we need to divide by $\sqrt{2\pi}\sigma$ to make the area equal to 1. Therefore, the density function for a normal random variable is defined as follows:

$$f(x) = \frac{1}{\sqrt{2\pi}\sigma} e^{-1/2\left(\frac{x-\mu}{\sigma}\right)^2}$$

Note that it is a continuous, rather than a discrete, distribution. The normal distribution is important both because we'll be interested in a lot of random variables that have normal distributions and because the normal distribution can be used as an approximation for many other distributions, such as the binomial distribution.

Some important examples of random variables with approximately normal distributions are these:

- the IQ of a randomly selected person
- the result of a measurement of a physical quantity, such as the molecular weight of a chemical
- the total that appears if you toss many dice
- scores on an aptitude test
- the velocities of molecules in a gas

The normal density function often applies to quantities in situations where extreme values are less likely to occur. Also, we will show that if you add together a large number of independent random variables with identical distributions, the resulting random variable will have a normal distribution.

The parameter μ defines where the center, or peak, of the distribution will be. In fact, as you've probably guessed, μ turns out to be equal to the mean of the distribution. You can tell that just by looking at the form of the density function, since the function is symmetric about $x = \mu$. Figure 12–4 shows a density function for a typical normal distribution.

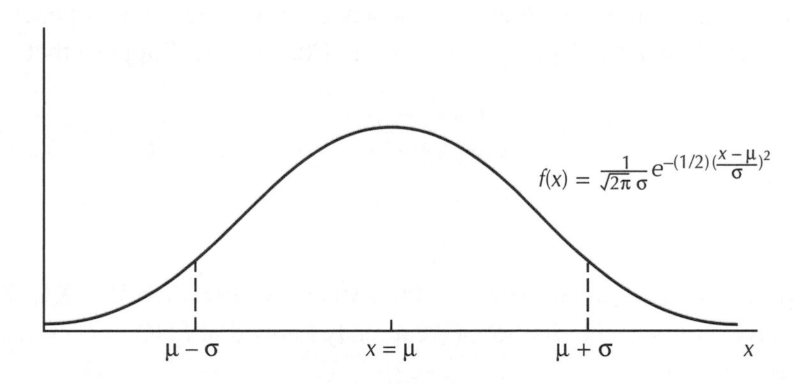

FIGURE 12–4

It turns out that σ^2 is the variance (see Exercise 26). By adjusting the value of σ^2, you can determine whether the distribution will be very spread out or whether most of the probability will be concentrated near the peak. Figure 12–5 shows four different normal density functions that have different values of σ^2.

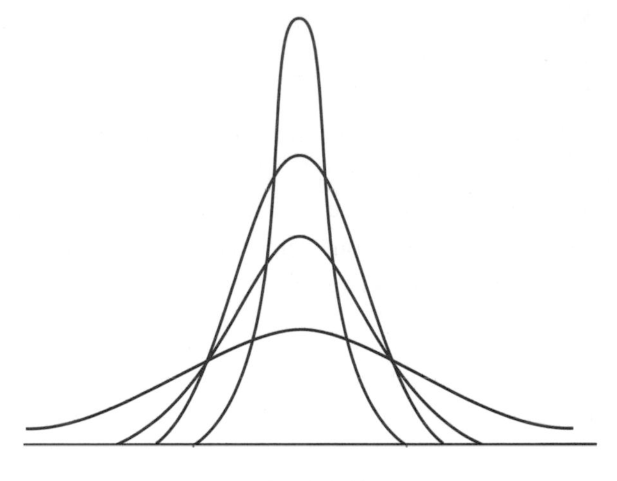

FIGURE 12–5

Specifically, the value of σ tells you how far away from the mean you have to go until you reach the *points of inflection*. These are the points where the curve stops facing downward and begins facing upward. The normal distribution has two points of inflection, one at $\mu - \sigma$ and the other at $\mu + \sigma$ (see Figure 12–4). (Calculus savvy readers will know that the points of inflection occur where the second derivation is zero; see Exercise 30.)

An important property of a normal random variable is the addition property. If X is a normal random variable with mean μ and variance σ^2, and $Y = aX + b$, where a and b are two constants, then Y has a normal distribution with mean $a\mu + b$ and variance $a^2\sigma^2$.

Also, suppose X and Y are two independent random variables with normal distributions. (Two random variables are independent if they don't affect each other. We'll define exactly what independence means in Chapter 14.) Suppose that

$$E(X) = \mu_x,$$
$$\mathrm{Var}(X) = \sigma_x^2,$$
$$E(Y) = \mu_y$$
$$\mathrm{Var}(Y) = \sigma_y^2.$$

If we form a new random variable by adding these two together, $V = X + Y$, then V will also have a normal distribution. (We already know that $E(V) = \mu_x + \mu_y$, and $\mathrm{Var}(V) = \sigma_x^2 + \sigma_y^2$.)

For example, suppose you decide to enter the hamburger business by opening restaurants at two different locations. The number of hamburgers that you sell each day at the downtown location is given by a normal random variable with mean 200 and variance 1,600. The number of hamburgers sold at the suburban location has a normal distribution with mean 100 and variance 400. Then the total number of hamburgers that you will sell at both restaurants has a normal distribution with mean

300 and variance 2,000 (assuming that the number of hamburgers sold at the two restaurants are independent from each other).

The Standard Normal Density Function

Suppose you with to know the probability that you will sell more than 200 hamburgers at your downtown location (with sales given by a normal distribution with $\mu = 200$, $\sigma = 40, \sigma^2 = 1{,}600$). We need to find the area under this normal curve to the right of the value 200.

From symmetry we can see that $\Pr(X > 200) = .5$, since the area to the left of the mean is a mirror image of the area to the right of the mean. In general, for any normal random variable X with mean μ, it is true that $\Pr(X > \mu) = .5$.

That is the only easy problem. Suppose now we need to find the probability that you will sell more than 230 hamburgers. We need to find the area under the density function curve to the right of the value 230 (see Figure 12–6). Or, to put it another way, we need to find this integral:

$$\int_{230}^{\infty} \left(\frac{1}{\sigma\sqrt{2\pi}} \right) e^{-(x-\mu)^2/(2\sigma^2)}$$

Unfortunately, there is no formula for the value of this integral. The situation looks grim when we turn to the back of the book and discover there is no table for a normal distribution with $\mu = 200$ and $\sigma = 40$. There is only one normal table: for $\mu = 0$ and $\sigma = 1$. Fortunately, it turns out that we can use this table to find the area for any normal distribution.

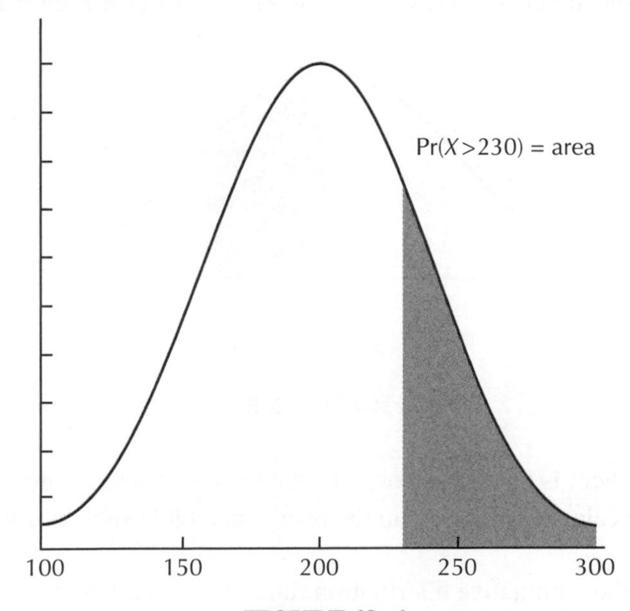

FIGURE 12–6

A normal random variable with $\mu = 0$ and $\sigma = 1$ is referred to as a *standard normal* random variable. Its density function is

$$f(x) = \frac{1}{\sqrt{2\pi}} e^{-(1/2)x^2}$$

Suppose that Z is a random variable with a standard normal density function. Figure 12–7 shows a graph of the density function for Z.

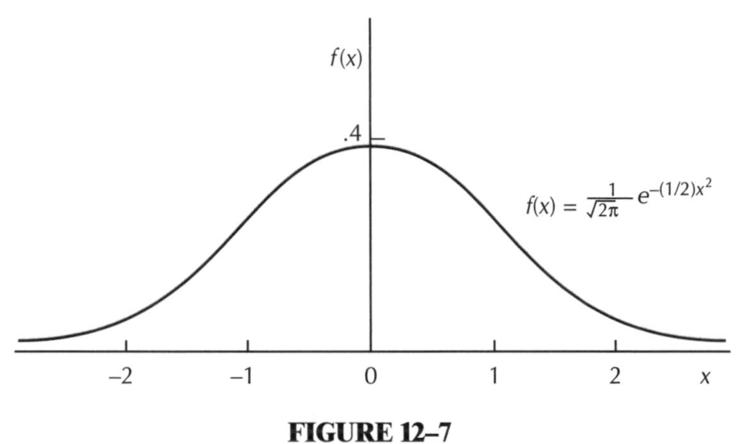

FIGURE 12–7

Since the density function is symmetric about $\mu = 0$, we can see that $\Pr(Z > 0) = 1/2$. Suppose that we need to know the probability that Z is between 0 and 1. Then we need to calculate the area under the curve between 0 and 1 (see Figure 12–8).

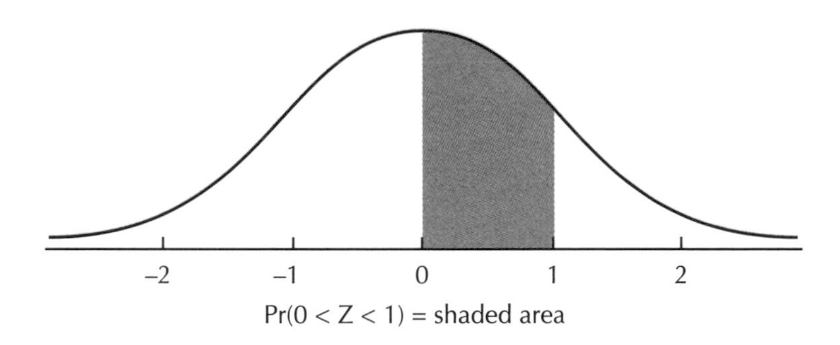

Pr(0 < Z < 1) = shaded area

FIGURE 12–8

Unfortunately, there is no simple formula that tells us what this area is. We have to use a computer or calculator to look up the results in a table such as Table A3–1 at the back of the book.

The table give the cumulative distribution function, which tells you the probability the Z will be less than a particular value. [The Greek letter Φ (phi) is often used to represent this function. $\Phi(z)$ means $\Pr(Z < z)$.]

The probability that Z will be between any two numbers a and b (with $a > b$) can be found from the formula

$$\Pr(a < Z < b) = \Pr(Z < a) - \Pr(Z < b) = \Phi(b) - \Phi(a)$$

We have already figured out the $\Phi(0) = .5$. We can see from Table A3–1 that $\Phi(1) = .8413$. Therefore, the probability that Z will be between 0 and 1 is $.8413 - .5000 = .3413$.

Because of the symmetry of the density function, we can see that there is also a .3413 probability that Z will be between –1 and 0 (see Figure 12–9). We can add these two probabilities together:

$$\Pr(-1 < Z < 0) + \Pr(0 < Z < 1) = .3413 + .3413$$
$$\Pr(-1 < Z < 1) = .6826$$

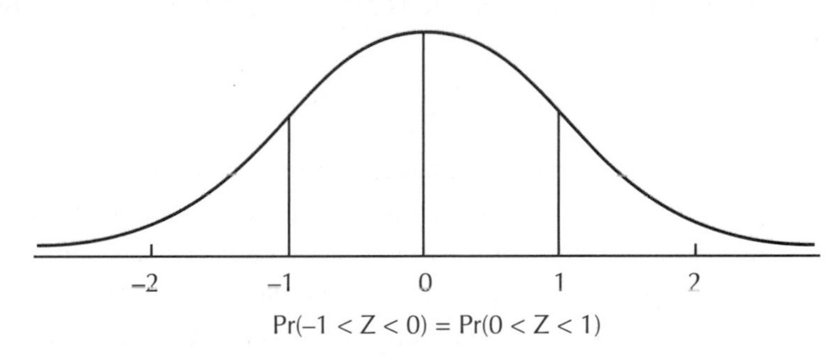

FIGURE 12–9

Therefore, there is a 68 percent chance that a standard normal random variable will be between –1 and 1. Putting it another way, there is a 68 percent chance that the value of a standard normal random variable will be within one standard deviation of its mean. (In this case the mean is 0 and the standard deviation is 1.)

This particular property also holds for *any* normal random variable, regardless of its mean and standard deviation: There is a 68 percent chance that any normal random variable will be within one standard deviation of its mean. For example, if X is a normal random variable with mean 200 and standard deviation 30, then there is a 68 percent chance that X will be between 170 and 230.

We can also use the table to show that there is a 95 percent chance that Z will be between –1.96 and 1.96. In general, we can say that any normal random variable has a 95 percent chance of being less than about 2 standard deviations away from its mean.

It will often be helpful to know the probability that a standard normal random variable will be between $-a$ and a, where a is a particular number. So, to make things more convenient, Table A3–2 at the back of the book lists these values. For example, the table shows that there is a .3830 probability that Z will be between –0.5 and 0.5.

The value of the standard normal random variable could conceivably be anything, since the density function never quite touches the axis. There is no number k such that $\Pr(Z > k) = 0$. However, you can see from Table A3–1 that there is only a .0002 probability that Z will be greater than 3.5. Larger values are even less likely, so we don't have to worry too much about the likelihood that Z might take on extreme values.

Here are three properties that hold for a standard normal random variable Z and any value of a:

$$(1) \qquad \Pr(Z > a) = 1 - \Pr(Z < a)$$

(This property holds true for any continuous random variable, but it does not hold true for discrete random variables.)

$$(2) \qquad \Pr(Z > a) = \Pr(Z < -a)$$

(This property holds true for any continuous random variable that is symmetric about zero.)

$$(3) \qquad \Pr(Z < -a) = 1 - \Pr(Z < a)$$

(This property is found by combining the two previous properties.)

It wouldn't be possible to produce a different table for every single possible value of μ and every single possible value of σ. However, we can use the standard normal tables to find the probabilities for any normal random variable by using the following trick. Suppose Y has a normal distribution with means 6 and variance 9, and we need to know the probability that Y will be between 5 and 8. We can create the random variable Z:

$$Z = \frac{Y - 6}{3}$$

Then Z will have a normal distribution with mean 0 and variance 1 because of the addition property. It should be clear that if Y is between 5 and 8, Z will be between $-1/3$ and $2/3$. Now we can look up the probability in the tables:

$$\begin{aligned} \Pr(5 < Y < 8) &= \Pr(-1/3 < Z < 2/3) \\ &= \Phi(.6667) - \Phi(-.3333) \\ &= .7486 - (1 - .6293) \\ &= .7486 - .3707 \\ &= .3779 \end{aligned}$$

In general, if X is a normal random variable with mean μ and variance σ^2, then $(X - \mu)/\sigma$ is a standard normal random variable. For example, suppose we would like to know the probability that you will sell more than 230 hamburgers at your downtown hamburger store.

Let X represent the number of hamburgers sold. We have $\mu = 200$ and $\sigma = 40$. Then we create a standard normal random variable Z as follows:

$$\Pr(X > 230) = \Pr\left[\frac{X - 200}{40} > \frac{230 - 200}{40}\right]$$
$$= \Pr(Z > .75)$$
$$= 1 - \Pr(Z < .75)$$
$$= 1 - .7734 \qquad \text{(from Table A3–1)}$$
$$= .2266$$

Now, suppose we would like to know the probability that you will sell a total of more than 330 hamburgers at the two locations. Let X be the total number of hamburgers. Then $\mu = 300$, $\sigma^2 = 2{,}000$, and $\sigma = 44.72$. Set up the standard normal random variable:

$$Z = \frac{X - 300}{44.72}$$

If $X > 330$, then $Z > .671$, and the probability of this happening is .2514.

The Excel function NORMDIST(*a, mean, sigma,* TRUE) will calculate the probability that a normal random variable with the given mean and standard deviation (sigma) will be less than the value A1. For example, NORMDIST(330,300,44.72,TRUE) will be .749, which is the probability you will sell fewer than 330 hamburgers at the two locations. (Changing the last TRUE to FALSE in the function will give you the height of the curve rather than the area under the curve.)

There also are many inexpensive scientific calculators that will calculate the normal distribution probabilities.

NOTE TO CHAPTER 12

The cumulative distribution function for the standard normal random variable can be approximated with this series:

$$\Pr(Z < x) = .5 + \int_0^x \frac{1}{\sqrt{2\pi}} e^{-y^2/2} dy$$
$$= .5 + \frac{1}{\sqrt{2\pi}}\left[x - \frac{x^3}{6} + \frac{x^5}{40} - \frac{x^7}{336} + \frac{x^9}{3456} - \frac{x^{11}}{42240} + \frac{x^{13}}{599040} - \cdots\right]$$

The ith term in the series can be found in this formula:

$$\frac{x^{2i-1}}{2^{i-1}(i-1)!(2i-1)}$$

(This is an example of a type of series known as a Taylor series. See a book on calculus).

There are many inexpensive scientific calculators now on the market that can do the work of Table A3–1, i.e. calculate Pr $(Z < x)$ when Z is a standard normal variable.

NORMAL DISTRIBUTION SUMMARY

If X has a normal distribution with mean μ and standard deviation σ, its density function is as follows:

$$f(x) = \frac{1}{\sigma\sqrt{2\pi}} e^{-(x-\mu)^2/(2\sigma^2)}$$

If X and Y are independent normal random variables, then $X + Y$ will also be a normal random variable (with mean $\mu_x + \mu_y$ and variance $\sigma_x^2 + \sigma_y^2$).

If X is a normal random variable and a and b are constants, then $aX + b$ will also be a normal random variable (with mean $a\mu_x + b$ and variance $a^2\sigma_x^2$).

In particular, $(X - \mu)/\sigma$ will be a normal random variable with mean 0 and standard deviation 1 (called a standard normal random variable, or Z random variable).

$$\Pr(X < a) = \Pr\left(\frac{X-\mu}{\sigma} < \frac{a-\mu}{\sigma}\right) = \Pr\left(Z < \frac{a-\mu}{\sigma}\right)$$

$$\Pr(Z > a) = 1 - \Pr(Z < a)$$
$$\Pr(Z > a) = \Pr(Z < -a)$$
$$\Pr(Z < -a) = 1 - \Pr(Z < a)$$

EXERCISES

1. Suppose the annual rainfall in a city has a normal distribution with mean 40 and standard deviation 5. What is the probability that the city will get less than 33 inches of rain next year? What is the probability that the city will get more than 38 inches of rain?

2. Suppose that the score the a student will get on an entrance exam is a random variable selected from a normal distribution with mean 550 and variance 900.

If you need a score of 575 to get into a certain college, what is the probability that you will get in? If instead you need a score of 540, what is the probability that you will get in?

3. You're coach of a football team that faces a third-down situation with four yards needed for a first down. If you select a play involving a runoff tackle, the number of yards you will gain on the play is given by a normal random variable with mean 2.5 and standard deviation 1. What is the probability that you will make the first down if you run this play?

4. Consider the same situation as in Exercise 3. Another play you might run is a tricky end-around double reverse. The results of that play are given by a normal random variable with mean 3 and variance 6. What is the probability that you will make the first down if you run that play?

5. Suppose you are measuring the speed of light. The result of your measurement is given by a normal random variable whose mean is the true value and whose standard deviation is 5×10^9 centimeters per second. What is the probability that your measurement will be within 2×10^9 centimeters per second of the true value?

6. Suppose you are running a lemonade stand. The number of glasses of lemonade that you sell each day is given by a normal random variable with mean 15 and standard deviation 10. What is the probability that you will sell at least 120 glasses of lemonade in a week (seven days)? What is the probability that you will sell at least 100 glasses? Is it all right to represent a variable such as the number of glasses of lemonade as a normal random variable?

7. List four other quantities that you think have approximately normal distributions.

In the following exercises, let X be a normal random variable with parameters μ and σ^2, density function $f_x(x)$, and cumulative distribution function $F_x(a)$. Use Table A3–1 to calculate $\Phi(x)$.

8. If $\mu = 0$ and $\sigma^2 = 100$, what is $\Pr(5 < X < 10)$?

9. If $\mu = -3$ and $\sigma^2 = 9$, and $F_x(a) = .6$, what is a?

10. If $\mu = 0$ and $F_x(5) = .8$, what is σ^2?

11. If $\mu = 3$, why can't $F_x(4) = .4$?

12. If $\mu = 73$ and $\sigma^2 = 81$, what is $\Pr(|X| > 100)$?

13. If $\mu = 25$ and $\sigma^2 = 100$, what is $\Pr(X = 25)$?

14. If $\mu = 1$ and $\sigma^2 = 64$, for what values of a is $.1 < F_x(a) < .3$?

15. If $f_x(x)$ takes a maximum value of 5 at $x = 10$, what are μ and σ^2?

16. On the same graph, plot $f_x(x)$ for $\mu = 0$ and (a) $\sigma^2 = 1$, (b) $\sigma^2 = 4$, and (c) $\sigma^2 = 9$.

17. If X is a normal random variable with mean μ_1 and standard deviation σ_1, and Y is a normal random variable with mean μ_2 and standard deviation σ_2, what is $\Pr(Y > X)$?

18. Suppose you have your choice between two jobs. Your annual earnings from an industrial job will have a normal distribution with mean \$45,000 and standard deviation \$6,000. Your annual earnings from a traveling sales job will have a normal distribution with mean \$36,000 and standard deviation \$30,000. What is the probability that you would earn more from the traveling sales job?

19. Show that $(X - \mu)/\sigma$ is a standard normal random variable.

20. Show that, for any normal random variable, there is a probability of .68 that the value of the random variable will be within one standard deviation of the mean.

21. Show that $\Phi(-x) = 1 - \Phi(x)$.

22. The median of a continuous random variable is the number x^{Φ} such that $\Pr(X < x^{\Phi}) = 1/2$. What is the median of a random variable with a normal distribution?

23. The mode of a continuous random variable is the point where the density function reaches its maximum value. What is the mode for a random variable with a normal distribution?

24. Show that, if three normal random variables are added together, the resulting random variable has a normal distribution.

☆ 25. Show that μ is the mean for a random variable with a normal distribution by calculating the integral.

☆ 26. Show that σ^2 is the variance for a normal random variable.

☆ 27. Calculate the area under the standard normal density function by evaluating the integral $\int_{-\infty}^{\infty} e^{(-1/2)^{x^2}} dx$. (HINT: Multiply by $\int_{-\infty}^{\infty} e^{(-1/2)y^2} dy$ and then convert the result to polar coordinates.)

28. Suppose that the grade that a student receives in an individual course is not very accurate, since there are many random factors that could cause the grade to be higher or lower than the student's true abilities would indicate. Suppose that the grade is given by a random variable with a normal distribution with mean 3.5 and variance 1/16. What is the probability that the student's grade for an individual course will be between 3.4 and 3.6? Now, suppose a student takes 36 courses, whose grades all have the same normal distribution. What is the probability that the average grade for all of the courses will be between 3.4 and 3.6?

29. Suppose the X is a random variable made up of the sume of 10 random variables with the following means and variances:

Mean	Variance
–5	25
–4	16
–3	9
–2	4
–1	1
1	1
2	4
3	9
4	16
5	25

What is the probability that X will be greater than 1?

30. Find the second derivative of the normal distribution density function and verify that its value is zero at $\mu - \sigma$ and $\mu + \sigma$.

Moment Generating Functions

WARNING: You are now entering a difficult mathematical area. We will need these results for some important theoretical proofs later on, but you are free to skip this section if you'd like.

Often you will want to know the value of $E(X)$, $E(X^2)$, $E(X^3)$, etc. (Take our word for it—you will.) These are called the *moments* of X. Specifically, $E(X^n)$ is called the nth moment of X. These can be calculated from a function $\psi(t)$ that is called the *moment generating function* (ψ is the Greek letter *psi*). The definition of the moment generating function (mgf for short) is this:

$$\psi(t) = E(e^{tX})$$

Here X is the random variable whose moment generating function we are calculating, and t is an ordinary variable that the mgf is a function of.

We can find the moments in the following way:

$$\begin{aligned}\psi'(t) &= \frac{d}{dt}E\left(e^{tX}\right)\\ &= E\left(\frac{d}{dt}e^{tX}\right)\\ &= E\left(Xe^{tX}\right)\end{aligned}$$

Notice that we have assumed that you can take the d/dt inside the parentheses (which you can for nice distributions, and certainly all of the ones used in this book are nice).

In general,

$$\begin{aligned}\psi^{(n)}(t) &= \left(\frac{d}{dt}\right)^n E\left(e^{tX}\right)\\ &= E\left(X^n e^{tX}\right)\end{aligned}$$

Then, if we calculate the value of the mgf when $t = 0$, we get

$$\psi^n(0) = E\left(X^n\right)$$

Therefore, if you know the moment generating function for a distribution, you can easily calculate all of the moments. To find the nth moment, just take the nth derivative and evaluate it at the point where $t = 0$.

Here are some examples. Suppose X has a binomial distribution with parameters n and p. Then

$$\begin{aligned}\psi(t) &= \sum_{i=0}^{n} e^{it}\binom{n}{i} p^{i}(1-p)^{n-i}\\ &= \sum_{i=0}^{n}\binom{n}{i}\left(pe^{t}\right)^{i}(1-p)^{n-i}\\ &= \left(pe^{t}+1-p\right)^{n}\end{aligned}$$

(by the binomial theorem).

If we take the derivative $\psi'(t)$, we get

$$\begin{aligned}\psi'(t) &= npe^{t}\left(pe^{t}+1-p\right)^{n-1}\\ \psi'(0) &= E(X)\\ &= np(\text{which is what we found earlier})\end{aligned}$$

Now find the second derivative:

$$\psi''(t) = np[e^{t}][(n-1)(pe^{t}+1-p)^{n-2}pe^{t}] + np(e^{t})(pe^{t}+1-p)^{n-1}$$

Evaluate $\psi''(0) = np[(n-1)(p+1-p)p + np(p+1-p) = np^{2}(n-1) + np$. This expression give us $E(X^2)$. Now, to find Var (X), calculate $E(X^2) - E(X))^2 = np^2(n-1) + np - n^2p^2 = np[p(n-1) + 1 - np] = np[pn - p + 1 - np] = np(1-p)$ again agreeing with the result we previously found for the binomial distribution.

Now, suppose X has a normal distribution with mean μ and variance σ^2. Then

$$\begin{aligned}\psi(t) &= E\left(e^{tx}\right)\\ &= \frac{1}{\sigma\sqrt{2\pi}}\int_{-\infty}^{\infty} e^{tx}e^{-(x-\mu)^2/2\sigma^2}dx\\ e^{tx}e^{-(x-\mu)^2/2\sigma^2} &= \exp\left[\frac{2\sigma^2 tx-(x-\mu)^2}{2\sigma^2}\right]\\ &= \exp\left[\frac{-x^2+2(\sigma^2 t+x)-\mu^2}{2\sigma^2}\right]\\ &= \exp\left[\frac{\left((\sigma^2 t+\mu)^2-\mu^2\right)-\left(x-(\sigma^2 t+\mu)\right)^2}{2\sigma^2}\right]\\ &= \exp\left[\frac{\left((\sigma^2 t+\mu)^2-\mu^2\right)}{2\sigma^2}\right]\exp\left[\frac{-\left(x-(\sigma^2 t+\mu)\right)^2}{2\sigma^2}\right]\\ &= \exp\left[\frac{\sigma^2 t^2}{2} t\mu\right]\exp\left[\frac{-\left(x-(\sigma^2 t+\mu)\right)^2}{2\sigma^2}\right]\end{aligned}$$

$$\psi(t) = \exp\left[\frac{\sigma^2 t^2/2}{\sigma\sqrt{2\pi}} + t\mu\right]\int_{-\infty}^{\infty} \exp\left[\frac{-\left(x-\left(\sigma^2 t+\mu\right)\right)^2}{2\sigma^2}\right]dx$$

(To save us from having to write complicated expressions as exponents, we write up exp [*expression*] to mean $e^{(expression)}$.)

If we let $v = \sigma^2 t + \mu$, then that second integral becomes equal to

$$\frac{1}{\sigma\sqrt{2\pi}}\int_{-\infty}^{\infty} \exp\left[\frac{-(x-v)^2}{2\sigma^2}\right]dx$$

which is equal to 1, since it is just the total area under the density function for a normal random variable. Therefore,

$$\psi(t) = \exp\left[\frac{\sigma^2 t^2}{2} + t\mu\right]$$

If we calculate the derivative $\psi'(t)$, we get

$$\psi'(t) = \left(\sigma^2 t + \mu\right)\exp\left[\frac{\sigma^2 t^2}{2} + t\mu\right]$$

We can see that $\psi'(0) = \mu$.

If X is a standard normal random variable, the moment generating function is even simpler:

$$\psi(t) = \exp\left(\frac{t^2}{2}\right)$$

The moment generating function contains a great deal of information about its random variable. (In fact, each random variable has its own unique mgf.) Moment generating functions will be very useful in showing some important properties of random variables, and they will be used in the proof of the central limit theorem.

EXERCISES

☆ 31. Let X and Y be two independent random variables, and let $Z = X + Y$. Show that $\psi_Z(t) = \psi_X(t)\psi_Y(t)$.

☆ 32. Use the mgf to calculate $E(X^2)$ when X is the number appearing on one die.

☆ 33. Calculate the moment generating functions of the following random variables, and use the moment generating function to calculate the mean and the variance:

Poisson variable with parameter $\lambda > 0$.

Geometric random variable with parameter $p, 0 < p \leq 1$.

☆ 34. If X has a normal distribution, show that $Y = aX + b$ also has a normal distribution when a and b are constants. Use moment generating functions.

☆ 35. If X and Y are independent random variables with normal distributions, show that $Z = X + Y$ has a normal distribution. Use moment generating functions.

13

Other Continuous Distributions

In this chapter we will discuss come probability distributions that seem rather esoteric at first, but turn out to be essential tools in statistics.

The Chi-square Distribution

Suppose Z is a standard normal random variable (that is, it has mean 0 and variance 1). Then suppose the $Y = Z^2$. This means that Y will also be a continuous random variable. We'd like to know what its probability density function will be. It's obvious that Y can't be less than 0, so its density function will look different from the standard normal density function. We'll give a name to this type of random variable: We'll say that it is a *chi-square* random variable. Since χ is the Greek letter chi, this distribution is usually symbolized by χ^2. You're probably wondering where the name chi-square comes from. (If you find out, let us know.)

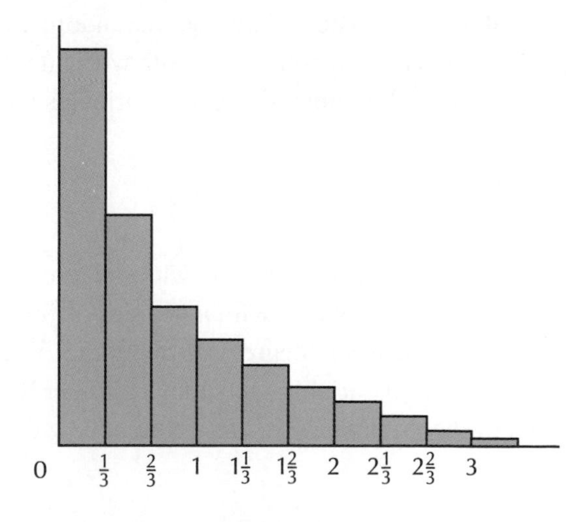

FIGURE 13–1

To give you an idea what the density function looks like, Figure 13–1 shows the frequency histogram for the squares of a group of numbers chosen from a standard normal distribution.

Some mathematical gymnastics show that the density function for the random variable $Y = Z^2$ is

$$f(y) = (2\pi)^{-1/2}y^{-1/2}e^{-y/2} \qquad \text{for } y \geq 0; f(y) = 0 \text{ if } y < 0$$

This function is drawn in Figure 13–2.

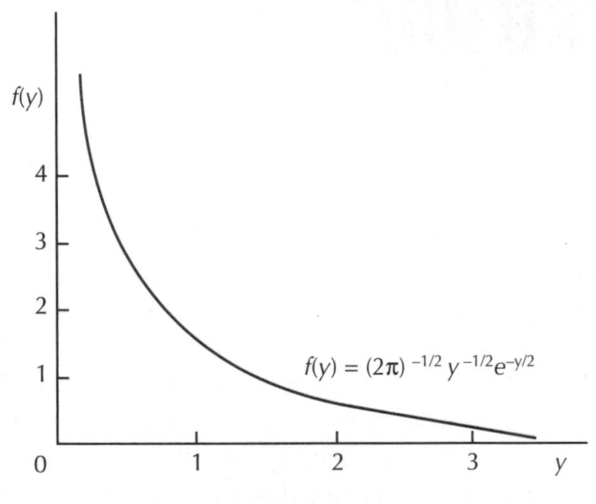

FIGURE 13–2

The mean of this random variable is easy to calculate, Since Z has a standard normal distribution, by definition $E(Z) = 0$ and $\text{Var}(Z) = 1$. Since $\text{Var}(Z) = E(Z^2) - E(Z)^2 = E(Z^2)$, it follows that $E(Z^2) = E(Y) = 1$. The variance of Y turns out to be 2.

There actually are several different types of chi-square random variables. The variable Y is strictly speaking called a chi-square variable with one *degree of freedom*. Suppose we square a lot of independent normal random variables and then take their sum, calling the result Y_n:

$$Y_n = Z_1^2 + Z_2^2 + \cdots + Z_n^2$$

Then Y_n is said to have the chi-square distribution with n degrees of freedom. (We'll write χ_n^2 to stand for "chi-square distribution with n degrees of freedom." You're probably wondering why we use the term "degrees of freedom." You can think of it this way. Each of the normal random variables acts like a number that you can choose freely, so since you have n of these numbers, it's as if you have n different free choices that you can make.

The general chi-square density function is

$$f(y) = \frac{1}{c} y^{n/2-1} e^{-y/2}$$

In this formula c is a constant number that has the appropriate value so that the total area under the curve is 1 (as we know must be the case if the function is to be a legitimate probability density function). The note at the end of the chapter tells how to calculate the value of c.

We can easily calculate the expectation and variance of a χ^2 random variable with n degrees of freedom. Since Y_n is the sum of n random variables, each with expectation 1, $E(Y_n) = n$. Each χ^2 random variable is independent, so the variance is just the sum of all of the individual variances, and therefore

$$\text{Var}(Y_n) = 2n$$

Figure 13–3 shows several different chi-square distributions with different degrees of freedom.

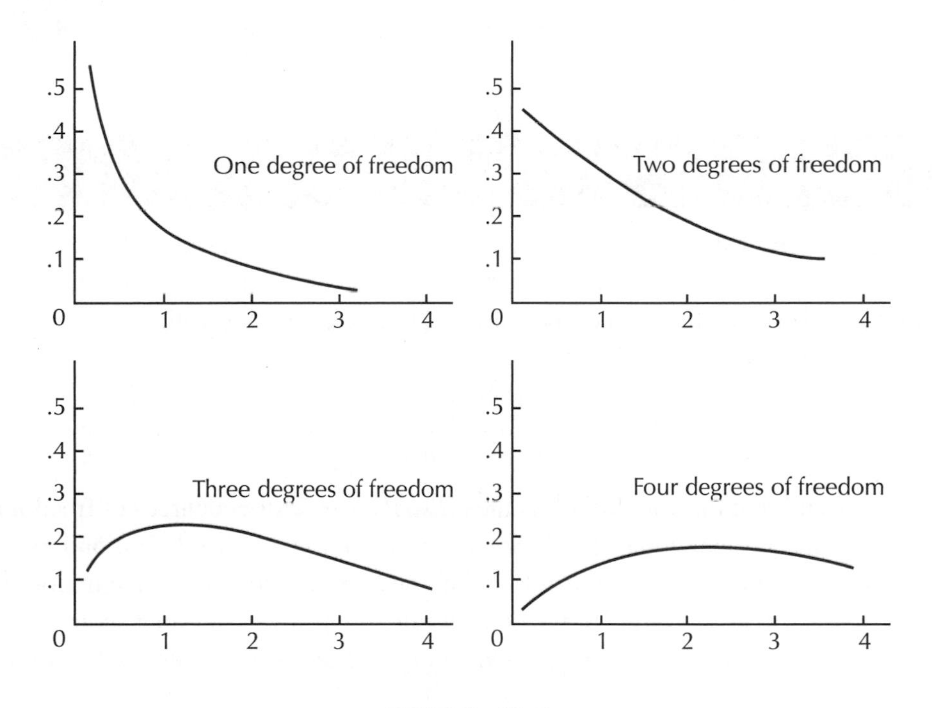

FIGURE 13–3

When the number of degrees of freedom is small, the density function is severely asymmetric. As the number of degrees of freedom increases, the density function gradually becomes more and more symmetric. As n becomes very large, the chi-square distribution begins to resemble a normal distribution. (We know this must be true because of the central limit theorem, which will be discussed in Chapter 15.)

Unfortunately, there is no simple expression for the cumulative distribution function for a χ^2 random variable. The only way to calculate values of $F(y)$ is to use a computer. Table A3–3 lists some results.

The Excel function CHIDIST(a, *df*) will give the probability that a chi-square distribution with *df* degrees of freedom will be greater than the value a (which is 1 minus the probability of being less than the value of a). However, in statistics you often know the probability in advance, and you need to determine the value of a. Typically you will want to know the value of a such that there is a 5 percent probability that the chi-square random variable will be greater than a, which can be found with the function CHIINV(.05, *df*). For example, CHIINV(.05, 12) gives the value 21.0, meaning there is a 5 percent chance that a chi-square random variable with 12 degrees of freedom will be greater than 21.0 (and a 95 percent chance it will be less than that value, as shown in Table A3–3).

The χ^2 random variable is very important in statistical estimation. For example, the distribution of the sample variance of a random sample drawn from the normal distribution will be closely related to a χ^2 distribution (see Chapter 17). Also, this distribution provides the basis for an important statistical test known as the χ^2 test (see Chapter 18).

The *t* Distribution

Another important distribution related to the normal distribution is the t distribution. Suppose that Z and Y are independent random variables. Let Z be a standard normal random variable (mean 0, variance 1) and Y a chi-square variable with m degrees of freedom. Let us make the definition

$$T = \frac{Z}{\sqrt{Y/m}}$$

Then it is said that the variable T has the *t distribution* with m degrees of freedom. (The t distribution is sometimes called *Student's distribution*.) This definition looks somewhat mystifying and pointless at first, but the t distribution does have important uses in statistics. The derivation of the density function is a very arduous, complicated process. (The function itself is complicated enough). The resulting density function looks like this:

$$g(x) = c\left(1 + \frac{x^2}{m}\right)^{-(m+1)/2}$$

Once again, c is an appropriate constant whose nature is described in the note at the end of the chapter.

Looking at the form of the density function provides some clues about the nature of the *t* distribution. Since $g(x) = g(-x)$, it follows that the density function is symmetric about $x = 0$. You can also see that the maximum value of $g(x)$ occurs when $x = 0$. Figure 13–4 shows two sample *t* density functions.

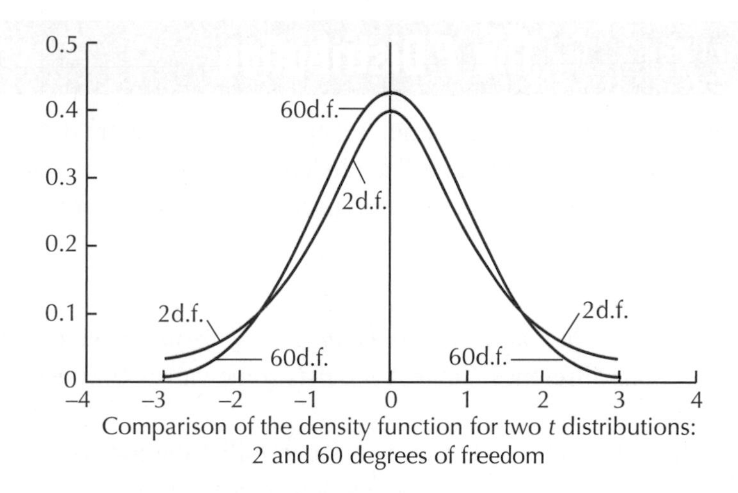

Comparison of the density function for two *t* distributions: 2 and 60 degrees of freedom

FIGURE 13–4

The density function has a bell shape that is roughly similar to the standard normal distribution. In general, though the *t* distribution has thicker tails than the normal distribution does. In other words, a *t* random variable has a higher chance of being far from 0 than does a standard normal random variable. However, as the number of degrees of freedom (m) increases, the *t* distribution approaches very close to the standard normal distribution. Table A3–4 contains some values for the cumulative distribution function for *t* distributions with various values for the degrees of freedom.

The Excel function TDIST(*a, df, 1*) will calculate the probability that a *t* distribution random variable with *df* degrees of freedom will be greater than *a*. For example, TDIST(2.228, 10, 1) will be .025 because $\Pr(T_{10} > 2.228) = .025$, where T_{10} is a random variable with a *t* distribution with 10 degrees of freedom. Another version of the function is TDIST(*a, df, 2*) which will give the probability that $-a < T < a$, where T is a *t*-distribution random variable with *df* degrees of freedom. For example, TDIST(2.228, 10, 2) will be .05 because $\Pr(-2.228 < T_{10} < 2.228) = .05$.

The most common calculation result you will need to know for the *t* distribution is this: what is the value *a* such that $\Pr(-a < T < a) = .95$? These values are found in the middle column of Table A3–5, and they can be found from the Excel function TINV(.05, *df*), where *df* is the number of degrees of freedom and .05 is used because $.05 = 1 - .95$. For example, TINV(.05,10) gives the result 2.228, meaning $\Pr(-2.228 < T_{10} < 2.228) = .95$. (TINV, standing for *t*-inverse, is the inverse of the TDIST function.)

If $m = 1$ the mean of the t distribution doesn't exist. (The t distribution with $m = 1$ is also called the *Cauchy distribution*.) If $m > 1$, then the mean does exist, and it is equal to 0 because the distribution is symmetric about zero. The variance of the t distribution does not exist if $m = 1$ or $m = 2$, but if $m > 2$ the variance is $m/(m - 2)$.

The *F* Distribution

Now we will mention one other distribution related to the chi-square distribution that has important uses in statistics. If X and Y are independent chi-square random variables with degrees of freedom m and n, respectively, then the random variable

$$F = \frac{X/m}{Y/n}$$

is said to have the *F distribution with m and n degress of freedom*. Note that the order of m and n makes a big difference. Table A3–6 lists some values for the cumulative distribution function.

Figure 13–5 illustrates the density function for an F distribution with 3 and 30 degrees of freedom. Note that an F random variable can never be negative. In practice, you are usually most interested in the value a such that $\Pr(F < a) = .95$ (or, alternatively, that $\Pr(F > a) = .05$, so those values are given in the table. We will see application of the F distribution in later chapters.

The Excel function FINV(.05, *df1, df2*) calculates the value a such that $\Pr(F < a) = .95$, where F has the F distribution with *df1* and *df2* degrees of freedom and $.05 = 1 - .95$. For example, FINV(.05, 3, 30) gives the value 2.92.

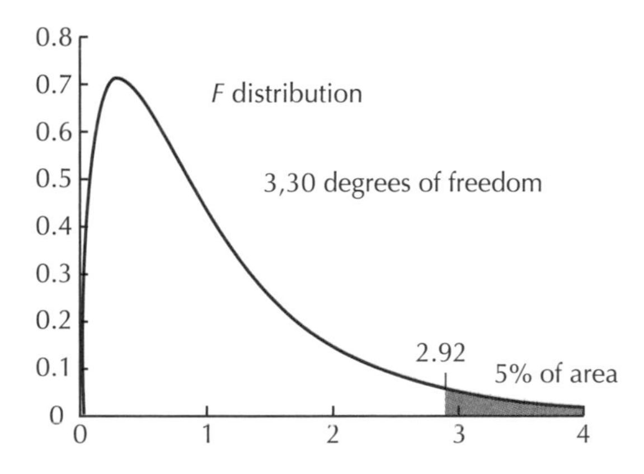

FIGURE 13–5

The Exponential Distribution

Light bulbs have a tendency to burn out unpredictably, so the length of time that a light bulb burns before it burns out is a good example of a continuous random variable. The type of distribution that is often appropriate in this situation is that of the *exponential random variable*. The density function for an exponential random variable is

$$f(x) = \lambda e^{-\lambda x} \quad \text{if } x \geq 0$$
$$f(x) = 0 \quad \text{if } x < 0$$

The Greek letter lambda (λ) is used as a parameter of the distribution. An exponential distribution is completely determined once you know the value of X. Figure 13–6 illustrates a sample exponential density function.

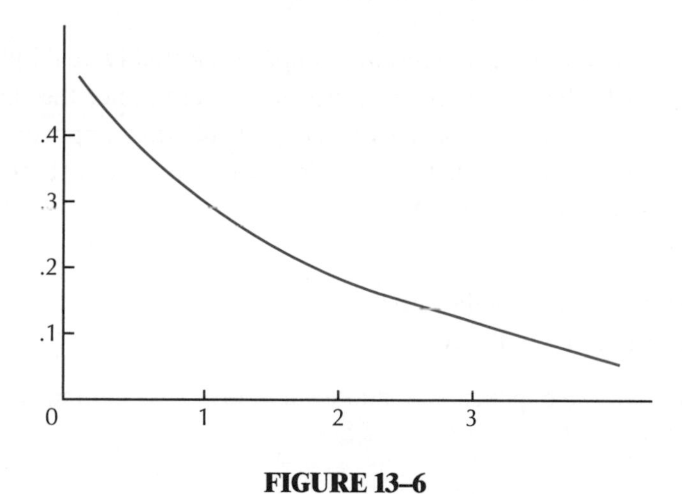

FIGURE 13–6

We can use calculus to derive the cumulative distribution function:

$$F(a) = \Pr(X < a) = 1 - e^{-\lambda x}$$

and therefore

$$\Pr(X > a) = e^{-\lambda x}$$

The exponential distribution has an interesting and unique property. Suppose we are interested in the probability that the light bulb will survive a particular day. If time is measured in hours, then the probability that the light bulb will survive the first day that we install it is $\Pr(X > 24) = e^{-24\lambda}$. Suppose that the light bulb does indeed survive the first day, and in fact it survives the first week. Then we want to know the probability that it will survive the next day. That means we want to know $\Pr(X > 168 + 24) = \Pr(X > 192)$. However, we know that the light bulb has already survived 168 hours (there are 168 hours in one week), so we are really interested in the conditional probability that X will be greater than 192, given that we know that it is already greater than 168. We can use the formula for conditional probability:

$$\begin{aligned}\Pr(X > 192 \mid X > 168) &= \frac{\Pr[(X > 192) \text{ and } (X > 168)]}{\Pr(X > 168)} \\ &= \frac{\Pr(X > 192)}{\Pr(X > 168)} \\ &= \frac{e^{-192\lambda}}{e^{-168\lambda}} \\ &= e^{-24\lambda}\end{aligned}$$

Therefore, the probability that the bulb will survive the eighth day, given the fact that it has survived the first seven days, is exactly the same as the probability that it will survive the first day. In general, for an exponential distribution,

$$\Pr\big[[X > (t + s)] \mid (X > t)\big] = \Pr(X > s)$$

This property is called the *lack of memory property*. (Actually, real light bulbs probably do not satisfy the lack of memory property exactly, since the chances that a particular light bulb will burn out on a given day probably do go up with time.)

Any continuous random variable that has this property for s and t both positive has an exponential distribution for some λ.

Examples include:

- the time that it takes for a radioactive particle to decay.
- the distance a light ray will travel through air before being refracted.

The expectation of the exponential random variable is $1/\lambda$, and the variance is $1/\lambda^2$ (see Exercise 10).

NOTE TO CHAPTER 13

The constant factors in the chi-square density function and the t density function can be defined by using an unusual function called the *gamma function*, symbolized by Γ. The Γ function has these properties:

$$\begin{aligned}\Gamma(0) &= 1 \\ \Gamma(1/2) &= \sqrt{\pi} \\ \Gamma(n) &= (n-1)\Gamma(n-1)\end{aligned}$$

Therefore, if n is an integer,

$$\Gamma(n) = (n-1)!$$

The chi-square density function can be written

$$f(x) = \frac{1}{2^{n/2}\Gamma(n/2)} x^{n/2-1} e^{-x/2}$$

The t density function can be written

$$g(x) = \frac{\Gamma[(m+1)/2]}{\sqrt{m\pi}\,\Gamma(m/2)}\left(1 + \frac{x^2}{m}\right)^{-(m+1)/2}$$

The gamma function can be defined by the integral

$$\Gamma(n) = \int_0^{\infty} x^{n-1}e^{-x}dx$$

It also is used in the definition of the *gamma distribution* with parameters a and b:

$$f(x) = \frac{b^a}{\Gamma(a)}x^{a-1}e^{-bx} \qquad (\text{if } x > 0)$$

There are two other quantities (in addition to the mean and the variance) that we can calculate to learn about the shape of the distribution of a random variable. The *skewness* of a distribution measures whether or not the distribution is symmetric. The skewness is defined to be

$$\frac{E\left[(X-\mu)^3\right]}{\sigma^3}$$

If a distribution is symmetric, then its skewness is 0. If it has a long tail in the positive direction, then it has positive skewness; it if has a long tail in the negative direction then it has negative skewness.

The *kurtosis* of a distribution measures the thickness of the tails of a distribution. It is defined to be

$$\frac{E\left[(X-\mu)^4\right]}{\sigma^4}$$

EXERCISES

1. What is the probability that a chi-square random variable with three degrees of freedom will be greater than 4?

2. What is the probability that a chi-square random variable with five degrees of freedom will be less than 2?

3. What is the probability that a random variable with a t distribution with five degrees of freedom will be between 0 and 1?

4. What is the probability that a random variable with a t distribution with 10 degrees of freedom will be between –2 and –1?

5. Suppose someone tells you that she observed the value 16 as the result of a chi-square random variable with 12 degrees of freedom. Do you believe her?

6. If someone tells you that he observed the value 5 for a chi-square random variable with 16 degrees of freedom, do you believe him?

7. If someone tells you that he observed the value –2.3 for a random variable that was selected from a *t* distribution with 3 degrees of freedom, do you believe him?

8. If someone tells you that she observed the value 0.7 for a random variable selected from a *t* distribution with 12 degrees of freedom, do you believe her?

9. Suppose X has a *t* distribution with n degrees of freedom. How can you characterize the distribution of X^2?

☆ 10. Calculate the expectation and variance for an exponential random variable with parameter λ.

☆ 11. Derive the cumulative distribution function for an exponential random variable.

☆ 12. Calculate the moment generating function for an exponential random variable.

☆ 13. If X has an exponential distribution, show that

$$\Pr\left[X > (a+b)\right] = \Pr(X > a)\Pr(X > b)$$

☆ 14. Derive the density function for a chi-square random variable with one degree of freedom.

☆ 15. Calculate the mean and variance for a chi-square random variable with one degree of freedom.

☆ 16. Calculate the moment generating function for a chi-square random variable with n degrees of freedom.

☆ 17. Suppose Y_1 has a chi-square distribution with n_1 degrees of freedom, and Y_2 has a chi-square distribution with n_2 degrees of freedom. Show that $Y_1 + Y_2$ has a chi-square distribution with $n_1 + n_2$ degrees of freedom. (Use moment generating functions.)

☆ 18. Show that the *t* distribution approaches the normal distribution as the number of degrees of freedom becomes very large. (To make things easier, assume that the value of the constant c approaches $1/\sqrt{2\pi}$ as m becomes large.)

☆ 19. Calculate the skewness and kurtosis for the normal, chi-square, and *t* distributions. Compare the results as the number of degrees of freedom for the chi-square and *t* distributions increases.

14 Distributions with Two Random Variables

Joint Density Functions

Suppose we have two discrete random variables X and Y, and we are interested in the probabilities that they will take on particular values. Just as in the case of one random variable, we can define a probability density function and a cumulative distribution function. Let's say that X has six possible values: X_1, X_2, X_3, X_4, X_5, and X_6. Also, we'll say that Y has six possible values: Y_1, Y_2, Y_3, Y_4, Y_5, and Y_6. Now, let's conduct a random experiment in which we observe values for X and Y. The result of the experiment consists of two numbers: the observed value of X and the observed value of Y. Then there are 36 possible outcomes for the experiment. (In general, if there are m possible values for x and n possible values for Y, there will be mn possible results.) To characterize the experiment completely, we need to calculate the probability for each one of the 36 possible results. We can arrange our results in a table.

For example, suppose that X is the number that shows up on the top of a die when it is rolled and Y is the number on the bottom of the die. Then the probability table is

Y	$X = 1$	$X = 2$	$X = 3$	$X = 4$	$X = 5$	$X = 6$
1	0	0	0	0	0	1/6
2	0	0	0	0	1/6	0
3	0	0	0	1/6	0	0
4	0	0	1/6	0	0	0
5	0	1/6	0	0	0	0
6	1/6	0	0	0	0	0

(The table looks like this because dice are marked so that the numbers on any two opposite faces add up to 7.)

For another example, suppose that X and Z are the numbers that appear on two different dice. Then the probability table looks like this:

Z	X = 1	X = 2	X = 3	X = 4	X = 5	X = 6
1	1/36	1/36	1/36	1/36	1/36	1/36
2	1/36	1/36	1/36	1/36	1/36	1/36
3	1/36	1/36	1/36	1/36	1/36	1/36
4	1/36	1/36	1/36	1/36	1/36	1/36
5	1/36	1/36	1/36	1/36	1/36	1/36
6	1/36	1/36	1/36	1/36	1/36	1/36

(In this case, obviously, each of the 36 outcomes is equally likely.)

When there are many possible values for either X or Y it will be too cumbersome to make a table. In general, though, we can define a *joint probability density function* $f(x, y)$ like this:

$$\begin{aligned} f(x, y) &= \Pr[(X = x) \text{ and } (Y = y)] \\ &= \Pr[(X = x) \cap (Y = y)] \end{aligned}$$

In this case f is a function of two variables. (Note that an ordinary probability density function is a function of only one variable.) Note that $f(x,y) = 0$ if (x,y) is not a possible result for X and Y. The sum of all of the possible values of the function $f(x, y)$ must be 1.

(Note: previously we have used the term "probability function" for discrete random variables and "density function" for continuous random variables. In this chapter we will use the term "density function" for both because the properties we develop here apply either to the probability function of a discrete random variable or the density function of a continuous random variable.)

We can also define the *joint cumulative distribution function*

$$\begin{aligned} F(x, y) &= \Pr[(X \leq x) \text{ and } (Y \leq y)] \\ &= \Pr[(X \leq x) \cap (Y \leq y)] \end{aligned}$$

The joint cumulative distribution function for two continuous random variables can be defined in the same way. The joint density function for two continuous random variables can be pictured as follows. Imagine a little hill spread out across a flat plane with an x axis and a y axis marked on it. The total volume under the hill must be 1. Then the probability that the values of X and Y will be within any specified rectangle of the plane is equal to the volume of the hill above that rectangle.

Interpreting a joint density function for two continuous random variables requires the use of a double integral:

$$\Pr[(a < X < b) \text{ and } (c < Y < d)] = \int_{x=a}^{x=b} \int_{y=c}^{y=d} f(x, y)dy\, dx$$

Marginal Density Functions for Individual Random Variables

Suppose that we are interested in the value of X but don't care about the value of Y. (For example, most people don't really care what number shows up on the bottom of a die they have just tossed.) In that case, the joint probability density function gives us far more information than we want. What we'd like to do is to figure out some way to derive the regular density function for X alone. (We will write $f_X(x)$ to stand for the density function of X.) Let's look at the joint density function for X and Y shown above. We should be able to use this information to find the probability that X will equal 1. As you can see from the table on page 171, there are six possible outcomes of the experiment that have X equal to 1:

$$(X=1, Y=1);\quad (X=1, Y=2); (X=1, Y=3);$$
$$(X=1, Y=4);\quad (X=1, Y=5); (X=1, Y=6)$$

To get the probability that X will equal 1, we need to add all these probabilities. We get $0+0+0+0+0+1/6=1/6$.

Now, look at the table that gives us the joint density function for X and Z. Once again we can get the probability that X will equal 1 by adding all of the terms in the first column: $1/36+1/36+1/36+1/36+1/36+1/36=1/6$. Likewise, by adding up each of the other columns in the tables we can find $\Pr(X=2)$, $\Pr(X=3)$, and so on.

In general, if there are n possible values for Y, then

$$\begin{aligned}\Pr(X=x) &= \Pr[(X=x) \text{ and } (Y=y_1)] \\ &\quad + \Pr[(X=x) \text{ and } (Y=y_2)] + \ldots \\ &\quad + \Pr[(X=x) \text{ and } (Y=y_n)]\end{aligned}$$

We can write this relation in terms of the joint density function:

$$\begin{aligned}fx(x) &= f(x,y_1)+f(x,y_2)+\cdots+f(x,y_n) \\ &= \sum_{i=1}^{n} f(x,y_i)\end{aligned}$$

An individual density function derived from a joint density function in this way is sometimes called a *marginal density function*.

We can also find the marginal density function for y:

$$f_Y(y)=f(x_1,y)+f(x_2,y)+\ldots+f(x_m,y)$$

If $f(x, y)$ is a continuous joint density function for X and Y, then the marginal density function of X can be found in a similar manner, with the summation over the discrete values of y replaced with an integral:

$$fx(x) = \int_{-\infty}^{\infty} f(x,y)dy$$

That formula will not mean anything to you unless you know something about calculus. If you don't, then you're free to ignore that formula.

Conditional Density Functions

There will often be times when we know the value of one of two random variables and would like to know the value of the other. Sometimes knowing the value of one random variable will help us a lot when we try to guess the value of the other one. For example, if we know that the variable Y discussed above has the value 4, then we know for sure that X has the value 3. However, knowing that Z (the number on another die) is 4 does not tell us anything about the value of X.

The *conditional density function* for X tells us what the density function for X is, given that Y has a specified value. Note that this concept is very much like the conditional probabilities that we discussed in Chapter 7. We will write "the conditional probability of X given that Y has the value y^*" like this:

$$f|x|(Y = y^*) \quad \text{or} \quad f(x|y^*)$$

(Remember that the vertical line | means "given that.") Then we can use the definition of conditional probability to find

$$f(x \mid Y = y^*) = \frac{\Pr[(X = x) \text{ and } (Y = y^*)]}{\Pr(Y = y^*)}$$

$$= \frac{f(x,y^*)}{f_Y(y^*)}$$

In other words, the conditional density function is equal to the joint density function $f(x,y^*)$ divided by the marginal density function of y. For example, suppose $Y = 3$. Then $f_Y(3) = 1/6$, so

$$f(x \mid Y = 3) = \begin{cases} \frac{0}{1/6} & \text{when } x = 1 \\ \frac{0}{1/6} & \text{when } x = 2 \\ \frac{0}{1/6} & \text{when } x = 3 \\ \frac{1/6}{1/6} & \text{when } x = 4 \\ \frac{0}{1/6} & \text{when } x = 5 \\ \frac{0}{1/6} & \text{when } x = 6 \end{cases}$$

Independent Random Variables

It would be nice if we could tell whether two random variables affect each other. We will say that two random variables X and Y are independent if knowing the value of Y does not tell you anything about the value of X, and vice versa. That means that the conditional density for X given Y is equal to the regular marginal density for X (since knowing the value of Y does not change any of the probabilities for X.) Therefore, if X and Y are independent,

$$f(x \mid Y) = f_X(x)$$

We know from the definition of conditional probability that

$$f(x \mid y) = \frac{f(x,y)}{f_Y(y)}$$

Therefore, when X and Y are independent,

$$f_X(x) = \frac{f(x,y)}{f_Y(y)}$$
$$f(x,y) = f_X(x)\, f_Y(y)$$

When the two variables are independent, the joint density function can be found simply by multiplying together the two marginal density functions. This means that independent random variables are much easier to deal with. In general, it is very difficult to reconstruct the joint density function for two individual random variables if you are given their marginal density functions. (In fact, it is impossible unless you have some additional information about how the two random variables are related.)

In the examples above, X, Y, and Z all have identical marginal density functions. However, the joint density function for X and Y is much different from the joint density function for X and Z, since X and Z are independent but X and Y are not.

Here is an example of the joint density function for two independent random variables U and V:

V	$U=1$	$U=2$	$U=3$	$U=4$	$f(v)$
1	.02	.04	.06	.08	.2
2	.02	.04	.06	.08	.2
3	.06	.12	.18	.24	.6
$f_U(u)$	.1	.2	.3	.4	1

The marginal density function for U, $f_U(u)$, can be found by adding each column, and the marginal density function for V, $f_V(v)$, can be found by adding each row. You can see that for each entry in the table $f(u, v)$ is equal to $f_U(u) f_V(v)$, so the two random variables are independent.

Here is another useful result that works if two random variables are independent. If U and V are independent, then

$$E(UV) = E(U)\,E(V)$$

In other words, the expectation of their product is just equal to the product of their individual expectations. That is another reason why life is much simpler when we have independent random variables.

Covariance and Correlation

Suppose X and Y are two random variables that are not independent. We still would like to be able to measure how closely they are related. If X and Y are very closely related, learning the value of X tells you a lot about the value of Y. If they are only slightly related, then knowing the value of X helps you a little bit, but not much, when you try to guess the value of Y.

The quantity that measures the degree of dependence between two random variables is called the *covariance*. The covariance of X and Y [written $\text{Cov}(X, Y)$] is defined as follows:

$$\text{Cov}(X, Y) = E[[X - E(X)][Y - E(Y)]]$$

Suppose that, when X is larger than $E(X)$, you know that Y is also larger than $E(Y)$. In that case, $E[[X - E(X)][Y - E(Y)]]$ will be positive. In general, when two random variables tend to move together, their covariance is positive. And if two random variables tend to move in the opposite direction (for example, if X tends to be big at the same time Y is small, and vice versa) then their covariance is negative. There is a short-cut formula for calculating that is much easier to use than the defining formula.

$$\begin{aligned}
\text{Cov}(X,Y) &= E[[X - E(X)][Y - E(Y)]] \\
&= E[XY - Y\,E(X) - X\,E(Y) + E(X)E(Y)] \\
&= E(XY) - E[Y\,E(X)] - E[X\,E(Y)] + E(X)\,E(Y) \\
&= E(XY) - E(X)\,E(Y) - E(X)\,E(Y) + E(X)\,E(Y) \\
\text{Cov}(X,Y) &= E(XY) - E(X)\,E(Y)
\end{aligned}$$

For example, when X and Y are the numbers on the top and bottom of a die, respectively, we can find $E(XY)$ as follows:

x	y	xy	$\Pr[(X = x)$ and $(Y = y)]$	$(xy) \times \Pr[(X = x)$ and $(Y = y)]$
1	6	6	1/6	6/6
2	5	10	1/6	10/6
3	4	12	1/6	12/6
4	3	12	1/6	12/6
5	2	10	1/6	10/6
6	1	6	1/6	6/6
			6/6	**56/6**
			Sum $= E(XY) = 56/6 = 9.333$	

Therefore,

$$\text{Cov}(X, Y) = 9.333 - 3.5 \times 3.5 = -2.917$$

The covariance is negative, because larger values of X are associated with smaller values of Y.

Here is another example. Suppose the joint density function for two discrete random variables S and T is given by

	$S = 5$	$S = 10$	$S = 20$
$T = 2$	0.20	0.30	0.02
$T = 3$	0.10	0.15	0.23

We can find the marginal probabilities for S and T by adding up the columns and the rows:

	$S = 5$	$S = 10$	$S = 20$	$f_T(t)$
$T = 2$	0.20	0.30	0.02	0.52
$T = 3$	0.10	0.15	0.23	0.48
$f_S(s)$	**0.30**	**0.45**	**0.25**	**1**

We can find $E(TS)$:

t	s	ts	$\Pr[(T = t)$ and $(S = s)]$	$ts \times \Pr[(T = t)$ and $(S = s)]$
2	5	10	0.20	2.0
2	10	20	0.30	6.0
2	20	40	0.02	0.8
3	5	15	0.10	1.5
3	10	30	0.15	4.5
3	20	60	0.23	13.8
				Total $= E(TS) = 28.6$

We can find that $E(S) = 11$ and $E(T) = 2.48$, so the covariance is $28.6 - 11 \times 2.48 = 1.32$.

From the short-cut formula you can see immediately that $\text{Cov}(X, Y) = 0$ if X and Y are independent. (Use the result $E(XY) = E(X)E(Y)$ if X and Y are independent.) Therefore $\text{Cov}(X, Z) = 0$ when X and Z are the numbers on two different dice. (However, it unfortunately works out that just showing that $\text{Cov}(X, Y) = 0$ is not enough to ensure that X and Y are independent.)

If Cov(X,Y) is nonzero, then we know that X and Y are not independent. However, the size of $\text{Cov}(X, Y)$ does not tell us much, because it depends mostly on the size of X and Y. We define a new quantity, called the *correlation*, that we can use directly to tell how strong the relation between X and Y is.

The most important property of the correlation is that its value is always between –1 and 1. If X and Y are independent, then clearly their correlation is zero. If the correlation coefficient is positive, you know that when X is big, Y is also likely to be big. They are then said to be *positively correlated*. The more closely related X and Y are, the closer the correlation coefficient is to 1. On the other hand, if the correlation is negative, then Y is more likely to be small when X is big. They are *negatively correlated*, and the negative relationship is stronger if the correlation coefficient is closer to –1.

The correlation between two random variables X and Y can be written $r(X, Y)$ or $\rho(X, Y)$. The symbol ρ is the Greek letter rho; it is often used for the correlation when you have population data, while r is used for the correlation when you have sample data. In this chapter we know all the possible values for our random variables, because we are dealing with simple random situations such as dice. In reality, we will often be limited to sample data. Chapter 21 shows how *regression analysis* is used to investigate the relationship between two variables based on your observations.

The definition of the correlation coefficient is the following:

$$\rho(X,Y) = \frac{\text{Cov}(X,Y)}{\sqrt{\text{Var}(X)\text{Var}(Y)}} = \frac{\text{Cov}(X,Y)}{\sigma_X \sigma_Y}$$

This can also be written

$$\frac{E(XY) - E(X)E(Y)}{\sqrt{[E(X^2) - (E(X))^2][E(Y^2) - (E(Y))^2]}}$$

In the preceding example we found $\text{Cov}(S, T) = 1.32$, $E(S) = 11$, and $E(T) = 2.48$. We can find $E(S^2) = 152.5$, $\text{Var}(S) = 152.5 - 11^2 = 31.5$, $E(T^2) = 6.4$, and $\text{Var}(T) = 6.4 - 2.48^2 = .2496$. The correlation is therefore $1.32/\sqrt{31.5 \times .2496} = .4708$. The correlation is not zero, so the variables are not independent. It is positive, so larger values of S are associated with larger values of T. It is less than 1, so there is not a perfect linear relationship between S and T.

EXAMPLE Suppose you flip three pennies and three nickels. Let U be the total number of heads, and let V be the number of heads on the pennies. Then the joint probability table looks like this:

V	$U=0$	$U=1$	$U=2$	$U=3$	$U=4$	$U=5$	$U=6$	$f_V(v)$
0	1/64	3/64	3/64	1/64	0	0	0	1/8
1	0	3/64	9/64	9/64	3/64	0	0	3/8
2	0	0	3/64	9/64	9/64	3/64	0	3/8
3	0	0	0	1/64	3/64	3/64	1/64	1/8
$f_U(u)$	**1/64**	**6/64**	**15/64**	**20/64**	**15/64**	**6/64**	**1/64**	**1**

We can calculate that $E(UV) = 5.25$. We already know that $E(U) = 3$ and $E(V) = 3/2$. Therefore, $\text{Cov}(U, V) = 5.25 - 3 \times 3/2 = .75$. We also have found that $\sigma_U = 1.225$, and $\sigma_V = .8660$. That means that $r(U, V) = .75/(1.225 \times .8660) = .7070$. The correlation is positive, because larger values of U are associated with larger values of V.

Two random variables are *perfectly correlated* when there is a relationship between them of the form

$$Y = aX + b$$

where a and b are two constants, and a is greater than 0. For example, suppose the joint density function for X and Y is this:

Y	$X=0$	$X=1$	$X=2$	$X=3$	$X=4$	$X=5$	$X=6$	
0	1/7	0	0	0	0	0	0	1/7
4	0	1/7	0	0	0	0	0	1/7
8	0	0	1/7	0	0	0	0	1/7
12	0	0	0	1/7	0	0	0	1/7
16	0	0	0	0	1/7	0	0	1/7
20	0	0	0	0	0	1/7	0	1/7
24	0	0	0	0	0	0	1/7	1/7
	1/7	**1/7**	**1/7**	**1/7**	**1/7**	**1/7**	**1/7**	**1**

In this case it is clear that Y is always equal to $4X$. We can calculate that $E(X) = 3$, $\sigma_X = 2$, $E(Y) = 12$, $\sigma_Y = 8$, and $E(XY) = 52$.

Therefore, $\text{Cov}(X, Y) = 52 - 3 \times 12 = 16$, and the correlation is $16/(2 \times 8) = 1$, which confirms what we suspected.

Variance of a Sum

We found earlier that $\text{Var}(X + Y) = \text{Var}(X) + \text{Var}(Y)$ if X and Y are two independent random variables (see page 98).

We can now derive a general formula for $\text{Var}(X + Y)$.

$$\begin{aligned}
\text{Var}(X+Y) &= E[(X+Y)^2] - [E(X+Y)]^2 \\
&= E[X^2 + 2XY + Y^2] - [E(X) + E(Y)]^2 \\
&= E(X^2) + 2E(XY) + E(Y^2) - (E(X))^2 - 2E(X)E(Y) - (E(Y))^2 \\
&= E(X^2) - (E(X))^2 + E(Y^2) - (E(Y))^2 + 2[E(XY) - E(X)E(Y)] \\
&= \text{Var}(X) + \text{Var}(Y) + 2\,\text{Cov}(X, Y)
\end{aligned}$$

For example, suppose you own stock in Worldwide Fastburgers, Inc. Your profit from the stock is a random variable (W) with mean 1,000 and variance 400. You would like to buy some more stock, and you are trying to decide between Have It Your Way Burgers, Inc., and Fun and Good Times Pizza, Inc. Both stocks also have profits that are random variables with mean 1,000 and variance 400. Which one should you choose? (Let H represent your profit if you choose Have It Your Way Burgers, and let F represent your profit if you choose Fun and Good Times Pizza.)

The expected value of your profit will be the same regardless of which one you choose. However, you would also like the variance of your profit to be as small as possible. A smaller variance means that your stock holding is less risky. In order to calculate the total variance of your portfolio, you need to look at the covariance between Worldwide Fastburgers and the other two companies. Suppose that $\text{Cov}(W, H) = 380$. The covariance is positive, meaning that both firms will prosper if the hamburger market is strong, but both firms will suffer if the hamburger market is weak. Suppose also that $\text{Cov}(W, F) = -200$. The negative covariance means that the pizza market is booming when the hamburger market is in a slump, and vice versa.

If you buy the Have It Your Way Burger stock, then the total variance of your profit will be

$$\begin{aligned}
\text{Var}(W+H) &= \text{Var}(W) + \text{Var}(H) + 2\,\text{Cov}(W, H) \\
&= 400 + 400 + 2 \times 380 \\
&= 1{,}560
\end{aligned}$$

If you buy the pizza stock, then the variance is

$$\begin{aligned}
\text{Var}(W+F) &= \text{Var}(W) + \text{Var}(F) + 2\,\text{Cov}(W, F) \\
&= 400 + 400 + 2 \times (-200) \\
&= 400
\end{aligned}$$

Clearly, it is much less risky to buy the pizza stock. In general, you can lower the risk of your stock holdings by diversifying and buying stocks that have negative covariances with each other. Or, to put it another way, don't put all your eggs in one basket.

The Multinomial Distribution

It is possible to have a distribution for more than two random variables. An important example is called the *multinomial distribution*.

The multinomial distribution is a generalization of the binomial distribution. With the binomial distribution we had two outcomes, which we called successes and failures. (In the example of the tossed coin there were heads and tails.) We now assume that there are m different outcomes for each trial. Let $X_1, X_2, \dots X_m$, be the random variables representing the number of times each outcome occurs. For each of the m outcomes let p_i be the probability that outcome number i will be the result of any one trial. (So we must have $p_1 + p_2 + p_3 + \dots + p_m = 1$.) Then the joint probability density function is

$$\Pr(X_1 = n_1, X_2 = n_2, \cdots X_m = n_m)$$
$$= \frac{n!}{n_1!n_2!\cdots n_m!} p_1^{n_1} \times p_2^{n_2} \times p_3^{n_3} \times \cdots \times p_m^{n_m}$$

where $n_1 + n_2 + n_3 + \dots + n_m = n$.

This formula follows by the same reasoning as the binomial distribution, since the probability of any single combination of n_1 occurrences of outcome number 1, n_2 occurrences of outcome number 2, etc., is $p_1^{n_1} \times p_2^{n_2} \times p_3^{n_3} \times \cdots \times p_m^{n_n}$, and there are

$$\frac{n!}{n_1!n_2!\cdots n_m!}$$

such outcomes.

As an example, what is the probability of rolling 3 ones, 4 twos, and 1 three out of 10 rolls of a die? Let X_1 equal the number of ones rolled, X_2 equal the number of twos rolled, X_3 equal the number of threes rolled, and X_4 equal the number of rolls that are not one, two, or three. Then $p_1 = p_2 = p_3 = 1/6$, and $p_4 = 3/6 = 1/2$. Also, $n_1 = 3$, $n_2 = 4$, $n_3 = 1$, $n_4 = 10 - 3 - 4 - 1 = 2$. So we have

$$\Pr(X_1 = n_1, X_2 = n_2, X_3 = n_3, X_4 = n_4)$$
$$= \frac{10!}{3!4!1!2!}\left(\frac{1}{6}\right)^3\left(\frac{1}{6}\right)^4\left(\frac{1}{6}\right)^1\left(\frac{1}{2}\right)^2$$
$$= 1.875 \times 10^{-3}$$

SUMMARY

If X and Y are discrete random variables, then their joint probability density function $f(x,y)$ is defined as:

$$f(x,y) = \Pr[(X = x) \text{ and } (Y = y)]$$

If X and Y are two continuous random variables, then the volume under their joint density function over a rectangle gives the probability the random variables will take on values in that rectangle.

The marginal density function $f_X(x)$ can be derived from the joint function:

Discrete:

$$f_X(x) = \sum_{i=1}^{n} f(x,y_i)$$

(where y_1 to y_n are the possible values of Y)

Continuous:

$$f_X(x) = \int_{-\infty}^{\infty} f(x,y)dy$$

Conditional density function for X, given that Y has the value y^*:

$$f[x \mid (Y = y^*)] = \frac{f(x,y)\,*}{f_Y(y^*)}$$

If X and Y are independent random variables:

$$f(x,y) = f_X(x) \times f_Y(y)$$

$$E(XY) = E(X)\,E(Y)$$

Covariance:

$$\text{Cov}(X,Y) = E(XY) - E(X)E(Y)$$

Correlation:

$$\rho(X,Y) = \frac{\text{Cov}(X,Y)}{\sigma_x \sigma_y}$$

If X and Y are independent, then $\text{Cov}(X, Y) = \rho(X,Y) = 0$.

$$\text{Var}(X + Y) = \text{Var}(X) + \text{Var}(Y) + 2\,\text{Cov}(X, Y)$$

EXERCISES

1. Roll three dice (one red, one blue, and one green.) Let X equal the sum of the numbers on the red and blue dice and let Y equal the sum of the numbers on the blue and green dice. Find the joint distribution of X and Y.

2. Roll two dice (one red and one blue). Let X be the sum of the numbers on two dice and let Y be the number on the red die. Find the joint distribution of X and Y.

3. Let X, Y be discrete random variables with $p_{X,Y}(x, y)$ their joint density function and $p_X(x)$ the density function of X. For a given x, which is larger: $p_X(x)$ or $p_{X,Y}(x, y)$?

4. Let X and Y be discrete random variables such that their joint density function $p_{X,Y}(x,y) = 1/15$ or 0. Suppose that X has only five possible values. What is the least number of possible values that Y can have?

For the next four exercises, calculate the marginal density functions for X and Y, the covariance between X and Y, and the correlation between X and Y. In each case the table gives the joint density function.

5.

X	**Y = 0**	**Y = 2**	**Y = 4**	**Y = 6**
–2	.1	0	.1	.2
4	0	.1	0	.1
5	.1	0	.1	.2

6.

X	**Y = 3**	**Y = 5**	**Y = 9**
–2	.15	.05	0
–1	.05	0	.2
0	0	.1	0
1	.15	0	.2
2	0	.1	0

7.

X	Y = 1	Y = 4	Y = 5	Y = 6	Y = 9
0	.1	0	.1	0	0
5	0	.1	0	0	.15
9	0	.05	0	0	0
11	0	.3	0	.05	0
13	.15	0	0	0	0

8.

X	Y = -9	Y = -4	Y = -3	Y = -1	Y = 0
-3	0	.1	0	.05	.1
2	.05	0	0	0	0
4	0	0	.1	0	.05
5	0	0	0	.05	0
7	0	.25	.05	0	.2

9. Suppose the joint density function for X and Y is given by the following table. Find $f_X(x)$ and $f_Y(y)$

X	Y = 1	Y = 2	Y = 3	Y = 4
1	.2	.1	.1	0
2	0	0	0	0
3	0	.1	.1	.2
4	0	0	.1	.1

10. Roll two dice. Let X represent the sum of the numbers on the top of the dice, and let Y represent the sum of the numbers on the bottom of the dice. Determine the joint density function.

11. Let X be the number of the month (January is 1, February is 2, etc.), and let Y be the day of the month for a randomly selected day in a regular (non-leap) year. Determine the joint density function for X and Y.

12. Give an example of two random variables X and Y such that $f_{X,Y}(x, y) = f_X(x)f_Y(y)$. Give another example such that

$$f_{X,Y}(x, y) \neq f_X(x)f_Y(y)$$

13. If $f_{X,Y}(x_1, y_1) > 0$, then we call (x_1, y_1) a *possible combined value* for X and Y. If $f_{X,Y}(x, y) \geq .13$ for all possible combined values (x,y), what is the largest number of possible combined values?

14. In a lottery based on a randomly selected three-digit number, let X be the first two digits and Y be the last two. Describe $f_{X,Y}(x, y)$.

15. Suppose the following table gives the joint density function for two random variables X and Y.

X	Y = 7	Y = 10	Y = 13	Y = 14
1	.1	0	.1	0
2	0	0	0	.15
5	0	.25	0	.1
9	.1	.1	.1	0

Find the joint cumulative distribution function.

16. Why can't

$$g(x,y) = \begin{cases} .1 & x = 2,\ y = 3/2 \\ .3 & x = 1/2,\ y = 4 \\ .5 & x = 2,\ y = 1 \\ .2 & x = 4,\ y = 1 \\ 0 & \text{everywhere else} \end{cases}$$

be a joint density function?

17. Calculate the covariance and correlation between X and Y when X represents the number on the top of a die and Y represents the number on the bottom of the die.

18. Calculate the covariance and correlation between X and Y when they represent the numbers that appear when two separate dice are tossed.

The next four exercises refer to the random variables U and V whose joint density function is given in Chapter 14, page 179.

19. Calculate the covariance and correlation for U and V.

20. Calculate the marginal density functions for U and V.

21. Find the conditional density function for U given that $V = 2$.

22. Find the conditional density function for V given that $U = 3$.

☆ 23. Show that $E(XY) = E(X)\,E(Y)$ if X and Y are independent.

☆ 24. Show that $[\text{Cov}(X, Y)]^2 \leq \text{Var}(X)\,\text{Var}(Y)$ for any two random variables X and Y.

25. Show that the correlation coefficient between any two random variables is always between –1 and 1. Use the result of the preceding exercise.

26. Let $Y = aX + b$, where X and Y are two random variables and a and b are two constants (with $a > 0$). Show that the correlation coefficient for X and Y is 1.

☆ 27. Given that X and Y are two continuous random variables whose joint density functions $f(x, y)$ is as follows:

$$\frac{1}{2\pi\sqrt{1-\rho^2}\sigma_x\sigma_y} \times$$

$$\exp\left[-\frac{1}{2(1-\rho^2)}\left[\left(\frac{x-\mu_z}{\sigma_x}\right)^2 - 2\rho\left(\frac{x-\mu_z}{\sigma_x}\right)\left(\frac{y-\mu_y}{\sigma_y}\right)\left(\frac{y-\mu_y}{\sigma_y}\right)^2\right]\right]$$

(Note: $\exp[u]$ means e^u.) Find the marginal density function for X by evaluating the integral

$$\int_{y=-\infty}^{y=\infty} f(x,y)dy$$

(When you evaluate this integral, treat x as a constant.) Warning: this is hard, but the result is very elegant.

☆ 28. Work with the density function in the previous exercise, but simplify matters by setting $\mu_x = \mu_y = 0$ and $\sigma_x = \sigma_y = 1$. Find the conditional density function for Y for a given value x^*. What is the nature of this function if $\rho = 0$? Investigate how it changes with other values of ρ.

15 Law of Large Numbers and the Central Limit Theorem

In this chapter we are going to derive the most important results in probability theory. We've already indicated that if you measure a random variable many times, then the average of all those observations will be close to the expectation value. Now we will prove that proposition. Also, we'll show some amazing features of the normal distribution.

Markov's Inequality

We'll start by stating a proposition called *Markov's inequality*. Let's suppose that X is a random variable that you are absolutely sure will always have a positive value. (In other words, $F(0) = \Pr(X \leq 0) = 0$.) Then Markov's inequality states that if $a > 0$,

$$\Pr(X \geq a) \leq \frac{E(X)}{a}$$

(For the proof, see the answer to Exercise 13.)

This result is true for *any* positive value of a, and it is true regardless of the nature of the distribution X. Notice that, as a becomes very large $E(X)/a$ becomes small, so the probability that X will be greater than a becomes small. Markov's inequality is valuable because it applies to any random variable, even if you know nothing about its distribution (other than that the random variable is always positive). However, the inequality doesn't always give you very much useful information. For example, suppose $a < E(X)$. Then $E(X)/a > 1$, so all that Markov's inequality tells you is that $\Pr(X \geq a)$ is less than some number greater than one. However, you already knew that, since any probability is always less than one.

Here is an example of the use of Markov's inequality. Let's say that Y is a chi-squared random variable with two degrees of freedom. Y will always be positive, so we can use Markov's inequality. We know $E(Y) = 2$. Let's calculate the probability that Y will be greater than 5. From Markov's inequality:

$$\Pr(Y \geq 5) \leq \frac{E(Y)}{5}$$
$$\Pr(Y \geq 5) \leq \frac{2}{5}$$

In this case we know from a chi-square table that the true value is about .1.

Here is an example where the limit set by Markov's inequality is closer to the actual value. Suppose X is a random variable with the following density function:

$$\Pr(X = 2) = .308$$
$$\Pr(X = 3) = .308$$
$$\Pr(X = 4) = .308$$
$$\Pr(X = 50) = .076$$

Then $E(X) = 6.572$. Markov's inequality says that $\Pr(X \geq 49) \leq 6.572/49$, or $\Pr(X \geq 49) \leq .134$.

Chebyshev's Inequality

Markov's inequality can be used to prove another important proposition known as *Chebyshev's inequality* (also spelled *Tchebysheff*). Chebyshev's inequality works for *all* random variables, not just positive ones as does Markov's inequality. Chebyshev's inequality tells us the answer to an important question: How likely is it that the value of a particular random variable will be very far away from its mean?

We'll say that X is a random variable with mean μ and variance σ^2. Chebyshev's inequality states:

$$\Pr(|X - \mu| \geq k) \leq \frac{\sigma^2}{k^2}$$

To prove this, let's define the random variable Y like this:

$$Y = (X - \mu)^2$$

Since Y is the square of a number, it will always be positive, so it must obey Markov's inequality. So we can write

$$\Pr(Y \geq k^2) \leq \frac{E(Y)}{k^2}$$

Now Y will be greater than k^2 only if $|X - \mu| \geq k$. Therefore,

$$\Pr(|X - \mu| \geq k) \leq \frac{E(Y)}{k^2}$$
$$\Pr(|X - \mu| \geq k) \leq \frac{E[(X - \mu)^2]}{k^2}$$

The expression $E[(X - \mu)^2]$ is a familiar old friend (or nemesis); it's just the definition of the variance of X (or σ^2). So

$$\Pr(|X - \mu| \geq k) \leq \frac{\sigma^2}{k^2}$$

Therefore, what Chebyshev's inequality does is put an upper limit to the likelihood that any random variable will be very far from its mean. Notice that if σ^2 is smaller, then σ^2/k^2 is smaller, so there is less chance that X will be very far from its mean. As k becomes bigger, σ^2/k^2 becomes smaller, which means that as you go farther and farther away from the mean there is a smaller and smaller chance that you will find the value of X there.

EXAMPLE Let X be a random variable with a binomial distribution, with parameters p and n. Then we know that $E(X) = np$, and $\text{Var}(X) = np(l - p)$. Since X will always be positive, we can use Markov's inequality:

$$\Pr(X \geq a) \leq \frac{np}{a}$$

From Chebyshev's inequality we have

$$\Pr(|X - np| \geq k) \leq \frac{np(1-p)}{k^2}$$

Suppose $p = .4$ and $n = 10$. Then Markov's inequality tells us that $\Pr(X \geq a) \leq 4/a$. For example, if $a = 9$, Markov's inequality states that $\Pr(X \geq 9) \leq 4/9$. In reality, the probability that $X \geq 9$ is .001.

Chebyshev's inequality tells us that

$$\Pr[(X \geq 4 + k) \text{ or } (X \leq 4 - k)] \leq \frac{2.4}{k^2}$$

For example, if $k = 3$, then Chebyshev's inequality says that there is less than a 27 percent chance that X will be less than 1 or greater than 7. The exact value of the probability is .071.

Let X be the uniform random variable on the interval from 0 to 10. Then $\mu = 5$, $\sigma^2 = 100/12 = 25/3$, and $f(x)1/10$ if $0 < x < 10$, and

$$\begin{aligned} \Pr(X \geq a) &= 1 && \text{if } a \leq 0 \\ &= 1 - \frac{a}{10} && \text{if } 0 < a < 10 \\ &= 0 && \text{if } a \geq 10 \end{aligned}$$

Markov's inequality states that, if $a > 0$, then

$$\Pr(X \geq a) \leq \frac{E(x)}{a} = \frac{5}{a}$$

From Chebyshev's inequality:

$$\Pr(|X - \mu| \geq k) \leq \frac{35}{3k^2}$$

Law of Large Numbers

Now we can prove a very important result in probability theory. We have already said that if we flip a coin a large number of times the number of heads that appear will be close to $n/2$. If we toss a die many times and take the average of all of the results, we expect that the average will be near $3\frac{1}{2}$. Both of these situations are examples of a more general law called the *law of large numbers*. (Strictly speaking, we're going to talk about the weak law of large numbers first, and then we'll talk about the strong law of large numbers later.)

Let us start with a random variable X. As is traditional, we will say $E(X) = \mu$ and $\text{Var}(X) = \sigma^2$. We'll observe X many, many times, and we'll call the observations X_1, X_2, X_3, and so on, up to X_n. We'll assume that all of these random variables are independent of each other. Take the average of all the X's (call it $\overline{x}$):

$$\overline{x} = \frac{X_1 + X_2 + \cdots + X_n}{n}$$

As you've probably guessed by now, $\overline{x}$ will be close to μ if n is very large. You might be skeptical and say, "But there still must be a very small probability that X might be a little way away from μ." The law of large numbers says that, as n becomes very, very large (that is, as n approaches infinity) the probability that $\overline{x}$ will not equal μ approaches zero.

To prove this law, we will use Chebyshev's inequality. First, note that X depends on the values of a bunch of random variables, so $\overline{x}$ is itself a random variable. We can calculate the mean and variance of $\overline{x}$:

$$\begin{aligned} E(\overline{x}) &= E\left[\frac{X_1 + X_2 + \cdots + X_n}{n}\right] \\ &= \frac{1}{n}E[X_1 + X_2 + \cdots + X_n] \\ &= \frac{1}{n}[E(X_1) + E(X_2) + \cdots + E(X_n)] \\ &= \frac{1}{n}(\mu + \mu + \cdots + \mu) \\ &= \frac{1}{n}n\mu \\ &= \mu \end{aligned}$$

We can use the properties of variances to calculate $\text{Var}(\overline{x})$:

$$\begin{aligned} \text{Var}(\overline{x}) &= \text{Var}\left[\left(\frac{1}{n}\right)(X_1 + X_2 + \cdots + X_n)\right] \\ &= \frac{1}{n^2}[\text{Var}(X_1 + X_2 + \cdots + X_n)] \\ &= \frac{1}{n^2}[\text{Var}(X_1) + \text{Var}(X_2) + \cdots + \text{Var}(X_n)] \end{aligned}$$

(The last step works only because all of the X's are independent.)

$$\begin{aligned}\text{Var}(\bar{x}) &= \frac{1}{n^2}(\sigma^2 + \sigma^2 + \cdots + \sigma^2) \\ &= \frac{1}{n^2}(n\sigma^2) \\ &= \frac{\sigma^2}{n}\end{aligned}$$

This equation establishes an important result: The variance of the sample average is less than the variance of each random variable taken individually. And as you increase the number of items in the sample, the variance of $\bar{x}$ becomes less and less.

Now we can use Chebyshev's inequality, which says

$$\Pr(|\bar{x} - \mu| \geq \varepsilon) \leq \frac{\text{Var}(\bar{x})}{\varepsilon^2} \quad (\varepsilon \text{ can be any small positive number.})$$

Since $\text{Var}(\bar{x}) = \sigma^2/n$, we can substitute:

$$\Pr(|\bar{x} - \mu| \geq \varepsilon) \leq \frac{\sigma^2}{n\varepsilon^2}$$

As n goes towards infinity, we can see that $\sigma^2/n\varepsilon^2$ goes to zero. Therefore, as n goes to infinity, the distance from $\bar{x}$ to μ must go to zero.

One other feature of the law of large numbers is important to remember. Suppose that you measure the random variable many times and there is a stretch during which you see many values that are less than the average. That does not mean that in the future you are more likely to get values greater than the average in order to cancel out the earlier values. Each measurement is still independent.

EXAMPLE Let X_1 be the number that appears when you toss a die. We have previously found that $E(X_1) = \mu = 3.5$ and $\text{Var}(X_1) = \sigma^2 = 2.91667$ (see pages 94 and 97). Let X_2 be the number on the second die, X_3 the number on the third die, and so on. Each X has identical expectation and variance, and all the dice are independent. If we roll two dice and let $\bar{X}_2$ be the average of the two numbers, we have

$$E(\bar{X}_2) = 3.5 \quad \text{Var}(\bar{X}_2) = \frac{2.91667}{2} = 1.4583$$

If we roll 100 dice, you can see that the variance of the average (call it $\bar{X}_{100}$) becomes much less:

$$E(\bar{X}_{100}) = 3.5 \quad \text{Var}(\bar{X}_{100}) = \frac{2.91667}{100} = 0.0292$$

If you roll 1,000 dice, you can have a very high degree of confidence that the average number that appears will be very close to 3.5:

$$E(\bar{X}_{1000}) = 3.5 \quad \text{Var}(\bar{X}_{1000}) = \frac{2.91667}{1{,}000} = 0.0029$$

EXAMPLE You are trying to estimate the mean commute time for people in a city by choosing a random sample. Let X_i be the commute time for the ith person chosen in the sample. Then $E(X_i) = \mu$, where μ is the unknown mean commute time for the entire population and Var $(X_i) = \sigma^2$, where σ^2 is the unknown variance of the entire population. Let $\bar{x}$ be the mean commute time of the n people in your random sample. Now find the expected value and variance of $\bar{x}$: $E(\bar{x}) = \mu$ and Var $(\bar{x}) = \sigma^2/n$. You can see that as the sample size gets larger, the variance of x gets smaller, so the larger sample will give you greater confidence in your ability to accurately estimate μ. This result is the foundation for calculating confidence intervals (see Chapter 17).

The Central Limit Theorem

Now comes something you've undoubtedly been waiting for: a reason to study the normal distribution. From the definition alone, it does seem a pretty weird thing to study so thoroughly. However, it turns out that if you add together a lot of independent identically distributed random variables, the resulting sum will have a normal distribution. That's one of the reasons why the normal distribution pops up everywhere. This is the kind of theorem we will present now. It is called the *central limit theorem.*

Formally, the theorem states: Let $X_1, X_2, \ldots$ be an infinite sequence of independent random variables with identical distributions. (Each X has mean μ and variance σ^2.) Then let $\bar{x} = [X_1 + X_2 + \cdots + X_n]/n$. We already know that $E(\bar{x}) = \mu$ and Var $(\bar{x}) = \sigma^2/n$. The central limit theorem says that, in the limit as n goes to infinity, $\bar{x}$ has a normal distribution. The amazing thing is that $\bar{x}$ moves closer to the normal distribution with large n no matter what the original distribution of X looks like.

Suppose that you assign a number to each person alive today. Let X_i be the volume of air in the ith person's lungs at a given second. Since n is roughly 4×10^9 and the X's have identical distributions and are independent, $(X_1 + X_2 + \ldots + X_n)/n$ has a distribution very close to the normal distribution.

There are two very simple corollaries we can establish. If we just add all of the x's together, then the result will also have a normal distribution, this time with mean $n\mu$ and variance $n\sigma^2$. Or, if we calculate the random variable

$$Z = \frac{\bar{x} - \mu}{\sigma / \sqrt{n}} = \frac{\sqrt{n}(\bar{x} - \mu)}{\sigma}$$

then Z will have a standard normal distribution.

Since the binomial distribution can be thought of as the sum of a group of independent two-valued random variables, the central limit theorem says that the binomial density function begins to look like the normal distribution as n becomes large. Note that this works even though the binomial distribution is a discrete distribution and the normal distribution is a continuous distribution. Table 15–1 illustrates what happens as n increases, and the binomial density function is illustrated in Figure 15–1.

TABLE 15–1: COMPARISON OF NORMAL AND BINOMIAL PROBABILITY DENSITY FUNCTIONS

	Pr ($X = k$)				$p = 0.5$			
k	$n = 6$		$n = 10$		$n = 20$		$n = 30$	
	Binomial	Normal	Binomial	Normal	Binomial	Normal	Binomial	Normal
1	.0938	.0859	.0098	.0103	...	.0001	...	...
2	.2344	.2334	.0439	.0417	.0002	.0003	...	...
3	.3125	.3257	.1172	.1134	.0011	.0013	...	...
4	.2344	.2334	.2051	.2066	.0046	.0049	...	...
5	.0938	.0859	.2461	.2523	.0148	.0146	.0001	.0002
6			.2051	.2066	.0370	.0360	.0006	.0007
7			.1172	.1134	.0739	.0725	.0019	.0020
8			.0439	.0417	.1201	.1196	.0055	.0056
9			.0098	.0103	.1602	.1614	.0133	.0132
10					.1762	.1784	.0280	.0275
11					.1602	.1614	.0509	.0501
12					.1201	.1196	.0806	.0799
13					.0739	.0725	.1115	.1116
14					.0370	.0360	.1354	.1363
15					.0148	.0146	.1445	.1457

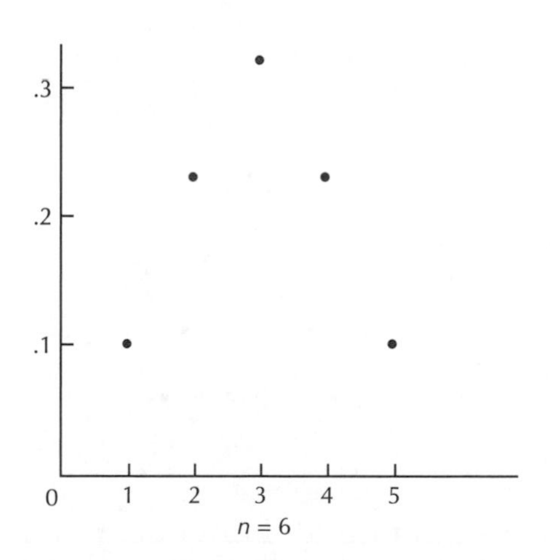

FIGURE 15–1

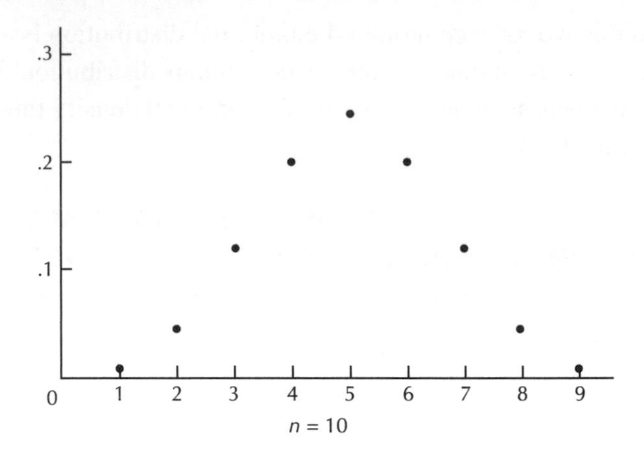
.3
.2
.1
0
1 2 3 4 5 6 7 8 9
n = 10

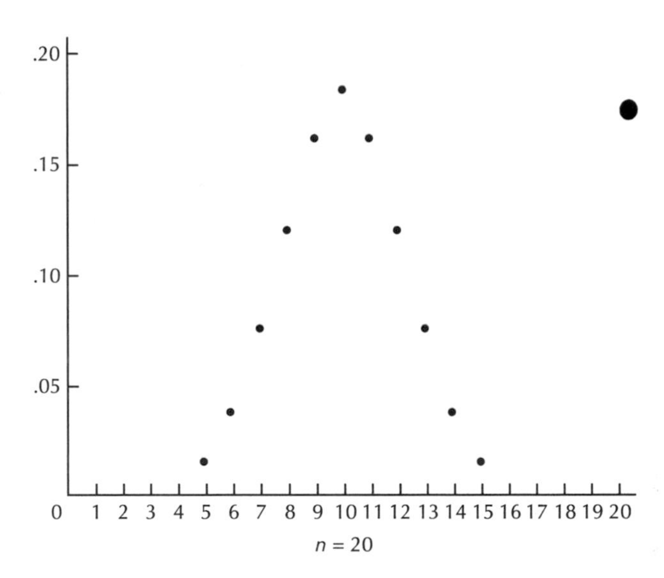
.20
.15
.10
.05
0
1 2 3 4 5 6 7 8 9 10 11 12 13 14 15 16 17 18 19 20
n = 20

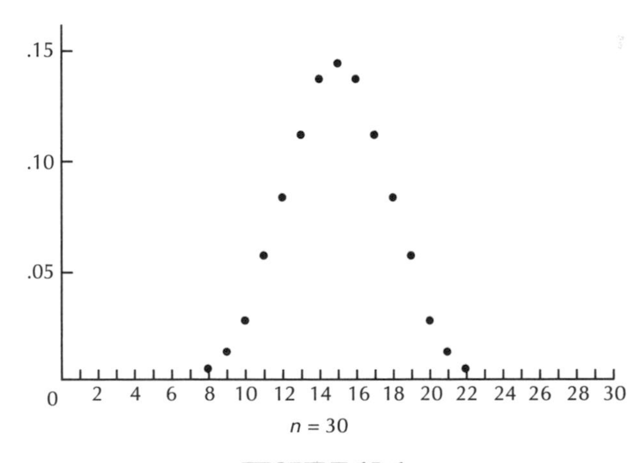
.15
.10
.05
0
2 4 6 8 10 12 14 16 18 20 22 24 26 28 30
n = 30

FIGURE 15–1

Of course, there are some limitations to the normal approximation. A normal random variable might be anywhere between minus infinity and plus infinity, whereas a binomial random variable must be between 0 and n. However, the probability that a normal random variable will take on such extreme values is very small, so there is no cause for worry. In practice, the normal approximation is all right when $n > 30$, provided that $np > 5$ and $n(1 - p) > 5$.

For another example, consider the probability function for the number that is the result of the roll of a die or dice (see Table 15–2 and Figure 15–2).

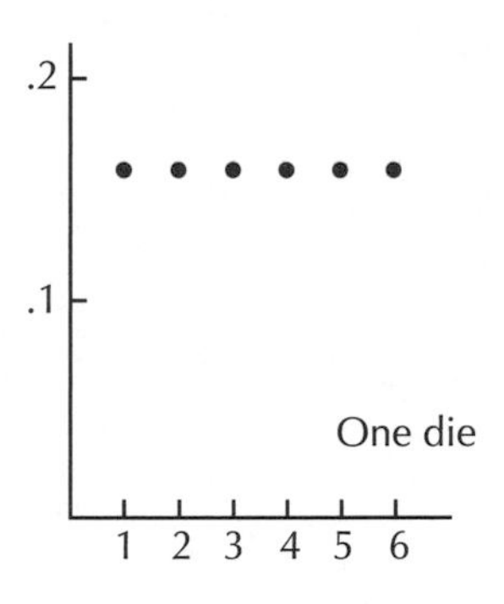

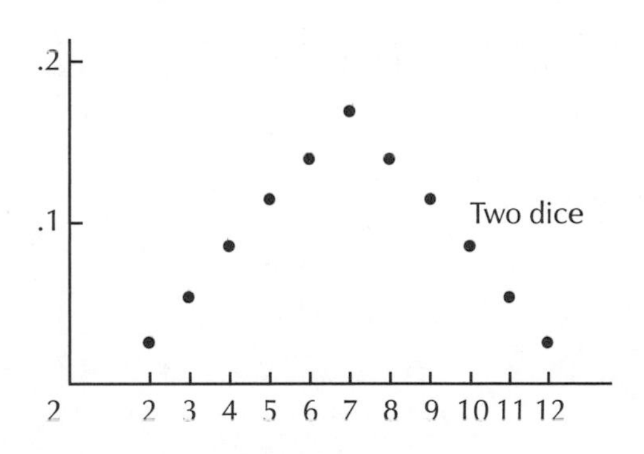

FIGURE 15–2

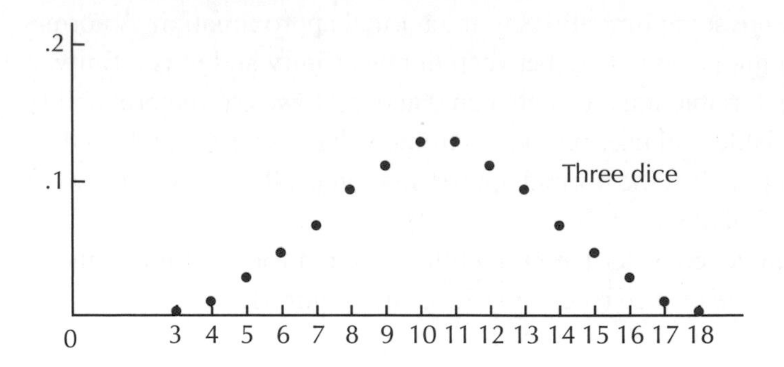

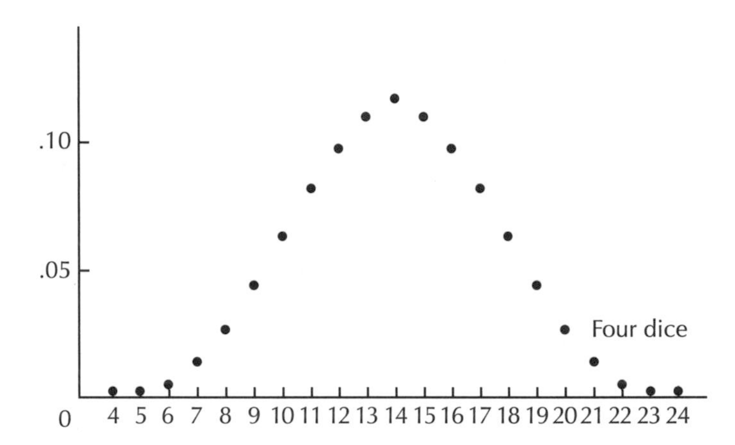

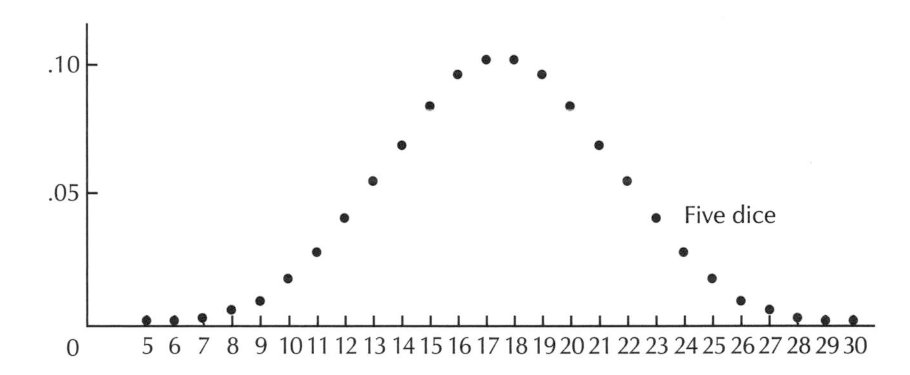

FIGURE 15–2

The probability function for one die doesn't look at all like the normal distribution. However, if you roll two dice, then the probability function for the result looks a bit more like the normal distribution, and if you increase the number of rolls then the probability function looks more and more like the normal density function.

TABLE 15–2

Number of Dice:	1		2		3		4		5	
Number on Dice	Number of Ways	Probability	Number of Ways	Probability	Number of Ways	Probability	Number of Ways	Probability	Number of Ways	Probability
1	1	.167								
2	1	.167	1	.028						
3	1	.167	2	.056	1	.005				
4	1	.167	3	.083	3	.014	1	.001		
5	1	.167	4	.111	6	.028	4	.003	1	…
6	1	.167	5	.139	10	.046	10	.008	5	.001
7			6	.167	15	.069	20	.015	15	.002
8			5	.139	21	.097	35	.027	35	.005
9			4	.111	25	.116	56	.043	70	.009
10			3	.083	27	.125	80	.062	126	.016
11			2	.056	27	.125	104	.080	205	.026
12			1	.028	25	.116	125	.096	305	.039
13					21	.097	140	.108	420	.054
14					15	.069	146	.113	540	.069
15					10	.046	140	.108	651	.084
16					6	.028	125	.096	735	.095
17					3	.014	104	.080	780	.100
18					1	.005	80	.062	780	.100
19							56	.043	735	.095
20							35	.027	651	.084
21							20	.015	540	.069
22							10	.008	420	.054
23							4	.003	305	.039
24							1	.001	205	.026
25									126	.016
26									70	.009
27									35	.005
28									15	.002
29									5	.001
30									1	…

EXAMPLE A scientist takes 100 measurements of pH in water samples chosen at random from a certain pond. If we assume that the average pH of the water (in general) is 6.8, and that the standard deviation for any single measurement is 0.1, what is the probability that the average of the 100 measurements will be between 6.79 and 6.81?

SOLUTION

We don't know much about the distribution of the water pH's, only the mean and standard deviation. Luckily the number of samples $n = 100$, is large enough that we can use the Central Limit Theorem. We know that if $X_1, X_2, \ldots, X_{100}$ are the random variables representing the water pH's, and $\overline{x}$ is their mean, then

$$\begin{aligned} Z &= \frac{\sqrt{n}(\overline{x} - \mu)}{\sigma} \\ &= \frac{\sqrt{100}(\overline{x} - 6.8)}{0.1} \end{aligned}$$

will be approximately a standard normal variable.

If $6.79 \le \overline{x} \le 6.81$, then

$$\begin{aligned} -0.01 &\le \overline{x} - 6.8 \le 0.01 \\ -0.01 &\le \sqrt{100}(\overline{x} - 6.8) \le 0.1 \\ -1 &\le \frac{\sqrt{100}(\overline{x} - 6.8)}{0.1} \le 1 \\ -1 &\le Z \le 1 \end{aligned}$$

Thus $\Pr(6.79 \le \overline{x} \le 6.81) = \Pr(-1 \le Z \le 1) = 0.6826$ (using Table A3–1 in Appendix 3).

The Proof of the Central Limit Theorem

WARNING: We are now entering very difficult mathematical terrain. If you don't intend to lose any sleep over the question "How do you prove the Central Limit Theorem?" then you can skip this section.

Now to prove the central limit theorem. We will need to use moment generating functions (described at the end of Chapter 12). First, we will prove a lemma concerning expectations.

Lemma: If X and Y are independent random variables, and g and h functions, then $E[g(X)\,h(Y)] = E[g(X)]\,E[h(Y)]$ (This is a generalization of the result $E(Y) = E(X)\,E(Y)$ proved earlier.)

Proof: If X and Y are continuous random variables and $f_{X,Y}(x, y) = f_x(x) f_Y(y)$ is their joint density function, then

$$\begin{aligned} E[g(X)h(Y)] &= \int_{-\infty}^{\infty} \int_{-\infty}^{\infty} g(x)h(y)f_{X,Y}(x,y)\,dx\,dy \\ &= \int_{-\infty}^{\infty} g(x)f_X(x)\,dx \int_{-\infty}^{\infty} h(y)f_Y(y)\,dy \\ &= E[g(X)]\,E[h(Y)] \end{aligned}$$

(The proof is similar in the discrete random variable case.)

To prove the central limit theorem we will have to use another lemma which we shall not prove but which should agree with your intuition.

Lemma: Let $X_1, X_2, \ldots$ be an infinite sequence of random variables. Let F_1 be the cumulative distribution function and ψ_i be the moment generating function of X_i. Let X be a random variable with cumulative distribution F_X and moment generating function ψ_X. Then, if $\lim_{n\to\infty} \psi_n(t) = \psi_X(t)$ for all t, then $\lim_{n\to\infty} F_n(t) = F_X(t)$ for all t where F_X is continuous. In other words, what this theorem says is that if the sequence of moment generating functions converges to a particular limit, then the corresponding cumulative distribution functions must converge to the corresponding limit.

Assume that this lemma is true. Then we need to show that if

$$Y_n = \frac{X_1 + X_2 + \cdots + X_n - n\mu}{\sigma\sqrt{n}}$$

then

$$\lim_{n\to\infty} \psi_{Y_n}(t) = e^{t^2/2}$$

(Remember that $e^{t^2/2}$ is the moment generating function for the standard normal distribution.)

First assume that $\mu = 0$ and $\sigma^2 = 1$. Then

$$Y_n = \frac{X_1 + X_2 + \cdots + X_n}{\sqrt{n}}$$

Since the X_i's all have identical distributions, we can let $\psi_X(t)$ be their common moment generating function. Then

$$\begin{aligned} \psi_{X_i/\sqrt{n}}(t) &= E(e^{tX_i/\sqrt{n}}) \\ &= \psi_X\left(\frac{t}{\sqrt{n}}\right) \end{aligned}$$

and

$$\begin{aligned}\psi_{Yn}(t) &= E[e^{tX_1/\sqrt{n}}e^{tX_2/\sqrt{n}}\ldots e^{tX_n/\sqrt{n}}]\\ &= E[e^{tX_1/\sqrt{n}}]E[e^{tX_2/\sqrt{n}}]\cdots E[e^{tX_n/\sqrt{n}}] \qquad \text{(because of the lemma above)}\\ &= \left[\psi_X\left(\frac{t}{\sqrt{n}}\right)\right]^n\end{aligned}$$

Now we expand $\psi_X(t)$ as a Taylor series. (If you haven't heard of Taylor series, you probably won't have made it into this section anyway.)

$$\psi_X(t) = \psi_0(0) + \psi'x(0)t + \psi''(0)t\frac{t^2}{2} + R_3(t)$$

$R_3(t)$ is the third-order remainder term, and $\lim_{t\to 0} R_3(t)/t^2 = 0$.

Then, since $\psi_X(0) = E[e^{0Xi}] = 1$, we have

$$\begin{aligned}\psi_X(t) &= 1 + tE(X) + \frac{t^2}{2}E(X^2) + R_3(t)\\ &= 1 + t\mu + \frac{t^2}{2}(\sigma^2 + \mu^2) + R_3(t)\\ &= 1 + \frac{t^2}{2} + R_3(t)\\ \psi_X\left(\frac{t}{\sqrt{n}}\right) &= 1 + \frac{t^2}{2} + R_3\left(\frac{t}{\sqrt{n}}\right)\\ \psi_{Yn}(t) &= \left[1 + \frac{t^2}{2} + R_3\left(\frac{t}{\sqrt{n}}\right)\right]^n\end{aligned}$$

Let $f(x) = \log(1 + x) - x$.

$$\begin{aligned}\lim_{x\to 0}\frac{\log(1+x) - x}{x} &= \lim_{x\to 0}\frac{f(x)}{x}\\ &= \lim_{x\to 0}\frac{1/(1+x) - 1}{1}\\ &= 0\end{aligned}$$

(by L'Hôpital's rule.)

Then

$$\begin{aligned}&\log\left[\left(1 + \frac{t^2}{2n} + R_3\left(\frac{t}{\sqrt{n}}\right)\right)^n\right]\\ &= n\left[\log\left(1 + \frac{t^2}{2n} + R_3\left(\frac{t}{\sqrt{n}}\right)\right)\right]\\ &= n\left[\frac{t^2}{2n} + R_3\left(\frac{t^2}{2n}\right) + f\left(\frac{t^2}{2n} + R_3\left(\frac{t}{\sqrt{n}}\right)\right)\right]\end{aligned}$$

Let $g(x) = f[x^2/2 + R_3(x)]$.

$$\begin{aligned}\lim_{x\to 0}\frac{g(x)}{x_2} &= \lim_{x\to 0}\frac{f[x^2/2 + R_3(x)]}{x^2}\\ &= \lim_{x\to 0}\left[\frac{f(x^2/2 + R_3(x))}{x^2/2 + R_3(x)}\right]\left[\frac{x^2/2 + R_3(x)}{x^2}\right]\\ &= 0 \times \frac{1}{2}\\ &= 0\end{aligned}$$

$$\begin{aligned}\log\left[\left(1 + \frac{t^2}{2n} + R_3\left(\frac{t}{\sqrt{n}}\right)\right)^n\right] &= n\left[\frac{t}{2n} + R_3\left(\frac{t}{\sqrt{n}}\right) + g\left(\frac{t}{\sqrt{n}}\right)\right]\\ &= \frac{t^2}{2} + nR_3\left(\frac{t}{\sqrt{n}}\right) + ng\left(\frac{t}{\sqrt{n}}\right)\end{aligned}$$

$$\begin{aligned}\lim_{n\to\infty} nR_3\left(\frac{t}{\sqrt{n}}\right) &= \lim_{n\to\infty}\frac{t^2[R_3(t/\sqrt{n})]}{(t/\sqrt{n})^2}\\ &= 0\\ \lim_{n\to\infty} ng\left(\frac{t}{\sqrt{n}}\right) &= \lim_{n\to\infty} t^2\left[\frac{g(t/\sqrt{n})}{(t/\sqrt{n})^2}\right]\\ &= 0\end{aligned}$$

and

$$\begin{aligned}\lim_{n\to\infty}\log\left[\left(1 + \frac{t^2}{2n} + R_3\left(\frac{t}{\sqrt{n}}\right)\right)^n\right] &= \lim_{n\to\infty}\log\left[\left(\psi_X\left(\frac{t}{\sqrt{n}}\right)\right)^n\right]\\ &= \frac{t^2}{2}\\ \lim_{n\to\infty}\left[\psi_X\left(\frac{t}{\sqrt{n}}\right)\right]^n &= \lim_{n\to\infty}\psi_{Y_n}(t)\\ &= e^{t^2/2}\end{aligned}$$

This is the moment generating function for the standard normal distribution. Voila! We did it.

If each X doesn't have $\mu = 0, \sigma^2 = 1$, set $\overline{X}_i = X_i - \mu/\sigma$; then

$$\frac{\overline{X}_1 + \overline{X}_2 + \cdots + \overline{X}_n}{\sqrt{n}} = \frac{X_1 + X_2 + \cdots + X - n\mu}{\sigma\sqrt{n}}$$

Strong Law of Large Numbers

There is an even stronger theorem (called, coincidentally enough, the *strong law of large numbers*) that says that the average of a large number of independent, identically distributed random variables is their common mean. In other words, if $X_1, X_2, \ldots$ are independent identically distributed random variables with mean μ, then

$$\Pr\left[\lim_{n\to\infty} \frac{X_1 + X_2 + \cdots + X_n}{n} = \mu\right] = 1$$

This theorem is even stronger than the weak law of large numbers. The weak law says that, for large n, $\overline{X}$ will probably stay close to μ, but for any $\varepsilon > 0$ there may be an infinite number of times that $|\overline{X} - \mu| > \varepsilon$. The strong law says that, with probability 1, this occurrence can happen only a finite number of times, which is more like the intuitive concept of limit.

The proof of this theorem is beyond the scope of this book.

EXERCISES

1. If you flip a coin 400 times, what is the probability that you will get more than 250 heads?

2. If you flip a coin 400 times, what is the probability you will get between 190 and 210 heads?

3. If you roll a die 50 times, what is the probability that the resulting sum will be greater than 200?

4. If you roll a die 50 times, what is the probability that the resulting number will be between 170 and 180?

5. The probability of getting a royal flush when you deal out a five-card hand is 1.54×10^{-6}. Suppose you deal out 8×10^6 hands. What is the probability that you will get 0 royal flushes? Make a list of the probabilities that you will get n royal flushes, for $n = 1$ to $n = 10$.

6. Suppose that there is only a 1/10,000,000 probability that a person will get struck by lightning. Out of 200 million people in the country, what is the probability that there are fewer than 15 people who are struck by lightning?

7. Suppose you roll a die 5,000 times. Use Chebyshev's inequality to estimate the probability that the average of the numbers that result will be between 3.4 and 3.6.

8. What is the maximum probability that a random variable might be 2 standard deviations from its mean?

9. What is the maximum probability that a random variable might be 4 standard deviations from its mean?

10. Let $X_1, \ldots, X_{25}$ be binomial random variables with identical parameters $n = 100$ and $p = .3$. Use the central limit theorem to calculate the probability that the sum of all of the X's will be less than 1,000.

11. Let $X_1, \ldots, X_{20}$ be geometric random variables with identical parameters $p = .7$. Use the central limit theorem to approximate the probability that the sum of the X's will be less than 10.

12. Suppose the density function for a random variable looks like the function shown in Figure 15–3.

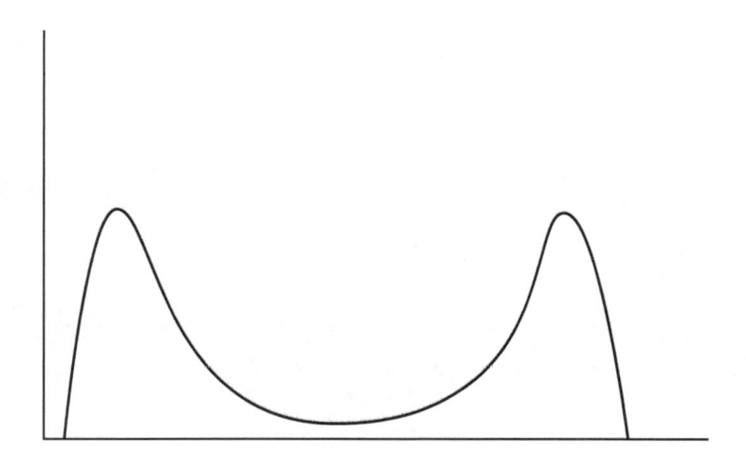

FIGURE 15–3

If you add up 100 random variables that all have this same density function, what will the density function for the sum look like?

☆ 13. Prove Markov's inequality.

☆ 14. Prove $E[g(X)\ h(Y)] = E[g(X)]\ E[h(Y)]$ when X and Y are independent discrete random variables.

16

Statistical Estimation

Up to now, in most of the problems we have done we have known in advance what all of the probabilities were. For example, when we draw cards or toss coins, we can calculate all of the probabilities explicitly. However, in most real problems we don't know the probabilities in advance. Instead, we have to use the methods of *statistical inference* to estimate them.

Suppose that the distribution of the heights of all of the people in the country can be described by a normal distribution. However, we don't know in advance what the mean (μ) of the distribution is. Instead, we will have to estimate it.

Suppose you are conducting a scientific experiment to measure the molecular weight of a chemical. On the average, you can expect that the results of the measurement will be the true value of the weight. However, any particular measurement is subject to random errors. It is often reasonable to suppose that the actual result of each measurement has a normal distribution whose mean is the true value of the quantity you are measuring.

Suppose we know that the number of successes of an experiment can be characterized by a binomial distribution, but that we don't know in advance the value of the probability-of-success parameter (p). We wish to use our observations of the results of the experiment to estimate the value of p.

You are trying to predict the results of an election. You don't know the proportion of people in the population who support your favorite candidate, but you will take a random sample in an attempt to estimate the population proportion.

Estimating the Mean

We will consider the general problem of trying to estimate the mean (μ) of a random variable X that has a normal distribution. Let's say that we are trying to estimate the average number of books read in the last month by the people in a large city. Assume that the number of books read has a normal distribution with mean μ and standard deviation σ. We know how we could calculate μ and σ: contact everyone in the city, ask them how many books they read, feed those numbers into a computer, and have the computer calculate the values. However, it will be too expensive to check with everyone. Instead, we need to check with a sample of randomly selected people. Suppose we wrote everyone's name on a slip of paper, put them all in a large drum and mixed them thoroughly, and then drew one name out. (It would be easier to use a computer random number generator to select a number, and then choose the name from the phone book that corresponded to that random number. The results would be similar.)

Let X_1 be the number of books read by the person whose name is drawn. The X_1 is a random variable with a normal distribution with mean μ and standard deviation σ. However, you would not have much confidence in guessing that μ is the same as X_1, since it would be only by coincidence if the number of books read by the person we randomly selected happened to be the same as the average number of books for all of the people in the city. Therefore, in order to obtain a more reliable estimate, we will start asking some more people. We will develop the theory that explains why a larger sample will be able to represent the population more accurately. You also should make sure that you can understand this intuitively. If we select just one person, there is a chance we might select one person who never reads at all, or someone who reads two books a day. Neither person is representative of the population. If we select 50 people in the sample, then it is highly unlikely that all 50 people in the sample never read, or that all 50 people read two books a day. Most likely some of the people in the sample will read very little and some will read a lot. The sample as a whole will begin to be representative of the population. If we make the sample larger, then it will tend to be more representative, provided that it is selected randomly.

Let X_2 be the number of books read by the second person in the sample, X_3 by the third person, and so on. For now, assume that the sample is selected with replacement, meaning that once a person is selected there is a chance they could be selected again. This means that we can treat each X as independent from the others. (In Chapter 19 we will discuss the difference between sampling with and without replacement.) Let n represent the number of people in the sample. Then the random variables X_1 to X_n all have the same distribution, and they are all independent. Therefore, they are said to form a random sample of size n chosen from the normal distribution with mean μ and standard deviation σ.

Intuitively, it is clear what our estimate for the mean should be

$$\text{estimate of mean} = \overline{x} = \frac{X_1 + X_2 + X_3 + \cdots + X_n}{n}$$

This quantity is just the average value of all of the Xs, which we called $\overline{x}$, or the sample average; $\overline{x}$ is an example of a *statistic*. A statistic is a particular function of the items in a random sample. When a statistic is used to estimate the value of some unknown quantity, it's called an *estimator*. In this case, $\overline{x}$ is used as an estimator for the quantity μ. Often a small hat sign ($\hat{\mu}$) is placed over a quantity to indicate that it is an estimator for a parameter. The statement "$\hat{\mu} = \overline{x}$" means, "We are using the sample average $\overline{x}$ as an estimator for the mean μ." An estimate is the value of an estimator in a particular circumstance. If the sample we observe consists of the numbers 5, 7, 4, 10, 12, and 4, then $\overline{x} = \hat{\mu} = 7$ is an estimate for the population mean.

What we'll do now is show that the sample average $\overline{x}$ has some appealing properties when it is used to estimate the mean.

Maximum Likelihood Estimators

Suppose that the true value of μ was 10,000, but the sample average turned out to be 7 for a particular sample. This occurrence is not likely to happen. On the other hand, if $\mu = 7$, then we are much more likely to get a value of 7 for the sample average. In fact, for every possible value of μ we can calculate the probability of getting a particular value $\overline{x}$ for the sample average. Intuitively, we would not estimate that μ has one of the values for which the probability of getting the observed sample average is very low. We would much rather choose our guess for it to be a value such that the probability of getting the observed sample average is high. In general, the value of μ that gives the highest probability of getting the actual observed value of $\overline{x}$ is called the *maximum likelihood* estimate of μ.

We certainly expect that the sample average $\overline{x}$ will be the maximum likelihood estimator for the mean μ, and we will in fact show this in the mathematical section at the end of the chapter.

The method of maximum likelihood can also be used for many other types of problems. In general, suppose that a is an unknown parameter in a particular probability distribution. Then in many cases we can calculate the maximum likelihood estimator for a.

EXAMPLE If $\overline{x_1}$ to $\overline{x_2}$ are from a random sample from a normal distribution with mean μ and variance σ^2, we can show that the maximum likelihood estimator of the variance ($\hat{\sigma}^2$) is

$$\hat{\sigma}^2 = \frac{(x_1 - \overline{x})^2 + (x_2 - \overline{x})^2 + \cdots + (x_n - \overline{x})^2}{n}$$

(We called this statistic s_1^2, the sample variance.) However, this estimator for the variance does not satisfy all of the criteria for a good estimator. In the next chapter we will discuss another estimate of the variance, s_2^2, which is equal to $s_1^2 \times n/(n-1)$.

EXAMPLE If we are trying to estimate the probability of success p for a random variable with a binomial distribution, then the maximum likelihood estimator is also the obvious one:

$$\frac{\text{number of successes}}{\text{number of attempts}}$$

EXAMPLE Suppose you have n observations each of two variables X and Y that come from a bivariate normal distribution (see Chapter 14, Exercise 27). Then the maximum likelihood estimators for the parameters μ_x, μ_y, σ_x, σ_y, and $\rho(X, Y)$ are the corresponding sample statistics, as you would expect:

- The maximum likelihood estimators for the means are $\bar{x}$ and $\bar{y}$.
- The maximum likelihood estimators for the variances are

$$\overline{x^2} - \bar{x}^2 \text{ and } \overline{y^2} - \bar{y}^2$$

- The maximum likelihood estimator for the correlation is

$$\frac{\overline{xy} - \bar{x}\,\bar{y}}{\sqrt{(\overline{x^2} - \bar{x}^2)(\overline{y^2} - \bar{y}^2)}}$$

Another important property of maximum likelihood estimators is called the *invariance property*. Suppose that $\hat{a}$ is the maximum likelihood estimator for a parameter a, but we really want to know the maximum likelihood estimator of $\sqrt{a}$. If we were forced to make a guess, we would probably estimate that $\sqrt{a}$ is equal to $\sqrt{\hat{a}}$, and fortunately we would be right. For example, the maximum likelihood estimator of the standard deviation ($\hat{\sigma}$,) is just the square root of the sample variance. In general, if $h(a)$ is any function of a parameter a, then the maximum likelihood estimator of $h(a)$ is just $h(\hat{a})$.

Consistent Estimators

Another important property that we would like our estimators to have is the consistency property. You've undoubtedly been confused by inconsistent people, and you can be just as confused by inconsistent estimators. Here is what we mean by the consistency property for estimators. Suppose we are able to increase our sample size by a lot and therefore acquire a lot more observations of the random variable X. In that case, do we know that the new value of $\bar{x}$ will be closer to the mean μ or is there a chance that it might be farther away? A consistent estimator is an estimator such that the probability is very high that the value of the estimator will move closer to the true value as you increase the number of elements in the sample.

Fortunately, $\bar{x}$ does satisfy the property of being a consistent estimator of the population mean.

Unbiased Estimators

Take a random sample of size n from a population and calculate the sample average $\overline{x}$:

$$\overline{x} = \frac{X_1 + X_2 + \cdots + X_n}{n}$$

Note that X_1 to X_n are random variables, because they depend on which items are chosen as part of the random sample. Therefore, $\overline{x}$ is also a random variable. As with any random variable, we can calculate its expectation. In this case we have already found that $E(\overline{x}) = \mu$ (the population mean). This seems to be a very desirable property. Since we intend to use $\overline{x}$ as an estimator for the unknown value of μ, it is good that the expected value of $\overline{x}$ is the same as the parameter we are trying to estimate. Because of this, $\overline{x}$ is said to be an *unbiased estimator* for the population mean.

In general, any estimator calculated from a random sample is a random variable. If the expected value of that estimator is equal to the true value of the parameter you are trying to estimate, then the estimator is said to be unbiased. For example, suppose you conduct an experiment n times, with a probability of success p each time (p is unknown). Let $\hat{p} = X/n$, where X is the number of successes. Then $E(\hat{p}) = E(X/n) = (1/n)E(X) = (1/n)(np) = p$. Therefore, the expected value of $\hat{p}$ is equal to p, which mean that $\hat{p}$ (the portion of successes that you observed) is an unbiased estimator of p (the unknown true probability of success).

However, suppose you are trying to estimate the value of the population standard deviation. You may calculate the first version of the sample variance:

$${s_1}^2 = \frac{(X_1 - \overline{x})^2 + (X_2 - \overline{x})^2 + \cdots + (X_n - \overline{x})^2}{n}$$

However, it turns out that this is not an unbiased estimator of the population variance. Instead, the second version of the sample variance, which has $n - 1$ in the denominator instead of n, is an unbiased estimator:

$${s_2}^2 = \frac{(x_1 - \overline{x})^2 + \cdots + (x_n - \overline{x})^2}{n - 1}$$

This will be discussed in the next chapter.

It is desirable for an estimator to be unbiased; however, the mere fact that an estimator is unbiased does not necessarily make it a good estimator. For example, suppose we are considering two alternative methods to estimate the mean (μ) of a population that has standard deviation σ:

1. Randomly choose one item; use the value of that item X as our estimate for the mean.
2. Choose 100 items; calculate their mean $\overline{x}$ and use that as our estimator for the mean.

Which of these methods is better? It should be obvious that we would have a better estimator if we have 100 observations than if we have just one. However, both of these estimators have expected value equal to μ, so they are both unbiased:

$$E(X) = \mu \quad E(\overline{x}) = \mu$$

The difference is in their variances; the variance of $\overline{x}$ is much smaller than the variance of X:

$$\text{Var}(X) = \sigma^2 \quad \text{Var}(\overline{x}) = \frac{\sigma^2}{n} = \frac{\sigma^2}{100}$$

In general, we can expect to have a good estimator if it is unbiased and it has a small variance. (You can usually decrease the variance of an estimator by observing a larger sample, but that may be too expensive to be worthwhile.) In some cases it may even be true that a biased estimator with a small variance is better than an unbiased estimator with a large variance.

Derivation of the Maximum Likelihood Estimator for the Mean

WARNING: mathematical area

We can show that $\overline{x}$ is the maximum likelihood estimator of μ. First, we show the density function of X_1, since X_1 is selected from a normal distribution:

$$fx_1(x_1) = \frac{1}{\sqrt{2\pi}\sigma} e^{-\frac{1}{2}\left[\frac{x_1-\mu}{\sigma}\right]^2}$$

Now we can calculate the joint density function of X_1 and X_2. Since X_1 and X_2 are independent, we can get the joint density function by multiplying the two individual density functions together:

$$f(x_1,x_2) = fx_1(x_1)fx_2(x_2) = \frac{1}{2\pi\sigma^2} e^{-\frac{1}{2}\left[\frac{x_1-\mu}{\sigma}\right]^2} e^{-\frac{1}{2}\left[\frac{x_2-\mu}{\sigma}\right]^2}$$

Exponents have the useful property that $e^a e^b = e^{a+b}$, so we can rewrite the joint density function as follows:

$$f(x_1,x_2) = \frac{1}{2\pi\sigma^2} e^{-\frac{1}{2\sigma^2}[(x_1-\mu)^2+(x_2-\mu)^2]}$$

We can follow this same procedure to get the joint density function of all n of the observations of X:

$$f(x_1, x_2, \cdots, x_n) = fx_1(x_1)fx_2(x_2)fx_3(x_3)\cdots fx_n(x_n)$$
$$= \frac{1}{(2\pi\sigma^2)^{n/2}} e^{-\frac{1}{2\sigma^2}[(x_1-\mu)^2+(x_2-\mu)^2+\cdots+(x_n-\mu)^2]}$$

Now, to find the maximum likelihood estimator of μ, we need to find the value of μ that makes this function as large as possible. (When a density function for a group of observations is expressed as a function of an unknown parameter like this, it is called a *likelihood function*.)

To make f as large as possible, we need to make

$$-\frac{1}{2\sigma^2}[(x_1 - \mu)^2 + (x_2 - \mu)^2 + \cdots + (x_n - \mu)^2]$$

as large as possible, which is the same as making

$$(x_1 - \mu)^2 + (x_2 - \mu)^2 + \cdots + (x_n - \mu)^2$$

as small as possible. We can rewrite the last expression as

$$x_1{}^2 - 2x_1\mu + \mu^2 + x_2{}^2 - 2x_2\mu + \mu^2 + \cdots + x_n{}^2 - 2x_n\mu + u^2$$
$$= (x_1{}^2 + x_2{}^2 + \cdots + x_n{}^2) - 2\mu(x_1 + x_2 + \cdots + x_n) + n\mu^2$$
$$= (x_1{}^2 + x_2{}^2 + \cdots + x_n{}^2) - 2\mu n\bar{x} + n\mu^2$$

We need to use a little calculus to show that the optimum value of μ (call it $\hat{\mu}$) will satisfy this equation:

$$-2n\bar{x} + 2n\hat{\mu} = 0$$
$$\hat{\mu} = \bar{x}$$

Therefore, just as we suspected all along, $\bar{x}$ is the maximum likelihood for μ.

EXERCISES

1. Suppose you are taking samples from a normal distribution with mean 7 and variance 9. If you observe 10 values, what is the probability that $\bar{x}$ will be between 6 and 8? What is the probability that $\bar{x}$ will be between 8 and 9? What is the probability that $\bar{x}$ will be between 9 and 10?

Estimate the mean and variance for these sets of numbers:

2. 20, 21, 12, 16, 24, 24, 23, 15, 20, 19, 25, 21, 19, 20, 14.

3. 10, 9, 10, 14, 8, 15, 6, 10, 7, 10, 11, 15, 8, 12, 10.

4. 15, 9, 18, 10, 14, 23, 14, 11, 15, 11, 18, 14, 18, 15, 13.

5. Assume that the score that your team makes in each game is selected from a normal distribution. Estimate the mean and variance.

6. For a particular game of your favorite football team, estimate the mean and variance for the yardage it gains on running plays and compare it with the mean and variance of the yardage it gains on passing plays.

7. Show that (successes)/(trials) is an unbiased estimator for the probability of success p for a random variable with a binomial distribution.

☆ 8. Show that S_1^2 is the maximum likelihood estimator for the variance when a random sample is taken from a normal distribution.

17

Accuracy of Estimates

In Chapter 16 we discussed how to use observations of a random variable to get information about an unknown parameter for the distribution that generates that variable. However, we still have to face one important question: Are these estimates likely to be very close to the true value?

For example, suppose you are trying to estimate the fraction of days that it rains in Florida. You naturally would use the estimator

$$\frac{\text{number of days you've been in Florida when it rained}}{\text{number of days you've been in Florida}}$$

However, if you've been in Florida one day and it rained that day, your estimate that it rains in Florida every day is not likely to be very accurate. If you've spent ten years in Florida you'll be able to estimate the fraction of rainy days much more accurately.

The value of a statistic that is used as an estimator depends on the values of a group of random variables, so that means that the estimator is itself a random variable. It would help if we could figure out what the distribution of the estimator looks like. For example, let's suppose we are using a sample average $\overline{x}$ to estimate the value of the mean of a normal random variable.

$$\overline{x} = \frac{X_1 + X_2 + \cdots + X_n}{n}$$

Each of these X's has a normal distribution, so because of the addition property for normal random variables $\overline{x}$ must also have a normal distribution ($\overline{x}$ is found by adding up a bunch of normal random variables). We have already found that $E(\overline{x}) = \mu$ and $\text{Var}(\overline{x}) = \sigma^2/n$ (see page 190).

Confidence Intervals

Now that we know the distribution of $\overline{x}$, we can be more precise about how good our estimate is. We know that the true value of μ is likely to be close to $\overline{x}$, but how close is close? Is $\overline{x}$ likely to be 1 unit away from μ? Or is it likely to be 50 units away? We'd like to know the probability that the distance from $\overline{x}$ to μ will be less than some specific value c.

In other words, we want to know the probability that the random interval $(\overline{x} - c)$ to $(\overline{x} + c)$ contains the unknown true value of μ. Obviously, the probability depends a lot on the value of c that we choose (see Figure 17–1). If we choose a large value of c, then we can be almost certain that the interval contains the true value of μ. For example, we could set c to be infinity. Then the probability that the interval ($\overline{x}$ – infinity) to ($\overline{x}$ + infinity) contains the true value of μ must be 100 percent. However, an interval that wide is not very useful. If we make the interval narrower by choosing a smaller value for c, then we can be more precise about the true value of μ. However, when we make the interval narrower there is a greater chance that the interval won't even contain the true value of μ.

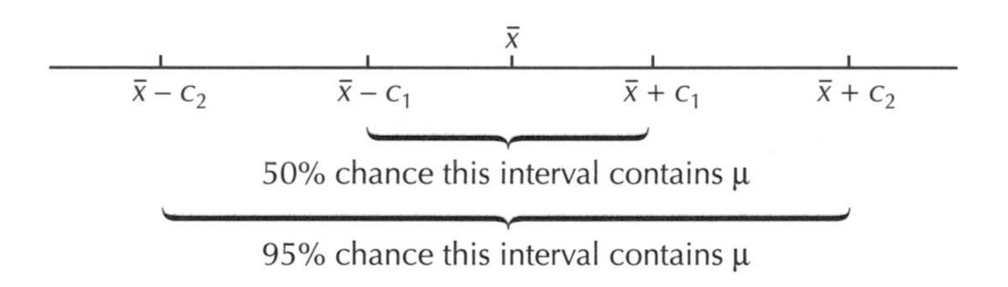

FIGURE 17–1

The normal procedure in statistics works like this: First, we choose the probability that we want—in other words, we set in advance the probability that the interval contains μ. Often this probability is set at 95 percent. Then we calculate how wide the interval must be so that there is a 95 percent chance that it contains the true value. This type of interval is called a *confidence interval*, and .95 is the *confidence level*.

Now we need to calculate the value of c that satisfies the equation

$$\Pr(\overline{x} - c < \mu < \overline{x} + c) = .95$$

Subtract $\overline{x}$ from all three parts:

$$\Pr(-c < \mu - \overline{x} < c)$$

Multiply by – 1:

$$\Pr(c > -\overline{x} - \mu > -c)$$

Reverse the order:

$$\Pr(-c < \bar{x} - \mu < c) = .95$$

Once we know the value of c, we know how wide the confidence interval must be. That means that our job is to calculate the value of c. Let's create a new random variable called Z:

$$Z = \frac{\bar{x} - \mu}{\sqrt{\sigma^2/n}}$$
$$= \sqrt{n}\frac{\bar{x} - \mu}{\sigma}$$

Using the properties we established for normal random variables, we know that Z has a standard normal distribution (mean 0 and variance 1; see Chapter 12). We can rewrite the equation:

$$\Pr\left(\frac{-c\sqrt{n}}{\sigma} < \sqrt{n}\frac{(\bar{x} - \mu)}{\sigma} < \frac{c\sqrt{n}}{\sigma}\right) = .95$$

$$\Pr\left(\frac{-c\sqrt{n}}{\sigma} < Z < \frac{c\sqrt{n}}{\sigma}\right) = .95$$

This problem calls for the standard normal probability table. Let's define a to be $a = c\sqrt{n}/\sigma$. Then

$$\Pr(-a < Z < a) = .95$$

Now we need to look in Table A3–2 until we find a value of a that satisfies this equation. Scanning down the columns, we can see that in this case the correct value for a is 1.96.

Now we can find the value for c:

$$c = \frac{a\sigma}{\sqrt{n}} = \frac{1.96\sigma}{\sqrt{n}}$$

Thus, we know how wide the confidence interval should be. There is a 95 percent chance that the true value of μ will be between $\bar{x} - 1.96\sigma/\sqrt{n}$ and $\bar{x} + 1.96\sigma/\sqrt{n}$.

There are two features of this result that are appealing to common sense. The confidence interval is wider (that is, more uncertain) if σ is bigger. If the variance of each individual observation is bigger, then it will be harder for us to pin down the true value of μ. On the other hand, the confidence interval is smaller if n is bigger. That means that as we take more and more observations we will be able to predict the true value of μ more accurately. However, note the square root sign. If we increase the number of observations by a factor of 4, the width of the confidence interval will only be cut in half.

We can be even more cautious if we want to. Suppose that we want to be 99 percent sure that our confidence interval contains the true value of μ. Then we will have to settle for a wider, less precise interval. Or, if we wanted to be less careful, we could have calculated a smaller confidence interval that had a lower probability of containing the true value.

Calculating Confidence Intervals by Using the *t* Distribution

There is one major difficulty with calculating confidence intervals this way. Often we don't know the true value of the population variance σ^2. Our first guess might be that we should use the known sample variance s_2^2 in place of the unknown population variance.

We found that the random variable

$$Z = \frac{\overline{x} - \mu}{\sqrt{\sigma^2 / n}}$$

has a standard normal distribution. So, if we instead consider the random variable

$$T = \frac{\overline{x} - \mu}{\sqrt{s_2^2 / n}}$$

we would expect it to have a distribution very similar to the standard normal distribution since the only change is that the sample variance has been substituted in place of the population variance. We did find one distribution that is very similar to the standard normal distribution: the *t* distribution. See Chapter 13. In particular, the random variable

$$T = \frac{\overline{x} - \mu}{\sqrt{s_2^2 / n}}$$

will come from a *t* distribution with $n - 1$ degrees of freedom. (See the mathematical section at the end of the chapter.)

Therefore, our confidence interval equation changes from

$$\Pr\left(\frac{-c\sqrt{n}}{\sigma} < \frac{\overline{x} - \mu}{\sqrt{\sigma^2 / n}} < \frac{-c\sqrt{n}}{\sigma}\right)$$

to

$$\Pr\left(\frac{-c\sqrt{n}}{s_2} < \frac{\overline{x} - \mu}{\sqrt{s_2^2 / n}} < \frac{c\sqrt{n}}{s_2}\right)$$

The term in the middle is our random variable T. Now make the definition $a = c\sqrt{n}/s_2$, and the equation becomes

$$\Pr(-a < T < a) = .95$$

If we have a sample size of $n = 20$, then we need to look in Table A3–5 to find a value for a t distribution with $n - 1 = 19$ degrees of freedom. Looking in the column for .95, we find the value $a = 2.093$. Once we know the value of a, we can find the value of the margin of error c (which is half of the width of the confidence interval):

$$c = \frac{as_2}{\sqrt{n}}$$

and the 95 percent confidence interval for the mean μ is from $\overline{x} - as_2/\sqrt{n}$ to $\overline{x} + as_2/\sqrt{n}$.

EXAMPLE In order to determine the average commute time for employees at a large company, you have surveyed a randomly chosen sample of 20 employees with these results. The commute times in minutes are

$$25, 18, 33, 42, 5, 12, 21, 18, 17, 23, 38, 15, 19, 22, 18, 17, 33, 29, 31, 9$$

SOLUTION

Enter these numbers into Excel and use the AVERAGE and STDEV functions to find

$$\overline{x} = 22.25$$
$$s_2 = 9.6019$$

The margin of error is

$$c = \frac{as}{\sqrt{n}} = \frac{2.093 \times 9.6019}{\sqrt{20}} = 4.4938$$

The confidence interval is

$$22.25 \pm 4.4938$$
$$17.76 \text{ to } 26.74$$

If our sample size had been n = 121, then we would have 120 degrees of freedom, and the value from Table A3–5 would be 1.98. Note that this is very close to the value 1.96 that we would have obtained from the standard normal distribution table. Since the t distribution approaches the standard normal distribution as the number of degrees of freedom increases, the t distribution table is needed only for relatively small samples. For larger samples, you can use the value $a = 1.96$ that comes from the standard normal distribution. If you are unsure if the sample is large enough to do this, look at Table A3–5 to see how close the value would be to 1.96 for the number of degrees of freedom that apply to your situation.

In general, here is the procedure to calculate confidence intervals using the t distribution if you have n observations of the random variable X.

CALCULATING CONFIDENCE INTERVALS FOR THE POPULATION MEAN WITH THE *t* DISTRIBUTION

1. Decide on the confidence level (CL). (.95 is the most common level. .90 and .99 are also used often.)
2. Calculate $\overline{x}$:

$$\overline{x} = \frac{x_1 + x_2 + \cdots + x_n}{n}$$

3. Calculate s_2:

$$s_2 = \sqrt{\frac{(x_1 - \overline{x})^2 + (x_2 - \overline{x})^2 + \cdots + (x_n - \overline{x})^2}{n - 1}}$$

$$= \sqrt{\frac{n}{n-1}(\overline{x^2} - \overline{x}^2)}$$

4. Look up the value of a in Table A3–5 for $n - 1$ degrees of freedom. (Note that if n is larger than 30, the t distribution is almost the same as the standard normal distribution.)

$$\Pr(-a < T < a) = \text{CL}$$

5. The confidence interval for μ is from

$$\overline{x} - s_2 a \,/\, \sqrt{n} \text{ to } \overline{x} + s_2 a \,/\, \sqrt{n}.$$

The Distribution of the Sample Variance

WARNING: mathematical section

Now we have to show that the distribution of the sample variance s_1^2 is related to the chi-square distribution. To start with, we'll assume that now we do know the true value of μ, the mean. We'll define the statistic s^2 like this:

$$s^2 = \frac{(X_1 - \mu)^2 + (X_2 - \mu)^2 + \cdots + (X_n - \mu)^2}{n}$$

The quantity s^2 is an estimator for the variance of X. We will rewrite the formula for s^2 as follows. (You may object that we shouldn't put σ into that formula, since we don't know the value of σ, but we're going to do it anyway.)

$$s^2 = \frac{\sigma^2}{n}\left[\frac{(X_1 - \mu)^2 + (X_2 - \mu)^2 + \cdots + (X_n - \mu)^2}{\sigma^2}\right]$$

$$= \frac{\sigma^2}{n}\left[\left(\frac{X_1 - \mu}{\sigma}\right)^2 + \left(\frac{X_2 - \mu}{\sigma}\right)^2 + \cdots + \left(\frac{X_n - \mu}{\sigma}\right)^2\right]$$

If we let $Z_1 = (X_1 - \mu)/\sigma$, $Z_2 = (X_2 - \mu)/\sigma$, and so on, then each Z has a standard normal distribution. That means that Z^2 has a chi-square distribution with one degree of freedom. Therefore, if we define Y^2 as

$$Y^2 = \left(\frac{X_1 - \mu}{\sigma}\right)^2 + \left(\frac{X_2 - \mu}{\sigma}\right)^2 + \cdots + \left(\frac{X_n - \mu}{\sigma}\right)^2$$

then Y^2 has a χ_n^2 distribution. Since

$$s^2 = \frac{\sigma^2}{n} Y^2$$

we can tell that $(n/\sigma^2)s^2$ has a χ_n^2 distribution.

However, in general we cannot use s^2 as an estimate of the variance, since we often don't know the true value of μ. However, we did discuss the statistic s_1^2, the sample variance, which is calculated using the sample average $\bar{x}$ as an estimator for the population mean:

$$s_1^2 = \frac{(X_1 - \bar{x})^2 + (X_2 - \bar{x})^2 + \cdots + (X_n - \bar{x})^2}{n}$$

Now we'll try to find the distribution of s_1^2. We know about the distribution of Y^2, so we'll start by looking at that:

$$Y^2 = \left(\frac{X_1 - \mu}{\sigma}\right)^2 + \left(\frac{X_2 - \mu}{\sigma}\right)^2 + \cdots + \left(\frac{X_n - \mu}{\sigma}\right)^2$$

We can change it around into a form that is more meaningful. Unfortunately, we have to make the expression worse before we can make it better.

$$\begin{aligned} Y^2 &= \frac{1}{\sigma^2}[(X_1 - \bar{x} + \bar{x} - \mu)^2 + \cdots + (X_n - \bar{x} + \bar{x} - \mu)^2] \\ &= \frac{1}{\sigma^2}[(X_1 - \bar{x})^2 + 2(X_1 - \bar{x})(\bar{x} - \mu) + (\bar{x} - \mu)^2 \\ &\quad + (X_2 - \bar{x})^2 + 2(X_2 - \bar{x})(\bar{x} - \mu) + (\bar{x} - \mu)^2 + \cdots \\ &\quad + (X_n - \bar{x})^2 + 2(X_n - \bar{x})(\bar{x} - \mu) + (\bar{x} - \mu)^2] \end{aligned}$$

All of those terms starting with 2 in the middle column quite conveniently add up to zero. Therefore,

$$Y^2 = \frac{1}{\sigma^2}[(X_1 - \bar{x})^2 + (X_2 - \bar{x})^2 + \cdots + (X_n - \bar{x})^2] + \frac{n}{\sigma^2}(\bar{x} - \mu)^2$$

We defined the statistic s_1^2 as follows:

$$s_1^2 = \frac{(X_1 - \bar{x})^2 + (X_2 - \bar{x})^2 + \cdots + (X_n - \bar{x})^2}{n}$$

Therefore, after all the dust clears, we can rewrite the expression for Y^2:

$$Y^2 = \frac{n}{\sigma^2} s_1^2 + \frac{n}{\sigma^2}(\bar{x} - \mu)^2$$

Remember that $\bar{x}$ has a normal distribution with mean μ and variance σ^2/n, so

$$Z = \sqrt{n}\left(\frac{\bar{x} - \mu}{\sigma}\right)$$

has a standard normal distribution. Therefore,

$$Z^2 = \frac{n(\bar{x} - \mu)^2}{\sigma^2}$$

has a χ_1^2 distribution. Then

$$\frac{n}{\sigma^2} s_1^2 = Y^2 - Z^2$$

where Y^2 has a chi-square distribution with n degrees of freedom, and Z^2 has a chi-square distribution with 1 degree of freedom. If you know something about chi-square random variables, you're likely to guess that ns_1^2/σ^2 has a chi-square distribution with $n - 1$ degrees of freedom. And, in fact, that guess turns out to be right.

Let $Y_{n-1}^2 = Y^2 - Z^2$ represent our chi-square random variable with $n - 1$ degrees of freedom. Then

$$\frac{n}{\sigma^2} s_1^2 = Y_{n-1}^2$$

Rewrite the formula for s_1^2:

$$s_1^2 = \frac{\sigma^2}{n} Y_{n-1}^2$$

Now we can find the expectation of s_1^2:

$$E(s_1^2) = E\left(\frac{\sigma^2}{n} Y_{n-1}^2\right)$$

Since σ and n are constants, they can be moved outside the expectation:

$$E(s_1^2) = \left(\frac{\sigma^2}{n}\right) E(Y_{n-1}^2)$$

Since the expectation of a chi-square random variable with $n - 1$ degrees of freedom is equal to $n - 1$, we have

$$\begin{aligned} E(s_1^2) &= \left(\frac{\sigma^2}{n}\right)(n - 1) \\ &= \frac{(n-1)\sigma^2}{n} \end{aligned}$$

Since the expected value of the sample standard variance s_1^2 is not the same as the population variance σ^2, this means that s_1^2 is *not* an unbiased estimator for the population variance. However, define s_2^2 as follows:

$$s_2^2 = \left(\frac{n}{n-1}\right) s_1^2$$

Now find the expectation of s_2^2:

$$E(s_2^2) = E\left(\left(\frac{n}{n-1}\right) s_1^2\right)$$

Move the constant $n/(n - 1)$ outside the expectation:

$$E(s_2^2) = \left(\frac{n}{n-1}\right) E(s_1^2)$$

We know $E(s_1^2) = (n - 1)\sigma^2/n$:

$$E(s_2^2) = \left(\frac{n}{n-1}\right)\left(\frac{(n-1)\sigma^2}{n}\right)$$
$$E(s_2^2) = \sigma^2$$

This verifies our claim that s_2^2 is an unbiased estimator of the population variance σ^2. Since s_2^2 is defined as

$$s_2^2 = \left(\frac{n}{n-1}\right)s_1^2$$

and s_1^2 is defined as

$$s_1^2 = \frac{(X_1 - \overline{x})^2 + (X_2 - \overline{x})^2 + \cdots + (X_n - \overline{x})^2}{n}$$

we can find the formula for s_2^2:

$$s_2^2 = \frac{(X_1 - \overline{x})^2 + (X_2 - \overline{x})^2 + \cdots + (X_n - \overline{x})^2}{n}\left(\frac{n}{n-1}\right)$$
$$s_2^2 = \frac{(X_1 - \overline{x})^2 + (X_2 - \overline{x})^2 + \cdots + (X_n - \overline{x})^2}{n-1}$$

Now we can see that the sample variance s_2^2 is defined with $n - 1$ in the denominator so that it will be an unbiased estimator for σ^2.

Next, we need to show that the random variable we called T will truly have the t distribution. The expression for T is

$$T = \frac{\overline{x} - \mu}{\sqrt{s_2^2/n}}$$

which we can rewrite as

$$T = \frac{\sqrt{n}(\overline{x} - \mu)}{\sqrt{s_2^2}}$$

Divide both the top and bottom of the fraction by σ:

$$T = \frac{\dfrac{\sqrt{n}(\overline{x} - \mu)}{\sigma}}{\sqrt{\dfrac{s_2^2}{\sigma^2}}}$$

The top part of the fraction represents a standard normal random variable Z:

$$T = \frac{Z}{\sqrt{\dfrac{s_2^2}{\sigma^2}}}$$

Multiply the fraction inside the square root sign by $(n-1)/(n-1)$:

$$T = \frac{Z}{\sqrt{\dfrac{(n-1)s_2^2}{(n-1)\sigma^2}}}$$

The expression $(n-1)\, s_2^2/\sigma^2$ that appears under the square root sign is the same as ns_1^2/σ^2, which is the same as our chi-square random variable Y_{n-1}^2:

$$T = \frac{Z}{\sqrt{\dfrac{Y_{n-1}^2}{(n-1)}}}$$

In words: the random variable T is equivalent to a standard normal random variable, divided by the square root of a chi-square random variable, divided by its degrees of freedom. That expression matches the definition of the t distribution (see Chapter 13, page 164), so this confirms that our random variable T has a t distribution with $n-1$ degrees of freedom.

NOTES TO CHAPTER 17

It is important to note that there is a subtle point involved with the calculation of confidence intervals. If you calculate a 95 percent confidence interval, that does not exactly mean that there is a 95 percent chance that μ is contained in the interval. The mean μ is a constant, even if we don't know its value. Instead, it is the interval $\overline{x} - c$ to $\overline{x} + c$ itself that is a random interval, so we are calculating the probability that this random interval will be chosen so that it happens to contain μ.

One more important point about the t distribution needs to be made. The definition of the t distribution requires that the standard normal random variable Z and the chi-square random variable Y^2 must be independent. But if we look at these expressions:

$$Z = \frac{\sqrt{n}(\overline{x} - \mu)}{\sigma}$$

$$Y_2^2 = \left(\frac{n-1}{\sigma^2}\right) s_2^2$$

we can see that Z depends on the sample mean and Y_2^2 depends on the sample variance. Since these two quantities are both calculated from the same sample, it doesn't seem as if they could be independent. However, when a sample is taken from a normal distribution it has the amazing property that the sample mean and the sample variance are independent. The proof of this fact is too hard and boring to include in this book, though. And you should note that, in general, when you are sampling from a distribution that is not a normal distribution, the sample mean and the sample variance will *not* be independent.

EXERCISES

1. Calculate a 95 percent confidence interval for the mean for the following random sample taken from a normal distribution with variance 11: 1.1, 0.2, 1.9, 4.1, 5.3, –6, .5, –.6

2–6. Calculate 95 percent confidence intervals for the means of the sets of numbers from Chapter 16, Exercises 2–6.

7. Calculate 95 percent confidence intervals for the test scores from the four subjects discussed in Chapter 3.

Calculate 99 percent confidence intervals for the mean if you have these observations for a random variable with a normal distribution with unknown mean and variance:

8. 18, 9, 15, 10, 16, 8, 7, 20, 13, 8, 12

9. 17, 13, 9, 8, 10, 13, 16, 17, 12, 11, 17

10. 15, 13, 13, 14, 15, 20, 15, 14, 9, 16, 12

11. 16, 12, 13, 12, 14, 11, 8, 13, 17, 15, 14

12. Show that s_2^2 is an unbiased estimator for the variance for a random sample.

13. Suppose a random sample of size 2 (X_1, X_2) is taken for a random variable with an unknown distribution. Calculate the expectation of

$$\frac{(X_1 - \overline{x})^2 + (X_2 - \overline{x})^2}{n - 1}$$

☆ 14. Show that $\sum_{i=1}^{n} 2(X_i - \overline{x})(\overline{x} - \mu) = 0$.

15. Create a spreadsheet that graphs the standard normal distribution and a t distribution with DF degrees of freedom.

16. Toss two dice 15 times. In this case we know that the mean of each observation is 7. However, pretend for now that you don't know the mean, and then calculate a 95 percent confidence interval for the mean, using the available data. Then, repeat the entire procedure 100 times. How many times did your estimated confidence interval contain the true value for the mean?

17. Estimate a 95 percent confidence interval for your grocery bill for several weeks.

18. Keep a record of the high temperature in your town every day this month. Then calculate a 95 percent confidence interval for the mean.

For each of the following values of $\overline{x}$, n, and s_2, calculate a 95 percent confidence interval for the mean.

	$\bar{X}$	n	s_2
19.	60	10	8
20.	60	20	8
21.	60	50	8
22.	60	100	8
23.	60	200	8
24.	60	500	8
25.	60	1,000	8
26.	60	50	0.5
27.	60	50	1
28.	60	50	5
29.	60	50	10
30.	60	50	50

18

Hypothesis Testing

In Chapter 3 we considered a specific hypothesis testing problem: If you toss a coin many times, how can you tell whether or not the coin is fair? Now we'll consider a more general treatment of the methodology that statisticians use when they formulate and test hypotheses.

Remember that the hypothesis that we want to test is called the null hypothesis (or H_0), and the hypothesis that says, "The null hypothesis is wrong" is called the alternative hypothesis. Examples of null hypotheses include:

- A coin is fair.
- The mean number of raisins in boxes of a particular brand of raisin cereal is 7.
- The difference in effectiveness between four cold medicines occurred entirely by chance.
- The rate of appointments to the U.S. Supreme Court fits the Poisson distribution.

If we decide to reject the null hypothesis, that means that we are almost sure the hypothesis is not true. More specifically, we usually design our test so that there is only a 5 percent chance that we would have rejected the hypothesis if it were really true. However, if we decide to accept the hypothesis, that does not mean for sure that the hypothesis is true. It just means that we have not yet found statistical evidence to reject it.

Test Statistics

The normal procedure in statistics is to calculate a specific quantity called a *test statistic*. There are several common test statistics. The one that you use depends on the problem you are facing. We will consider several examples in this chapter.

The test statistic is designed so that *if* the null hypothesis is true, you know exactly what the distribution of the test statistic is. Then you have to ask yourself: Suppose the null hypothesis is true. In that case, is the observed value of the test statistic a very plausible value? If the observed test statistic value is very unlikely to have occurred, then you figure that most likely the hypothesis is false.

For example, suppose you are testing a null hypothesis using a test statistic Z. Suppose you know that Z will have a standard normal distribution if the null hypothesis is true. Calculate the value of Z. If, for example, the value of Z turns out to be .878, then everything is fine. There is a reasonably good chance of drawing the number .878 from a standard normal distribution. Since the observed value is not particularly implausible, you have no grounds for rejecting the hypothesis.

However, suppose the observed value of the test statistic Z turned out to be 3. Then you should begin to get suspicious. You can see from the standard normal table that there is a probability of only .0026 that a standard normal random variable will be outside 3: (We will use the terms inside and outside in the following fashion. We will say that Z is *inside* a value c if $-c < Z < c$. We will say that Z is *outside* a value c if $Z < -c$ or if $Z > c$. In other words, Z is inside c if $|Z| < c$, and Z is outside c if $|Z| > c$.) You should say to the advocates of the null hypothesis, "You can't pull the wool over my eyes. I know that this test statistic value is very unlikely to have occurred if the null hypothesis were true, so I'm going to reject the hypothesis."

The null hypothesis advocates might respond, "If you reject the null hypothesis then you will be committing a type 1 error, since we think that the null hypothesis is really true. We admit that we had bad luck with our test statistic, and it turned out to have an implausible value. But it is still possible that you might draw the number 3 from a standard normal distribution."

Of course, there is no way that you can prove them wrong. There still is a slight possibility that the null hypothesis might be true, so you could commit a type 1 error by erroneously rejecting the hypothesis. But that is the risk that you will have to take. (Remember that a type 1 error occurs if you reject the null hypothesis when it is really true. A type 2 error occurs if you accept the null hypothesis when it is really false. See Chapter 3.) Normally, we design our test so that the risk of committing a type 1 error is less than 5 percent. The risk of committing a type 1 error is called the *level of significance* of the test. Therefore, we can say that our test is designed to be at the 5 percent significance level. The probability of a type 1 error is symbolized by α (alpha); the probability of a type 2 error is symbolized by β (beta).

From the standard normal tables we can see that there is a 95 percent chance that Z will be inside 1.96, so we will design our test so that the null hypothesis is accepted if Z is inside 1.96 and rejected otherwise. Therefore, we will call the region inside 1.96 the zone of acceptance and the region outside 1.96 the rejection region or critical region (see Figure 18–1). Sometimes the number that is the boundary between the rejection region and the zone of acceptance is called the *critical value* of the test statistic. In this case the critical values are 1.96 and –1.96.

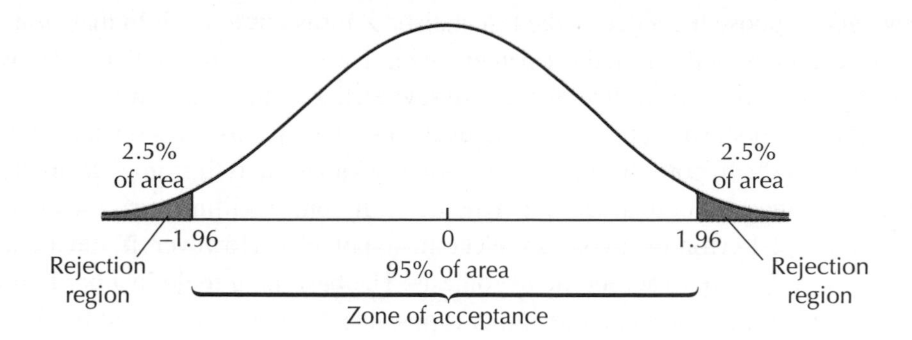

FIGURE 18–1

We can see that with this test:

$$\begin{aligned}\text{Level of significance} &= \Pr(\text{type 1 error}) \\ &= \Pr(\text{rejecting } H_0 \text{ if it is really true}) \\ &= \Pr[(Z > 1.96) \text{ or } (Z < 1.96)] \\ &= .025 + .025 \\ &= .05\end{aligned}$$

This is the result we want. If the observed value of the test statistic turns out to be outside 1.96, then we will say that we can reject the hypothesis at the 5 percent significance level.

However, suppose that we want to be more cautious. Suppose that it is very costly for us to reject the hypothesis erroneously, so we want to make sure that the probability of this event happening is only 1 percent. Then we need to widen the zone of acceptance (see Figure 18–2). There is a 99 percent chance that Z will be inside 2.58. Therefore, we can ensure that there is only a 1 percent chance of committing a type 1 error if we design our test so that the zone of acceptance runs from –2.58 to 2.58. If the value of the test statistic turns out to be –2.6, we can reject the hypothesis at the 1 percent significance level. (Unfortunately, this is confusing terminology, since a *more* significant test corresponds to a *lower* significance level.)

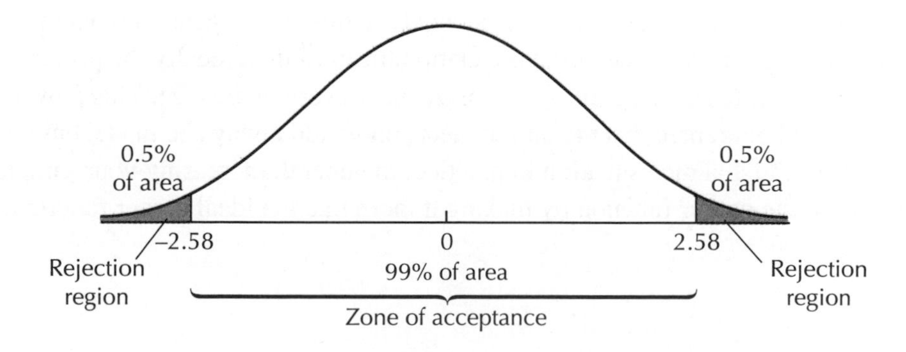

FIGURE 18–2

However, suppose the value of the test statistic Z turns out to be 2. In that case we cannot reject the hypothesis at the 1 percent level. If we want a test at that level, we must accept the hypothesis. However, as we saw earlier, with a test statistic of 2 we can reject the hypothesis at the 5 percent level. Test statistics like this one are in a sort of gray area. Is the hypothesis really true? Nobody knows, and this time we are not even sure whether or not to accept the hypothesis. If you're willing to risk a 5 percent chance of a type 1 error, then you can reject the hypothesis. However, if you are more cautious you will have to accept the hypothesis. The best thing to do in a borderline case is to collect more observations if that is possible, because then it will be easier to distinguish between a true hypothesis and a false hypothesis.

The situation is much more clear-cut when you get a test statistic such as 3 or larger. In that case you can reject the hypothesis at every significance level.

It will be helpful to remember these critical values: if Z has a standard normal distribution when the null hypothesis is true, then

- if $-1.96 < Z < 1.96$, accept the hypothesis at the 5 percent level.
- if $-2.58 < Z < 2.58$, accept the hypothesis at the 1 percent level.

When a hypothesis test is performed, the *p value* is the probability of getting a sample as extreme as the given one or worse, given that the null hypothesis is true. If the p-value is smaller than the significance level, the null hypothesis is rejected. The smaller the p value, the stronger is your evidence that the null hypothesis should be rejected.

For example, suppose you are performing a hypothesis test based on a standard normal (Z) random variable. If the test statistic value is $Z = 3$, then the p value is .0026. This is the area in both tails outside the range –3 to 3. From Table A3–1 we find $\Pr(Z < 3) = .9987$. From this we can tell that $\Pr(Z > 3) = 1 - .9987 = .0013$. Also $\Pr(Z < -3) = .0013$; therefore, $\Pr(Z > 3) + \Pr(Z < -3) = .0026$. A p value this small gives you strong evidence to reject the null hypothesis.

The *power* of a hypothesis test is the probability that the null hypothesis will be rejected, expressed as a function of the parameter being investigated. For example, if you were testing for the value of the population mean μ, then ideally the power function would equal 0 at the true value of μ, and 1 everywhere else. This power function would guarantee that the correct decision would always be made, but you can seldom expect such a nice situation in practice. In general, increasing your sample size will improve the power function by making it more like the ideal power function.

Testing the Value of the Mean

Now we'll see what test statistics arise in actual practice. Suppose we can observe a sequence of numbers drawn from a normal distribution. Suppose that we know the variance, but not the mean, of the distribution. We need to test the hypothesis that μ equals a particular value μ^*.

For example, suppose that we're quality control inspectors investigating the number of raisins in each (small) box of a raisin cereal. If there are too few raisins in the box, customers will complain. If there are too many, then the company will lose money on each box of cereal sold. The raisins are put into the boxes by an Automatic Raisin Packer. We know that the machine works in such a way that the number of raisins in each box has a normal distribution with variance 16.16 ($\sigma = 4.02$). On the average, the boxes are supposed to have 7 raisins. Our mission is to test the null hypothesis that the mean μ is equal to 7. We have $n = 13$ observations for the mean:

$$9, 11, 6, 10, 7, 4, 0, 7, 8, 6, 8, 2, 18$$

The sample average $\overline{x}$ is 7.38. Is that close enough to 7 so that we should accept the hypothesis? Or is it too far away? We know that if the hypothesis is true, then $\overline{x}$ will have a normal distribution with mean $\mu = 7$ and variance $16.16/n$. Therefore,

$$Z = \frac{\sqrt{n}(\overline{x} - 7)}{\sigma} = \frac{\sqrt{13}(7.38 - 7)}{4.02}$$

will have a standard normal distribution. Therefore, Z will be our test statistic. In our case, the computed value of Z is .341, which is well within the zone of acceptance. So we can accept the hypothesis that $\mu = 7$.

Suppose the mean number of raisins had been 9.3. Then the test statistic value would be

$$\frac{\sqrt{13}(9.3 - 7)}{4.02} = 2.06$$

This value is outside 1.96, so we reject the null hypothesis. Figure 18–3 illustrates the zone of acceptance and the rejection region.

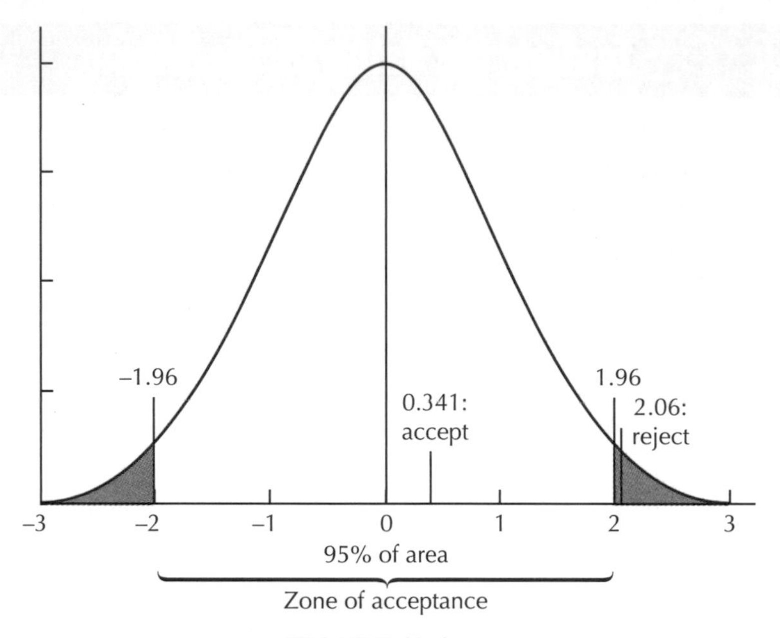

FIGURE 18–3

Of course, in general we cannot use the statistic $z = \sqrt{n}(\overline{x} - \mu^*)/\sigma$, because we ordinarily won't know the true value of σ. However, if the null hypothesis $\mu = \mu^*$ is true, then the test statistic

$$t = \sqrt{n}\left(\frac{\overline{x} - \mu^*}{s_2}\right)$$

will have a t distribution with $n - 1$ degrees of freedom (see Chapter 17).

For example, suppose that you have the following data points representing the weights of 27 sample players on a particular football team:

160, 185, 235, 208, 170, 185, 204, 180, 205, 215,
185, 188, 180, 220, 220, 221, 205, 235, 225, 190,
180, 205, 250, 210, 230, 210, 218

You want to test the hypothesis that these weights were selected from a normal distribution with mean 220. You need to calculate the two statistics $\overline{x} = 204.4$ and $s_2 = 22.1$. Then you can calculate the test statistic t:

$$t = \frac{(204.4 - 220)}{22.2}\sqrt{27} = -3.65$$

If the hypothesis is true, t will have a t distribution with 26 degrees of freedom. If you look up the results in a t table you can see that the critical value for a 1 percent test is 2.779. In other words, you can reject the null hypothesis at the 1 percent level if the value of the test statistic is outside 2.779. Since –3.65 is in the critical region, you have good statistical evidence to reject the hypothesis that the sample of football players was selected from a population with mean 220.

HYPOTHESIS TESTING FOR THE MEAN

General procedure to test the hypothesis that the mean $\mu = \mu^*$, when you have observed n values taken from a normal distribution:

Method 1. Use this method if you *know* the variance (σ^2) of the distribution.

1. Calculate the sample average $\bar{x}$.
2. Calculate the test statistic Z:

$$Z = \frac{\sqrt{n}(\bar{x} - \mu^*)}{\sigma}$$

3. If you want to test the hypothesis at the 5 percent significance level, then accept the hypothesis that $\mu = \mu^*$ if Z is between -1.96 and 1.96; otherwise reject the hypothesis.
4. If you want to test the hypothesis at another significance level, then look in Table A3–2 to find the critical value for Z.

Method 2. Use this method if you *don't know* the variance of the distribution.

1. Calculate the sample average $\bar{x}$.
2. Calculate the sample variance ${s_2}^2$:

$${s_2}^2 = \frac{(X_1 - \bar{x})^2 + (X_2 - \bar{x})^2 + \cdots + (X_n - \bar{x})^2}{n - 1}$$

3. Calculate the statistic t:

$$t = \sqrt{n}\frac{(\bar{x} - \mu^*)}{s_2}$$

4. The t statistic will have a t distribution with $n - 1$ degrees of freedom if the hypothesis is true. Look in Table A3–5 to find the critical value for the t distribution with the appropriate degrees of freedom.

One-tailed Tests

Suppose that you are a quality control inspector for a semiconductor firm that buys silicon wafers from a particular supplier. Each wafer has a certain number of defects. If there are too many defects, you must reject the wafer. The supplier tells you that, on the average, there are 14 defects per wafer. Your job is to find out whether or not the supplier is right. You have checked the number of defects for a sample of 17 wafers, with these results:

$$7, 16, 19, 12, 15, 9, 6, 16, 14, 7, 2,$$
$$15, 23, 15, 12, 18, 9$$

You want to test the hypothesis that the number of defects on each wafer has a normal distribution with mean $\mu = 14$. However, suppose it turns out that you can reject the hypothesis that $\mu = 14$ because the sample average is significantly less than 14. In that case you'll be totally happy—you surely won't complain to the supplier if the number of defects is less than is advertised. So you don't really want to test the null hypothesis that $\mu = 14$. Instead, you want to test the null hypothesis that $\mu \leq 14$. If you can reject this null hypothesis, then you will complain to the supplier. You can use the same t statistic again. The only difference is that, this time, you will only reject the null hypothesis if the value of the t statistic is in the region where the top 5 percent of the area is located. Figure 18–4 illustrates the rejection region and the zone of acceptance for this test. In our case, we will have a t distribution with 16 degrees of freedom. If we look in Table A3–4 we can see that the rejection region is located for values of t above 1.746.

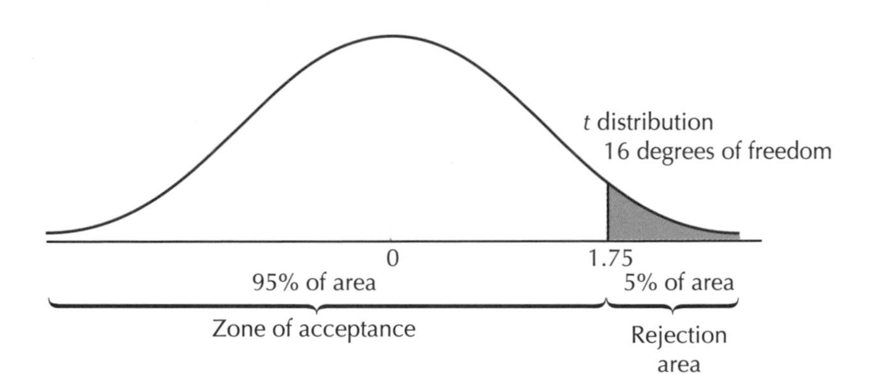

FIGURE 18–4

This type of test is called a *one-tailed test*, because the rejection region consists of only one tail of the distribution. In a one-tailed test the null hypothesis is rejected only if the test statistic has a value significantly greater than expected. (Or we could do a one-tailed test using only the left-hand tail, in which case the null hypothesis is rejected only if the test statistic value is significantly less than expected.)

The tests that we did before are called two-tailed tests. In a two-tailed test, you reject the null hypothesis if the test-statistic value is either very low or very high. Which type of test you should use depends on your situation. Normally, if the null hypothesis involves an inequality, such as $\mu \leq \mu^*$, you will want to use a one-tailed test. If the null hypothesis involves an equality, such as $\mu = \mu^*$, then you will use a two-tailed test.

In the semiconductor case, the sample average is $\bar{x} = 12.647$, so obviously you cannot reject the hypothesis that $\mu \leq 14$. The value of s_2 is 5.396, so the test statistic value is

$$T = \sqrt{n}\left(\frac{\bar{x} - 14}{s_2}\right) = -1.033$$

Testing Hypotheses About the Probability of Success

Now we return to the problem of attempting to tell whether or not a coin is fair. To make things easier, let's assume that we have performed enough tosses so that we can approximate the binomial distribution by the normal distribution. In that case, if the hypothesis $p = p^*$ is true, then the number of heads (X) will have a normal distribution with mean $\mu = p^*N$ and variance $\sigma^2 = Np^*(1 - p^*)$. The variable $Z = (X - \mu)/\sigma$ will have a standard normal distribution.

Now, suppose that we have tossed the coin 10,000 times, and we ended up with 5,056 heads. Should we accept the hypothesis that the coin is fair? The fair-coin hypothesis says that $p = .5$, so X will have a normal distribution with $\mu = 5{,}000$ and $\sigma = 10{,}000/4$. We need to calculate the test statistic Z:

$$Z = \frac{X - 5{,}000}{\sqrt{10{,}000/4}} = \frac{56}{50} = 1.12$$

This value is within the 95 percent zone of acceptance, so we can accept this hypothesis at the 5 percent significance level.

For another example, suppose there were 4,884 heads that resulted from the 10,000 tosses. In that case the value of the test statistic is –116/50 = –2.32. This value is outside 1.96, so we can reject the fair coin hypothesis at the 5 percent level (and the person who supplied the coin has some explaining to do). However, the value –2.32 is not outside the zone of acceptance at the 1 percent level, so if we want to be more cautious we cannot reject the fair-coin hypothesis.

Return to the first situation, in which 5,056 heads appeared in 10,000 flips. Suppose that we don't actually believe that the coin is fair. Instead, we think that the coin is unbalanced so that heads are slightly more likely to appear. In particular, we'll test the hypothesis that $p = .51$. If this null hypothesis is true, then X will have a normal distribution with mean 5,100 and variance $10{,}000 \times .51 \times .49 = 2{,}499$. In this case the value of our test statistic is –44/49.98 = –0.880. This value is well within the zone of acceptance, so we can't reject the hypothesis that $p = .51$. However, we have already found that we also can't reject the hypothesis that $p = .5$. There is no way that both of these hypotheses can be right, but we have no way to tell the two apart using only the information available to us. You probably could have guessed that it would be very difficult to tell the difference between these two hypotheses.

This fact illustrates that hypothesis testing methods can do a good job of proving hypotheses wrong, but they often can't do a very good job of proving hypotheses right. Even if you decide to accept the null hypothesis, that does not mean that there is not some other hypothesis that can also adequately account for the data. If you want to make a very convincing case that your hypothesis is true, then you will have to be able to reject all of the likely competing hypotheses. Since we haven't been able to do that in the coin example, we cannot say for sure that the coin is fair.

We have only one hope—if we flip the coin many, many times, then finally we will reach the point where we can tell the difference between the hypotheses $p = .5$ and $p = .51$. However, in the real world you often cannot increase your sample size by a lot. If you find yourself with two competing hypotheses neither of which can be rejected by the available data, then you're stuck.

Suppose you flip the coin 39,000 times, with 19,680 heads. If the null hypothesis is $p = .5$, our test statistic becomes

$$(19{,}680 - 39{,}000 \times .5)/\sqrt{39{,}000 \times .5 \times .5} = 180/\sqrt{9{,}750} = 1.82$$

This falls in the acceptance region. If the null hypothesis is $p = .51$, our test statistic becomes

$$(19{,}680 - 39{,}000 \times .51)/\sqrt{39{,}000 \times .51 \times .49} = -210/\sqrt{9{,}746.1} = -2.13$$

This falls in the rejection region. Therefore, with 39,000 tosses we are finally able to distinguish between the hypotheses $p = .5$ and $p = .51$.

Testing for the Difference Between Two Means

Suppose that you have two different populations whose means you want to compare. Assume that the random variables X_a (mean μ_a, variance σ_a^2) and X_b (mean μ_b, variance σ_b^2) have approximately normal distributions. If you take sample sizes of n_a and n_b, respectively, then the sample means $\overline{x_a}$ and $\overline{x_b}$ are normal random variables, and their difference is a normal random variable with mean $\mu_a - \mu_b$ and variance $\sigma_a^2/n_a + \sigma_b^2/n_b$.

Often we will want to test hypotheses about the difference between the population means $\mu_a - \mu_b$. For example, our null hypothesis might be that the population means are equal:

$$\mu_a = \mu_b \quad \text{or} \quad \mu_a - \mu_b = 0$$

More generally, let our null hypothesis be that

$$\mu_a - \mu_b = D$$

If the population variances σ_a^2 and σ_b^2 are known, we can form the test statistic Z:

$$Z = \frac{\overline{x_a} - \overline{x_b} - D}{\sqrt{\sigma_a^2/n_a + \sigma_b^2/n_b}}$$

This statistic will have a standard normal distribution if the null hypothesis is true.

For example, suppose we have two six-sided dice, A and B, which may be biased. We suspect that die A gives, on average, a value 0.7 larger than that of die B. Die A is rolled 20 times and gives the following values:

$$4, 5, 3, 6, 3, 5, 6, 3, 3, 6, 5, 1, 4, 2, 6, 1, 5, 5, 6, 2$$

Die B is rolled 15 times and gives the following values:

$$4, 3, 5, 4, 3, 2, 5, 1, 4, 1, 5, 6, 3, 6, 1$$

The manufacturer tells us that $\sigma_a^2 = 3.0$ and $\sigma_b^2 = 2.8$. We will assume that the manufacturer is statistically honest even though the dice may be shady. We can calculate our test statistic:

$$\overline{x_a} = 4.05 \quad \overline{x_b} = 3.53 \quad \sqrt{\frac{\sigma_a^2}{n_a} + \frac{\sigma_b^2}{n_b}} = .580$$

Therefore,

$$Z = \frac{4.05 - 3.53 - 0.7}{0.58} = -0.310$$

Since this is within the zone of acceptance, we will accept the null hypothesis.

If the population variances are unknown, we must turn once again to the t statistic. If the null hypothesis is true, and if we also know that $\sigma_a^2 = \sigma_b^2$, then

$$T = \frac{(\overline{x_a} - \overline{x_b}) - D}{\sqrt{s_p^2(1/n_a + 1/n_b)}}$$

has a t distribution with $n_a + n_b - 2$ degrees of freedom, where

$$s_p^2 = \frac{(n_a - 1)s_a^2 + (n_b - 1)s_b^2}{n_a + n_b - 2}$$

Upon sober reflection, we decide that the values for σ_a^2 *and* σ_b^2 given us by the dice manufacturer mentioned above were not reliable. We also decide that it is reasonable to assume that $\sigma_a^2 = \sigma_b^2$. Once again we will test the hypothesis that $\mu_a - \mu_b = 0.7$.

We can calculate that

$$s_a^2 = 2.892 \quad s_b^2 = 2.981 \quad s_p^2 = 2.930$$

and

$$T = \frac{4.05 - 3.53 - 0.7}{\sqrt{2.930(1/20 + 1/15)}} = -0.308$$

The random variable T will have a t distribution with 33 degrees of freedom if our null hypothesis is correct. Checking Table A3–5, we see that we will accept the null hypothesis at the 5 percent level, since –0.308 is inside the critical value of 2.030.

GENERAL PROCEDURE FOR HYPOTHESIS TESTING FOR THE DIFFERENCE BETWEEN TWO MEANS

Method 1 (if the values of σ_a^2 *and* σ_b^2 are known):

1. Calculate $\overline{x_a}$ and $\overline{x_b}$.

2. Calculate the test statistic Z:

$$Z = \frac{\overline{x_a} - \overline{x_b} - D}{\sqrt{\sigma_a^2/n_a + \sigma_b^2/n_b}}$$

3. If you want to test the hypothesis at the 5 percent significance level, then accept the hypothesis that $\mu_a - \mu_b = D$ if Z is between –1.96 and 1.96; otherwise reject the hypothesis.

4. If you want to test the hypothesis at another significance level, then look in Table A3–2 to find the critical value for Z.

Method 2: (if the values of σ_a^2 and σ_b^2 are unknown but assumed to be equal):

1. Calculate $\overline{x_a}$ and $\overline{x_b}$.

2. Calculate the sample variances (version 2, with $n - 1$ in the denominator) for both samples s_a^2 and s_b^2:

3. Calculate the pooled estimator s_p^2:

$$s_p^2 = \frac{(n_a - 1)s_a^2 + (n_b - 1)s_b^2}{n_a + n_b - 2}$$

4. Calculate the test statistic T:

$$T = \frac{(\overline{x_a} - \overline{x_b}) - D}{\sqrt{s_p{}^2(1/n_a + 1/n_b)}}$$

5. If the null hypothesis is true, then T will have a t distribution with $n_a + n_b - 2$ degrees of freedom. Look in Table A3–5 to find the critical values for the appropriate t distribution.

Paired Samples

Suppose you are investigating whether a random sample of students score higher on tests given on Friday or on Monday. The sample of 8 students gives this result:

Initials	Friday Test (*a*)	Monday Test (*b*)	Difference
M.Y.	98	90	8
B.K.	94	84	10
R.T.	91	90	1
G.A.	88	83	5
R.S.	86	80	6
T.J.	82	77	5
L.S.	80	76	4
J.B.	76	72	4

Your first temptation might be to use the test procedure for the difference of two means, using the null hypothesis $\mu_a - \mu_b = 0$, as in the preceding section:

$$s_p{}^2 = \frac{(n_a - 1)s_a^2 + (n_b - 1)s_b^2}{n_a + n_b - 2} = \frac{7 \times 7.3957^2 + 7 \times 6.50275^2}{8 + 8 - 2} = 48.491$$

$$T = \frac{(\overline{x_a} - \overline{x_b}) - 0}{\sqrt{s_p{}^2(1/n_a + 1/n_b)}} = 1.54$$

This result is inside the acceptance region for a t distribution with $8 + 8 - 2 = 14$ degrees of freedom, which could lead you to believe there is no statistically significant difference in test scores on the different days of the week.

However, glancing at the table, you can see that each student has a higher score on Friday than on Monday, which suggests there is a problem with the previous analysis. The problem is that there is so much variability in the scores between the different students that it tends to overwhelm the difference between the different days. The preceding analysis is inappropriate because we are not dealing with a random sample from Friday and an independent random sample from Monday. Instead, we have a paired sample; that is, a sample where we have pairs of values, one each for Friday and Monday, for each student. In this case we should perform the following t test:

1. Assume that the difference in scores between Monday and Friday for each student comes from a normal distribution with mean μ_D and variance σ_D^2.
2. The null hypothesis is that $\mu_D = 0$.
3. Let X_{Di} be the difference in scores for student i. Calculate the average $\overline{x_D} = (\Sigma X_{Di})/n$; this average will have a normal distribution with mean μ_D and variance σ_D^2/n.
4. Calculate the test statistic:

$$\frac{\overline{x_D}}{s_D/\sqrt{n}} = \frac{\sqrt{n}(\overline{x_D})}{s_D}$$

   Here s_D is the sample standard deviation (with $n - 1$ in the denominator) for the differences.

If the null hypothesis is true, this test statistic will come from a t distribution with $n - 1$ degrees of freedom.

For our example, we have $\overline{x_D} = 5.375$, $s_D = 2.722$, $n = 8$, and

$$T = \frac{\sqrt{8} \times 5.375}{2.722} = 5.58$$

This is well within the rejection region, so we reject the null hypothesis that there is no difference between Friday scores and Monday scores.

Testing for the Difference Between Two Proportions

Suppose now that we perform two series of tests. Suppose that test A is performed n_a times, with each test having an unknown probability p_a of success. If X_a is the number of successes, then $\hat{p}_a = X_a/n_a$ is an estimate of p_a. If n_a is large, then the central limit theorem says that $\hat{p}_a$ will have a normal distribution with mean p_a and variance $p_a(1 - p_a)/n_a$. Similarly, if X_b is the number of successes in n_b trials for test B, each with probability p_b of success, then for large n_b, $\hat{p}_b = X_b/n_b$ has a normal distribution with mean p_b and variance $p_b(1 - p_b)/n_b$. The difference $\hat{p}_a - \hat{p}_b$ also has a normal distribution with mean $(p_a - p_b)$ and variance equal to $[p_a(1 - p_a)/n_a + p_b(1 - p_b)/n_b]$.

If $p_a - p_b = D$, then the quantity

$$\frac{\hat{p}_a - \hat{p}_a - D}{\sqrt{p_a(1 - p_a)/n_a + p_b(1 - p_b)/n_b}}$$

has a standard normal distribution. If we hypothesize that $p_a - p_b = D$, then we want to test this hypothesis using the above statistic. Unfortunately, we need to know the values of p_a and p_b to calculate that statistic. But, if we knew these values, we wouldn't need to test the hypothesis in the first place.

How far off would we be if we substituted $\hat{p}_a$ for p_a and $\hat{p}_b$ for p_b? If n_a is large, then $|\hat{p}_a - p_a|$ will be small (by the law of large numbers), $|(p_a - \hat{p}_a) \times (1 - p_a - \hat{p}_a)|$ will be still smaller, and $|(p_a - \hat{p}_a)(1 - p_a - \hat{p}_a)/n_a|$ will be even smaller. The same reasoning applies to substituting $\hat{p}_b$ for p_b, and thus, if the null hypothesis is true and $p_a - p_b = D$, then the statistic Z,

$$Z = \frac{\hat{p}_a - \hat{p}_b - D}{\sqrt{\hat{p}_a(1 - \hat{p}_a)/n_a + \hat{p}_b(1 - \hat{p}_b)/n_b}}$$

has a standard normal distribution.

Suppose that two manufacturers, Abercrombie and Bayes, supply electric light bulbs to your corporation, which has many lamps. You suspect that Bayes's bulbs are less reliable then Abercrombie's and, in fact, that the probability of a Bayes bulb being defective is .001 greater than that of a defective Abercrombie bulb. Is your suspicion justified?

A random sample of 1,000 of Abercrombie's bulbs turns up 15 defective bulbs, while a random sample of 2,000 of Bayes's bulbs turns up 36 defective bulbs. Then

$$\hat{p}_a = \frac{15}{1{,}000} = .015 \quad \hat{p}_b = \frac{36}{2{,}000} = .018$$
$$\frac{\hat{p}_a(1 - \hat{p}_a)}{n_a} = .0000148 \quad \frac{\hat{p}_b(1 - \hat{p}_b)}{n_b} = .00000883$$

This means the value of Z is

$$Z = \frac{.015 - .018 - (-.001)}{\sqrt{.0000148 + .00000833}} = -.412$$

Thus, you would accept the hypothesis that $p_a - p_b = -.001$ at the 5 percent significance level, since $-1.96 < -.412 < 1.96$.

If $D = 0$, however, we can find a better estimator for p_a and p_b (which are assumed to be equal in this case). We will use the estimator $\hat{p} = (X_a + X_b)/(n_a + n_b)$. Substituting this for $\hat{p}_a$ and $\hat{p}_b$ in the denominator, we get

$$Z = \frac{\hat{p}_a - \hat{p}_b}{\sqrt{\hat{p}(1 - \hat{p})(1/n_a + 1/n_b)}}$$

Suppose we wish to test the null hypothesis that $p_a = p_b$. Then

$$\hat{p} = \frac{15 + 36}{1{,}000 + 2{,}000} = .017$$

$$\sqrt{\hat{p}(1 - \hat{p})\left(\frac{1}{n_a} + \frac{1}{n_b}\right)} = .005007$$

$$Z = \frac{.015 - .018}{.005007} = -.599$$

Thus, we would accept the null hypothesis at the 5 percent level, since $-1.96 < -.599 < 1.96$.

TESTING FOR THE DIFFERENCE BETWEEN TWO PROPORTIONS

General procedure for testing the hypothesis that $p_a - p_b = D$, given that test A has X_a successes in n_a trials, test B has X_b successes in n_b trials, and n_a and n_b are large:

Method 1: (If D does not equal zero):

1. Calculate $\hat{p}_a$ and $\hat{p}_b$:

$$\hat{p}_a = \frac{X_a}{n_a} \text{ and } \hat{p}_b = \frac{X_b}{n_b}$$

2. Calculate Z:

$$Z = \frac{\hat{p}_a - \hat{p}_b - D}{\sqrt{\hat{p}_a(1 - \hat{p}_a)/n_a + \hat{p}_b(1 - \hat{p}_b)/n_b}}$$

3. If you want to test the hypothesis at the 5 percent significance level, then accept the hypothesis if Z is between -1.96 and 1.96; otherwise reject the hypothesis.

4. If you want to test the hypothesis at another significance level, then look in Table A3–2 to find the critical value of Z.

Method 2 (if you are hypothesizing that $p_a = p_b$, that is, $D = 0$):

1. Calculate $\hat{p}$, $\hat{p}_a$, and $\hat{p}_b$:

$$\hat{p} = \frac{X_a + X_b}{n_a + n_b}, \quad \hat{p}_a = \frac{X_a}{n_a}, \quad \hat{p}_b = \frac{X_b}{n_b}$$

2. Calculate Z:

$$Z = \frac{\hat{p}_a - \hat{p}_b}{\sqrt{\hat{p}(1 - \hat{p})(1 / n_a + 1n_b)}}$$

3. If you want to test the hypothesis at the 5 percent significance level, then accept the hypothesis if Z is between –1.96 and 1.96; otherwise reject the hypothesis.

4. If you want to test the hypothesis at another significance level, then look in Table A3–2 to find the critical value of Z.

The Chi-square Test

Let's suppose that we are trying to test whether there is any difference between four competing cold-prevention medicines. None of the medicines is guaranteed to work—instead, they just promise to reduce your chances of getting a cold. Therefore, the number of people who try each kind of medicine and then get colds can be regarded as a random variable. Suppose we have checked with a sample of 495 people. We asked them what kind of medicine they used, and then whether or not they got colds. The results were as follows:

	Medicine 1	**Medicine 2**	**Medicine 3**	**Medicine 4**	**Total**
How many got colds	15	26	9	14	64
How many did not	111	107	96	117	431
Total	**126**	**133**	**105**	**131**	**495**

(This type of table is called a *contingency table*—in this case with two rows and four columns.) Each location in the table is called a *cell*. This table has 8 cells.

We can see from the table that medicine 3 seemed to be the most effective. Only 8.5 percent of the people who tried medicine 3 got colds. However, there are many other things that could have determined whether or not those people got colds. Maybe the people who used medicine 3 just happened to be exposed to fewer cold germs, so the fact that they got fewer colds is just a chance happening that has nothing to do with the fact that they used medicine 3.

Therefore, our null hypothesis will be as follows: There is basically no difference between the four medicines. In that case, the observed differences between the four medicines arose solely by chance.

Now we need to develop a test statistic to check this hypothesis. We can observe that in the total sample the fraction of people who got colds was .129 and the fraction who did not was .871. If there really was no difference between the medicines, then the fraction of people who got colds or did not get colds in each group should be close to these fractions.

Here is the procedure for performing the test in Excel. Enter the data:

	A	B	C	D	E	F
1	Actual	Med1	Med2	Med3	Med4	
2	w/cold	15	26	9	14	
3	w/o cold	111	107	96	117	

Next we need to calculate the totals in each row. Enter the formula =sum(B2:B3) into cell B4 and copy it to the right. Enter the fomula =sum(B2:E2) into cell F2 and copy it down:

	A	B	C	D	E	F
1	Actual	Med1	Med2	Med3	Med4	Total
2	w/cold	15	26	9	14	64
3	w/o cold	111	107	96	117	431
4	Total	126	133	105	131	495

Now we need to create a new table showing the predicted values if the null hypothesis were true. Enter the formula =B$4*$F2/F4 into cell B7. (The dollar signs need to be entered exactly as shown so the cells will copy correctly.) Copy this formula down to cell B8 and then across to fill the range B7:E8.

	A	B	C	D	E	F
1	Actual	Med1	Med2	Med3	Med4	Total
2	w/cold	15	26	9	14	64
3	w/o cold	111	107	96	117	431
4	Total	126	133	105	131	495
5						
6	Predicted	Med1	Med2	Med3	Med4	Total
7	w/cold	16.2909	17.1960	13.5758	16.9374	
8	w/o cold	109.7091	115.8040	91.4242	114.0626	

We want to base our test statistic on the difference between the observed frequencies and the frequencies that are predicted if there is in fact no difference between the medicines. If this difference is small, we can reasonably accept the no-difference hypothesis. If this difference is large, then we should reject the hypothesis.

Let f_i represent the observed frequency in cell i and let f_i^* represent the predicted frequency for cell i. Then, if there are n cells, we will use this test statistic:

$$S = \frac{(f_1 - f_1^*)^2}{f_1^*} + \frac{(f_2 - f_2^*)^2}{f_2^*} + \cdots + \frac{(f_n - f_n^*)^2}{f_n^*}$$

$$= \sum_{i=1}^{n} \frac{(f_i - f_i^*)^2}{f_i^*}$$

As it turns out, if the null hypothesis is true this test statistic has a chi-square distribution, so it is called the chi-square statistic. The degrees of freedom is

degrees of freedom

$$= (\text{number of rows} - 1) \times (\text{number of columns} - 1)$$

In our case, we have 2 rows and 4 columns, so our chi-square statistic has $(2 - 1) \times (4 - 1) = 3$ degrees of freedom. (In fact, the statistic described here has only approximately a chi-square distribution.)

Excel can save us a lot of work because we don't have to explicitly enter the formula for the test statistic. In cell A10 enter the formula =CHITEST(B2:E3,B7:E8). This will calculate the p value for the test, which for this example is .0538. Since this is slightly greater than .05, it means that we accept the null hypothesis at the 5 percent significance level. Accepting the null hypothesis indicates that it makes no difference which medicine a person takes—that is, the medicine taken is independent of whether or not the person got a cold. However, the p value is very close to .05, which means this is a close call and so we should investigate further.

To obtain the actual value of the chi-square statistic in Excel, enter the formula =CHIINV(A10,3). (In this example, A10 is the cell where we put the p value from the CHITEST function, and 3 is the number of degrees of freedom. CHIINV stands for the inverse of the chi-square distribution, meaning that you give the function a probability and it calculates the value of the distribution corresponding to that probability.) The resulting chi-square statistic is 7.7. A chi-square random variable with 3 degrees of freedom has a 5 percent chance of being greater than 7.8 (from Table A3–3), so the rejection region includes values of the test statistic above 7.8. Our observed test statistic (7.7) falls within the zone of acceptance, but it is very close to the boundary.

Goodness-of-fit Tests

The chi-square test can also be used to test whether or not a particular probability distribution fits the observed data very well. This type of test is called a *goodness-of-fit test*. Once again, we want to compare the observed frequencies f of a particular occurrence with the frequencies f^* that are predicted to occur if the alleged distribution really does fit the data well. Once again, we compute the statistic

$$\sum_{i=1}^{n} \frac{(f_i - f_i^*)^2}{f_i^*}$$

If the null hypothesis is true, this statistic will have approximately a chi-square distribution. If the value of the test statistic turns out to be too large, that means there is too much of a discrepancy between the actual results and the predicted results, so we can reject the hypothesis that the predicted distribution fits the data. The number of degrees of freedom for the chi-square statistic is

$n - 1$ – number of parameters that you have to estimate using the sample

For example, if you use the sample to estimate the mean of the distribution you are using, then the χ^2 statistic will have $n - 2$ degrees of freedom.

TABLE 18–1: APPOINTMENTS TO THE UNITED STATES SUPREME COURT

Period	Number of Appointments	Period	Number of Appointments
1790–94	3	1900–04	2
1795–99	4	1905–09	2
1800–04	2	1910–14	6
1805–09	2	1915–19	2
1810–14	2	1920–24	4
1815–19	0	1925–29	1
1820–24	1	1930–34	3
1825–29	2	1935–39	4
1830–34	1	1940–44	5
1835–39	5	1945–49	4
1840–44	1	1950–54	1
1845–49	3	1955–59	4
1850–54	2	1960–64	2
1855–59	1	1965–69	3
1860–64	5	1970–74	3
1865–69	0	1975–79	1
1870–74	4	1980–84	1
1875–79	1	1985–89	2
1880–84	4	1990–94	4
1885–89	3	1995–99	0
1890–94	4	2000–04	0
1895–99	2		

Let's perform a goodness-of-fit test to see if the Poisson distribution is appropriate for predicting the number of United States Supreme Court justices that will be appointed in a five-year period. Table 18–1 shows the number of Court appointments that have been made during each five-year period in U.S. history.

The mean is 2.465, so that on average 2.465 Supreme Court appointments are made in a 5-year period. Here is the frequency distribution of these observations. The upper figure is the number of appointments, the lower figure the number of periods in which there were that many appointments:

0	1	2	3	4	5	6
4	9	11	6	9	3	1

If the number of Supreme Court appointments is really given by a Poisson distribution with mean $\lambda = 2.465$, the predicted frequency is given by the formula $43 \times e^{-\lambda}\lambda^i/i!$ (since there are 43 five-year periods). Here are the values of the predicted frequencies:

0	1	2	3	4	5	6
3.655	9.010	11.105	9.125	5.624	2.773	1.139

We can now calculate the chi-square statistic:

$$\frac{(4-3.655)^2}{3.655} + \frac{(9-9.010)^2}{9.010} + \frac{(11-11.105)^2}{11.105} + \frac{(6-9.125)^2}{9.125} + \cdots$$
$$+ \frac{(9-5.624)^2}{5.624} + \frac{(3-2.773)^2}{2.773} + \frac{(1-1.139)^2}{1.139}$$

$$= 0.0326 + 0.0000 + 0.0010 + 1.0703 + 2.0270 + 0.0186 + 0.0170$$
$$= \text{test statistic} = 3.1666$$

It looks as though the observed frequencies match the predicted frequencies quite well. We have $n = 7$ categories, and we had to use the sample data to estimate the mean, so that leaves us with $7 - 1 - 1 = 5$ degrees of freedom. We can see from a chi-square table that a χ_5^2 random variable has a 95 percent chance of being less than 11.07, so the rejection region occurs for values of the test statistic above 11.07. The observed value is well within this limit, so we will accept the hypothesis that the rate of Supreme Court appointments can be described by the Poisson distribution. (Note that this was a one-tailed test, since we only wanted to reject the hypothesis if the test statistic was larger than expected. If the test statistic is very small, that means that the predicted frequencies are very close to the observed frequencies.)

Analysis of Variance

Suppose that we have observed scores on a particular aptitude test for three different groups of ten people each. The results were as follows:

Group a: 88, 92, 91, 89, 89, 86, 92, 86, 89, 89
Group b: 91, 92, 85, 94, 93, 87, 87, 92, 91, 89
Group c: 87, 88, 95, 88, 92, 87, 89, 88, 87, 88

The average scores for the three groups are close together. It seems reasonable to suppose that there is in fact no difference in aptitude between the groups, and that the observed difference in the average score has arisen solely by chance.

Suppose that we check the scores for three different groups and find these results:

Group d: 87, 94, 91, 89, 89, 84, 92, 86, 89, 89
Group e: 82, 76, 84, 79, 77, 84, 81, 69, 79, 74
Group f: 69, 79, 67, 64, 65, 69, 69, 64, 72, 66

In this case it seems clear that there is a real difference in aptitude between the three groups. In other words, we can reject the hypothesis that the observed differences between the groups arose solely by chance.

In both of these cases it was obvious whether or not there was a significant difference between the average scores for the groups. However, in general it will be more difficult to tell if the observed differences in scores are significant or random. We need to develop a new method that is called *analysis of variance*. For now, let's assume that we have $m = 3$ groups (call them group a, group b, and group c), and that there are n people in each group. Assume that we know that the aptitude scores for the people in each group are selected from a normal distribution, and assume that the variance of the distribution is the same for all three groups.

Let's say that μ_a is the unknown mean aptitude test score for group a, μ_b is the mean for group b, and μ_c is the mean for group c.

Our mission is to test the null hypothesis:

$$\mu_a = \mu_b = \mu_c = \mu$$

In other words, the null hypothesis states that the mean score for each group is the same. The alternative hypothesis simply states that the means are not all the same.

First, one obvious thing to do is calculate $\overline{a}, \overline{b}$, and $\overline{c}$ (the sample averages for each sample.) If $\overline{a}, \overline{b}$, and $\overline{c}$ are close to each other, we will be more willing to accept the hypothesis that μ_a, μ_b, and μ_c are all equal. We can calculate $\overline{x}$, the average for all the numbers:

$$\overline{x} = \frac{\overline{a} + \overline{b} + \overline{c}}{m} = \frac{\overline{a} + \overline{b} + \overline{c}}{3}$$

We can also calculate the sample variance (version 2) for these three averages (we'll call that variance S^{*2}):

$$S^{*2} = \frac{(\overline{a} - \overline{x})^2 + (\overline{b} - \overline{x})^2 + (\overline{c} - \overline{x})^2}{m - 1}$$

The larger S^{*2} is, the *less* likely we will be to accept the no-difference null hypothesis.

We should also look at the sample variance for each individual sample:

$$S_a^2 = \sum_{i=1}^{n} \frac{(a_i - \overline{a})^2}{n - 1}, \quad S_b^2 = \sum_{i=1}^{n} \frac{(b_i - \overline{b})^2}{n - 1}, \quad S_c^2 = \sum_{i=1}^{n} \frac{(c_i - \overline{c})^2}{n - 1}$$

It will turn out to be useful to calculate the average of the three sample variances (call it S^2):

$$S^2 = \frac{s_a{}^2 + s_b{}^2 + s_c{}^2}{3}$$

The larger these three variances are, the more likely we are to see $\overline{a}$, $\overline{b}$ and $\overline{c}$ spread out, even if they really do come from distributions with the same mean. For example, suppose that the observed values of $\overline{a}$, $\overline{b}$ and $\overline{c}$ are 500, 400, 450. If $s_a{}^2 = s_b{}^2 = s_c{}^2 = 1$, we know right away that it is extremely unlikely that $\overline{a}$, $\overline{b}$ and $\overline{c}$ could have the observed values if they really did come from distributions with the same mean. On the other hand, if $s_a{}^2 = s_b{}^2 = s_c{}^2 = 10{,}000$, then it would be quite likely to see $\overline{a}$, $\overline{b}$, and $\overline{c}$ spread out by this much even if the null hypothesis is true. Therefore, the larger S^2 is, the more likely we will be to accept the null hypothesis.

We will calculate the following statistic (call it F):

$$F = \frac{nS^{*2}}{S^2}$$

If the value of F is large, we will reject the null hypothesis. We can show that the F statistic will have an F distribution with $(m - 1)$ and $m(n - 1)$ degrees of freedom. (Here m is the number of groups and n is the number of items in each group.) The F distribution is described in Chapter 13, and Table A3–6 lists some values for the cumulative distributive function.

In the examples we discussed earlier, we had $m = 3$ and $n = 10$. Therefore, the F statistic will have 2 and 27 degrees of freedom. Table A3–6 shows that this type of F statistic has a 95 percent chance of being less than about 3.3. Therefore, if the observed F statistic value is greater than 3.3, we will reject the null hypothesis; otherwise we will accept the null hypothesis.

In the first example (groups a, b, and c), the F statistic is $4.133/6.693 = .6175$. Just as we suspected all along, we should accept the hypothesis. In the second example (groups d, e, and f), the F value is $1061/16.996 = 62.426$, so we should reject the hypothesis.

GENERAL PROCEDURE FOR AN ANALYSIS OF VARIANCE TEST

(Assume that you have *m* groups, each with *n* members.)

1. Calculate the sample average for each group:

$$\overline{a} = \frac{a_1 + a_2 + \cdots + a_n}{n}$$
$$\overline{b} = \frac{b_1 + b_2 + \cdots + b_n}{n}$$
$$\overline{c} = \frac{c_1 + c_2 + \cdots + c_n}{n}$$

and so on.

2. Calculate the average of all the averages:

$$\overline{x} = \frac{\overline{a} + \overline{b} + \overline{c} + \cdots}{m}$$

3. Calculate the sample variance of the averages:

$$S^{*2} = \frac{(\overline{a} - \overline{x})^2 + (\overline{b} - \overline{x})^2 + (\overline{c} - \overline{x})^2 + \cdots}{m - 1}$$

4. Calculate the sample variance for each group:

$$S_a{}^2 = \frac{(a_1 - \overline{a})^2 + (a_2 - \overline{a})^2 + \cdots + (a_n - \overline{a})^2}{n - 1}$$
$$S_b{}^2 = \frac{(b_1 - \overline{b})^2 + (b_2 - \overline{b})^2 + \cdots + (b_n - \overline{b})^2}{n - 1}$$
$$S_c{}^2 = \frac{(c_1 - \overline{c})^2 + (c_2 - \overline{c})^2 + \cdots + (c_n - \overline{c})^2}{n - 1}$$

and so on.

5. Calculate the average of all of the sample variances:

$$S^2 = \frac{s_a{}^2 + s_b{}^2 + s_c{}^2 + \cdots}{m}$$

6. Calculate the value of the *F* statistic:

$$F = \frac{nS^{*2}}{S^2}$$

7. Look in Table A3–6 to find the critical value for an *F* distribution with $(m - 1)$ and $m(n - 1)$ degrees of freedom.

8. If the observed value of the *F* statistic is greater than the critical value, reject the null hypothesis; otherwise accept it.

We can use Excel to perform the analysis of variance test. Enter the data in three different columns. Put the labels A, B, and C at the top of the appropriate columns. Then call for the Tools Menu, then Data Analysis, then ANOVA. Highlight the range with your data (including the labels, and check the "Labels in First Row" box). Then click OK, and you'll see the results:

ANOVA: Single Factor

SUMMARY

Groups	Count	Sum	Average	Variance		
A	10	891	89.1	4.544444444		
B	10	901	90.1	8.766666667		
C	10	889	88.9	6.766666667		
ANOVA						
Source of Variation	SS	df	MS	F	P-value	F crit
Between Groups	8.266666667	2	4.133333333	0.617598229	0.546682068	3.354131195
Within Groups	180.7	27	6.692592593			
Total	**188.9666667**	**29**				

This example shows the results for the first example on page 248.

EXAMPLE To illustrate, let's investigate the effects that two different surgical procedures have on the rate of growth of laboratory rats. The two procedures we will investigate are area postrenal lesion and ovariectomy. We will need four groups of rats: one group that has both procedures; one group that has neither procedure; and two groups each of which has only one of the procedures. Then we will investigate their rates of growth.

Here are the rates of growth from experiments performed by Liz Ashburn at Yale University:

Group 1 (both):

−0.088;	−0.165;	−0.099;	0.031;	0.030;	0.046;	−0.010;
−0.070;	−0.099;	0.028;	0.019;	0.059;	−0.037;	0.066;
−0.044;	−0.038;	0 ;	−0.034;	0 ;	−0.013;	0.014

Group 2 (ap lesion):

−0.051;	−0.079;	−0.120;	−0.017;	−0.019;	−0.065;	0.001;
−0.038;	0.027;	−0.030;	−0.013;	−0.012;	0.001;	0 ;
−0.074;	−0.104;	−0.050;	0.033;	0.009;	0.039;	0.022

Group 3 (ovariectomy):

.152;	−.120;	.018;	0 ;	.032;	.031;	−.020;
−0.143;	0.009;	0.012;	0.023;	−0.014;	0.047;	−0.007;
−0.061;	0.137;	0.010;	−0.136;	0.132;	0 ;	0.048

Group 4 (neither):

–0.006;	0.047;	0.030;	0.016;	0.060;	–0.045;	0.004;
0.018;	0.038;	0.006;	0.021;	0.006;	0.012;	0.031;
0.044;	0.009;	0.029;	–0.029;	0.05 ;	–0.027;	0.068

We need to calculate the sample average and the standard deviation s_2 for each group:

$$\text{Group 1: } \bar{x} = -0.0192, s_2 = 0.0596$$
$$\text{Group 2: } \bar{x} = -0.0257, s_2 = 0.0445$$
$$\text{Group 3: } \bar{x} = -0.0071, s_2 = 0.0786$$
$$\text{Group 4: } \bar{x} = -0.0182, s_2 = 0.0292$$

Now we can calculate the variance between the four averages: $s^{*2} = 0.000436$. The value of the test statistic is

$$\frac{21 \times 0.000436}{0.00314} = 2.92$$

If the null hypothesis is true, the test statistic will have an F distribution with $m - 1 = 3$ and $m(n - 1) = 80$ degrees of freedom. The F table shows that this random variable has a 95 percent chance of being less than about 2.7, so our observed statistic is just within the regection region and we can therefore reject the null hypothesis that the procedures have no effect on growth. However, these results consist of body-weight measurements within the first two weeks of the procedures—the differences between the different groups of rats become less when more time has elapsed since the procedures were done.

Sum of Squares

The analysis of variance approach can be further understood by looking at the sum of squared deviations (or the *sum of squares*). Again assume that we have m groups, each with n members. Each group has been treated differently in some manner, and our goal is to see if the treatments really made any difference. For example, we may have groups of people who were given different medicines. We will use x_{ij} to represent the ith item in the jth group. Note that we need two subscripts to identify each element uniquely. For example, look at the following $m = 3$ groups with $n = 5$ members:

Group 1	Group 2	Group 3
16	38	19
13	21	14
36	36	17
29	39	15
18	26	12

We can represent each like this:

$$\begin{array}{lll} x_{11} = 16 & x_{12} = 38 & x_{13} = 19 \\ x_{21} = 13 & x_{22} = 21 & x_{23} = 14 \\ x_{31} = 36 & x_{32} = 36 & x_{33} = 17 \\ x_{41} = 29 & x_{42} = 39 & x_{43} = 15 \\ x_{51} = 18 & x_{52} = 26 & x_{53} = 12 \end{array}$$

To represent the sum of all of the items, we need to use two sigmas:

$$T = \sum_{i=1}^{n}\sum_{j=1}^{m} x_{ij}$$

This expression can be broken down like this:

$$\begin{aligned} T &= \sum_{i=1}^{n}(x_{i1} + x_{i2} + \cdots + x_{im}) \\ &= x_{11} + x_{12} + \cdots + x_{1m} \\ &\quad + x_{21} + x_{22} + \cdots + x_{2m} \\ &\quad + \cdots \\ &\quad + x_{n1} + x_{n2} + \cdots + x_{nm} \end{aligned}$$

For our example:

$$\begin{aligned} T &= 16 + 38 + 19 \\ &\quad + 13 + 21 + 14 \\ &\quad + 36 + 36 + 17 \\ &\quad + 29 + 39 + 15 \\ &\quad + 18 + 26 + 12 \\ &= 349 \end{aligned}$$

We will use $\bar{\bar{x}}$ (*x* double bar) to represent the overall mean of all the elements (we could call it the *grand mean*):

$$\begin{aligned} \bar{\bar{x}} &= \frac{T}{mn} \\ &= \frac{\sum_{i=1}^{n}\sum_{j=1}^{m} x_{ij}}{mn} \\ &= \frac{349}{15} \\ &= 23.267 \text{ for our example} \end{aligned}$$

For each element in the list we can calculate how far away it is from the grand mean $\bar{\bar{x}}$:

$$\text{distance from } x_{ij} \text{ to } \bar{\bar{x}} = |x_{ij} - \bar{\bar{x}}|$$

As you may have begun to suspect, we would like to square this distance:

$$\text{squared distance from } x_{ij} \text{ to } \bar{\bar{x}} = (x_{ij} - \bar{\bar{x}})^2$$

Now add all of the squared distances:

$$\text{TSS} = \sum_{i=1}^{n} \sum_{j=1}^{m} (x_{ij} - \bar{\bar{x}})^2$$

We will call this quantity the *total sum of squares*, or *TSS*. For our example we can calculate the following:

$$\begin{aligned} \text{TSS} &= (16 - 23.267)^2 + (38 - 23.267)^2 + (19 - 23.267)^2 \\ &\quad + (13 - 23.267)^2 + (21 - 23.267)^2 + (14 - 23.267)^2 \\ &\quad + (36 - 23.267)^2 + (36 - 23.267)^2 + (17 - 23.267)^2 \\ &\quad + (29 - 23.267)^2 + (39 - 23.267)^2 + (15 - 23.267)^2 \\ &\quad + (18 - 23.267)^2 + (26 - 23.267)^2 + (12 - 23.267)^2 \\ &= 1358.93 \end{aligned}$$

We will look at several different sum of squares statistics. Note that a sum of squares is similar to a variance except that there is no division by n. The variance is calculated from the sum of squares about the mean, but we will see that there are several other types of sums of squares.

To analyze the total sum of squares, we need to break it into two parts. We know that, if the null hypothesis is true, the population mean is the same for each group and the deviation of any one element from the grand mean arises only by chance. On the other hand, if the population means are different, there are two reasons why an individual element may deviate from the grand mean: (1) because the mean of its own group is different from the overall population mean, and (2) because there is chance variation within its own group. Therefore, we will split the total sum of squares into two parts: the part arising from deviations of individual elements from their group mean, and the part arising from deviations of the group means from the grand mean. We can write the TSS formula like this:

$$\begin{aligned} \text{TSS} &= \sum_{i=1}^{n} \sum_{j=1}^{m} (x_{ij} - \bar{\bar{x}})^2 \\ &= \sum_{i=1}^{n} \sum_{j=1}^{m} [(x_{ij} - \bar{x}_j) + (\bar{x}_j - \bar{\bar{x}})]^2 \end{aligned}$$

We will use $\bar{x}_j$ to stand for the sample average of group j. We have not done anything to our original expression but add and subtract $\bar{x}_j$. Now we have

$$\text{TSS} = \sum_{i=1}^{n} \sum_{j=1}^{m} [(x_{ij} - \bar{x}_j)^2 + (\bar{x}_j - \bar{\bar{x}})^2 + 2(x_{ij} - \bar{x}_j)(\bar{x}_j - \bar{\bar{x}})]$$

It turns out quite conveniently that, when the double summation is taken over the last term, the result is always zero. Hence we are left with

$$\text{TSS} = \sum_{i=1}^{n} \sum_{j=1}^{m} [(x_{ij} - \bar{x}_j)^2 + (\bar{x}_j - \bar{\bar{x}})^2]$$

We can break that up into two double summations:

$$\text{TSS} = \sum_{i=1}^{n} \sum_{j=1}^{m} [(x_{ij} - \bar{x}_j)^2 + \sum_{i=1}^{n} \sum_{j=1}^{m} (\bar{x}_j - \bar{\bar{x}})^2$$

Since the last term does not contain anything that depends on i, we can replace the summation $\sum_{i=1}^{n}$ by simply multiplying by n:

$$\text{TSS} = \sum_{i=1}^{n} \sum_{j=1}^{m} [(x_{ij} - \bar{x}_j)^2 + n \sum_{j=1}^{m} (\bar{x}_j - \bar{\bar{x}})^2$$

We have now broken the total sum of squares into two components. The first part represents the deviation of each element from its group mean. Since we can think of these differences as arising because of unknown random factors (called statistical error), we will call this the *error sum of squares*, or *ERSS*. Then

$$\text{ERSS} = \sum_{i=1}^{n} \sum_{j=1}^{m} [(x_{ij} - \bar{x}_j)^2$$

The second part of the total sum of squares represents the deviation of each group mean from the grand mean. We can think of these deviations as arising because the groups were given different treatments, so we will call this the *treatment sum of squares* or *TRSS*. Then

$$\text{TRSS} = n \sum_{j=1}^{m} (\bar{x}_j - \bar{\bar{x}})^2$$

We can see that TSS = ERSS + TRSS. The larger the treatment sum of squares becomes, the less likely we will be to accept the null hypothesis that the treatments don't matter.

Each sum-of-squares statistic has an associated quantity called its degrees of freedom. The degrees of freedom of TRSS is $m - 1$. The degrees of freedom of ERSS is $m(n - 1)$. (We are adding together m different sum of squares for each sample. Each individual sum of squares has $n - 1$ degrees of freedom.) When a sum of squares is divided by its degrees of freedom, the result is called a *mean square variance*:

$$\text{treatment mean square variance} = \frac{\text{TRSS}}{m-1}$$

$$\text{error mean square variance} = \frac{\text{ERSS}}{m(n-1)}$$

When we take the treatment mean square variance divided by the error mean square variance, then the result is exactly the same F statistic we used before:

$$F = \frac{\text{TRSS} \,/\, (m-1)}{\text{ERSS} \,/\, [m(n-1)]}$$

For our example, we have

$$\text{TRSS} = 694.533 \quad m - 1 = 2$$

$$\text{ERSS} = 664.400 \quad m(n-1) = 12$$

$$F = \frac{\dfrac{694.533}{2}}{\dfrac{664.400}{12}} = 6.272$$

If the null hypothesis is true and the treatments do not matter, then the F statistic will have an F distribution with 2 and 12 degrees of freedom. The critical value for a 5 percent test is 3.9, so in this case we can reject the null hypothesis.

The information from the test can be summarized in a table known as an *ANOVA table*. (ANOVA is short for "analysis of variance.") The ANOVA table for our example looks like this:

Source of Variation	**Sum of Squares**	**Degrees of Freedom**	**Mean Square Variance**	***F* Ratio**
Between means (treatment)	694.533	2	347.267	6.272
Within samples (error)	664.400	12	55.367	
Total	**1,358.933**	**14**		

The table arranges the information about the sums of squares, their degrees of freedom, and the F statistic. Note that the total sum of squares has $mn - 1 = 14$ degrees of freedom, which is the sum of the degrees of freedom for the TRSS and ERSS.

The general formula for the ANOVA table looks like this:

TABLE 18–2: ANOVA TABLE FOR *m* DIFFERENT GROUPS EACH WITH *n* MEMBERS

Source of Variation	Sum of Squares	Degrees of Freedom	Mean Square Variance	*F* Ratio
Between means (treatment)	$\text{TRSS} = n\sum_{j=1}^{m}(\bar{x}_j - \bar{\bar{x}})^2$	$m-1$	$\frac{\text{TRSS}}{m-1}$	$\frac{\text{TRSS}/(m-1)}{\text{ERSS}/[m(n-1)]}$
Within samples (error)	$\text{ERSS} = \sum_{i=1}^{n}\sum_{j=1}^{m}(x_{ij} - \bar{x}_j)^2$	$m(n-1)$	$\frac{\text{ERSS}}{m(n-1)}$	
Total	**TSS = TRSS + ERSS**	***mn* - 1**		

Derivation of the *F* Statistic

WARNING: mathematical area

We now have to show the F statistic we used in the last section really does have an F distribution. Let's assume that we have $m = 3$ groups (call them group a, group b, and group c), and that there are n people in each group.

Let's say that μ_a is the unknown mean for group a, μ_b is the mean for group b, and μ_c is the mean for group c. Calculate the averages $\bar{a}, \bar{b}$, and $\bar{c}$. The average $\bar{a}$ will have a normal distribution with mean μ_a, and variance σ^2/n. If the null hypothesis is true, then $\bar{a}, \bar{b}$, and $\bar{c}$ act like three observations taken from a normal distribution with mean μ and variance σ^2/n.

We can estimate the variance σ^2 by using the sample variance of these three averages. (We'll call that variance S^{*2}.)

$$S^{*2} = \frac{(\bar{a} - \bar{x})^2 + (\bar{b} - \bar{x})^2 + (\bar{c} - \bar{x})^2}{m-1}$$

We know from Chapter 17 that the statistic

$$Z^{*2} = \frac{n(m-1)S^{*2}}{\sigma^2}$$

will have a chi-square distribution with $m - 1$ degrees of freedom. If the mean for each group really is different, then we could expect that the variance across $\bar{a}$, $\bar{b}$, and $\bar{c}$ should be greater than the variance within each sample. However, if the mean of each group is the same, then s_a^2, s_b^2, and s_c^2 can all be used as estimators for the variance of the whole population. Or we could use the average of these three variances (which we called S^2):

$$S^2 = \frac{s_a^2 + s_b^2 + s_c^2}{m} = \frac{s_a^2 + s_b^2 + s_c^2}{3}$$

We know that $[(n-1)/\sigma^2]S_a^2$ has a chi-square distribution with $n - 1$ degrees of freedom, so

$$Z^2 = \frac{m(n-1)}{\sigma^2} \times \frac{s_a^2 + s_b^2 + s_c^2}{m}$$

will have a chi-square distribution with $m(n - 1)$ degrees of freedom.

Now, we'll compare the two statistics Z^2 and Z^{*2} by calculating their ratio:

$$F = \frac{Z^{*2}/(m-1)}{Z^2/m(n-1)}$$

Because F equal to the ratio of two chi-square random variables divided by their degrees of freedom, it will have an F distribution.

Now all we have to do is show that this expression for the F statistic is really the same as the expression we used in the last section.

$$F = \frac{nS^{*2}\ /\sigma^2}{S^2\ /\ \sigma^2} = \frac{nS^{*2}}{S^2}$$

And therefore we have reached the result that we wanted.

EXERCISES

Given the following samples, test the hypothesis that the mean is as given:

1. 1, 5, 17, 9, 23, 17, 4, 3, 8, 8, 7, 8, 6, 0, –1: mean 7
2. 4, 30, –17, –29, 8, 7, –5, 4, 3, –6: mean –2
3. 15, 22, –19, 0, 1, 2, 4, 3, –3, 7: mean 14
4. 17, –9, –8, –10, 8, 5, 4, –7, 3, 4, –5, –7, –3, 2, 3: mean 0

5. Suppose four new pesticides are being tested in a laboratory, with the following results:

	Type 1	Type 2	Type 3	Type 4	Total
Insects killed	139	100	73	98	410
Insects surviving	15	50	80	47	192
Total tested	**154**	**150**	**153**	**145**	**602**

Is pesticide 1 significantly better than the rest?

6. Suppose five different meteorological theories are tested to see if they predict the weather correctly. The results are as follows:

	Theory 1	Theory 2	Theory 3	Theory 4	Theory 5	Total
Reports correct	50	48	53	47	46	244
Reports incorrect	76	74	75	76	77	378
Total reports	**126**	**122**	**128**	**123**	**123**	**622**

Is theory 3 significantly better than the rest?

7. Estimate the mean of the scores of your favorite football team last year. Test the hypothesis that the mean score is greater than 14.

8. Test the hypothesis that the mean number of pages in your favorite daily newspaper is greater than 50.

9. Estimate the mean price of a gallon of milk in your city over the past month. Test the hypothesis that the price this month is significantly greater than the price last month.

10. Consider a random variable with a binomial distribution with p unknown. How big does n have to be for you to be able to tell the difference (at the 5 percent significance level) between the hypothesis $p = .5$ and the hypothesis $p = .51$?

11. Perform a goodness-of-fit test to see if the normal distribution is appropriate for the heights of a sample of people you know.

12. Perform a goodness-of-fit test to see if the Poisson distribution is appropriate for the number of phone calls that you receive at your house.

13. Perform a goodness-of-fit test to see if the uniform distribution is appropriate for the numbers that you roll on a die.

14. Test the hypothesis that there is no significant difference between groups a, b, c, and d:

	Group *a*	Group *b*	Group c	Group *d*
Number of successes	16	12	7	13
Number of failures	54	94	66	49

15. Perform an analysis of variance test to see if the following sets of numbers were selected from distributions with the same mean:

Group 1: 18, 9, 15, 10, 16, 8, 7, 20, 13, 8, 12
Group 2: 17, 13, 9, 8, 10, 13, 16, 17, 12, 11, 17
Group 3: 15, 13, 13, 14, 15, 20, 15, 14, 9, 16, 12
Group 4: 16, 12, 13, 12, 14, 11, 8, 13, 17, 15, 14

16. Make a list of the first letters of the last names of 20 people you know and assign each letter a number ($A = 1, B = 2$, etc.). Test the hypothesis that the resulting numbers come from a distribution with mean 13.

17. Repeat the same procedure as in Exercise 16, only this time use the first names, and assume that the variance of the distribution is 25.

18. Divide your friends into categories by hair color and eye color. Perform a chi-square test to see if there is a significant difference in eye color for people with different hair colors.

19. Perform a goodness-of-fit test to see if the weights of a group of your friends fit the normal distribution.

20. Perform an analysis of variance test to see if people you know with different hair colors have the same height.

21. Look through old newspapers to find predictions from a well-known psychic. Test the hypothesis that the psychic could have made the predictions by pure guessing.

For each of the given values of $\bar{x}$, n, s_2, and μ^*, test the hypothesis that $\mu = \mu^*$.

	$\bar{x}$	n	s_2	μ^*
22.	60	10	8	62
23.	60	20	8	62
24.	60	50	8	62
25.	60	100	8	62
26.	60	500	8	62
27.	60	50	1	62
28.	60	50	5	62
29.	60	50	10	62
30.	60	50	20	62

19

Polling, Sampling, and Experimental Design

Opinion Polls

How can we find out how many people in a population have a particular characteristic? For example, we might want to know how many voters support our favorite presidential candidate. Or we might be interested in some general characteristics of people in a certain state—for example, how many are children, how many live in cities, how many are employed, and so on. It is usually too expensive to ask everyone, but occasionally this is done: an election is held every 4 years in the United States to find out the presidential preferences of all voters, and every 10 years a census is held to obtain detailed information about all of the people in the country.

Opinion polls typically ask a sample of about 1,000 people. At first glance, you should be skeptical that a sample of this size will be representative of all 300 million people in the United States. You probably think it is unlikely that each person in the poll has the same opinions as his or her 300,000 closest neighbors. However, polls do seem to be fairly accurate. The election results predicted by polls are usually close to the actual results of the election (with a few notable exceptions).

Now we will look at the theory that explains why these results tend to be accurate. We'll let N stand for the number of people in the population we are investigating. Assume that M of these people favor our candidate and $N - M$ favor the opposing candidate. Our goal is to estimate M/N, the fraction of people who support our candidate. Let $p = M/N$, which we'll call the population proportion. We'll ask a sample of n people. Let X represent the number of people in the sample who favor our candidate. If our poll is any good, X/n will be close to M/N.

As we decide on our process for selecting the sample, it is crucial that we avoid choosing a sample that is systematically biased in some way. We can't just start asking our friends since our friends might be more likely to be on our side. We can't select just one neighborhood and interview everyone there since people in one particular neighborhood are not likely to be representative of the diverse characteristics of all of the people in the town. If we conduct the survey over the internet, then we will systematically exclude those that lack internet access.

The way to prevent systematic bias in the sample is to select the sample completely at random. We should design the sampling system so that each person has an equal chance of being selected. Not only that, we should also design the system so that every single possible sample that we might conceive of has an equal chance of being the sample that we actually choose.

How are we going to do this? One way is to write everyone's name on a slip of paper, put the papers in little capsules, and then put all of the capsules in a large drum. If we mix up the capsules very thoroughly and then pull n capsules out of the drum, we will have a random sample of n people. However, this approach is difficult if N is very large. For one thing, we would need a very large drum. Another problem is that it is difficult to mix the capsules well if there is a large number of them. If the capsules are not mixed well, then the people whose names were put in last have a greater chance of being selected, and we will no longer have a pure random sample.

An easier method is to give everyone in the population a number. Then we can select a bunch of random numbers and interview the people whose numbers we've chosen. How do we select the random numbers? It's not as easy as it sounds. We can't just start making up numbers since it is hard for a person to make up a long string of numbers without falling into some sort of pattern. (Try it yourself sometime.) The best approach is to have a computer generate the random numbers. The numbers it generates are not true random numbers since they are generated according to a fixed rule. However, the rule is unpredictable enough that for all practical purposes the numbers seem to have been selected totally at random.

Random selection protects us from systematic bias, but we still need to worry about bad luck—that is, that our random sample may be unrepresentative of the population because our random selection procedure may have happened to choose a sample that doesn't reflect the population. Since all possible samples are equally likely, this could happen. However, you should have an intuitive feeling that it is unlikely that the sample will be extremely unrepresentative, for the same reason that it is unlikely that dealing 5 cards from a 52-card deck will result in 4 cards of the same rank.

A poll will give you an indication of how far off it might be by listing the *margin of error*. For example, the poll might report that 52 percent of the population support your candidate, with a margin of error of 3 percent. The results of the poll indicate that the true value of the population proportion is between 49 percent and 55 percent, but they cannot guarantee that with 100 percent accuracy. If the pollsters are honest, they will include fine print that says something like this: "The margin of error was 3 percentage points. This means that if this procedure were followed 20 times, we would expect that in 19 of these times the proportion found in this sample would be within 3 percentage points of the proportion that would be found if all members of the population were interviewed." In other words, the margin of error is just half of the width of the 95 percent confidence interval for the population proportion.

Confidence Interval for the Proportion

Fortunately, we can work out the probabilities when a sample is selected without replacement, using the formula for the hypergeometric distribution (see Chapter 6):

$$\Pr(X = k) = \frac{\binom{M}{k}\binom{N-M}{n-k}}{\binom{N}{n}}$$

where

$N =$ population size
$M =$ number of people in the population that support our candidate
$n =$ sample size
$X =$ number of people in the sample that support our candidate (call it a success when we select a person for the sample that supports our candidate)

Also, we define

$p = M/N =$ population proportion $=$ probability of success
$\hat{p} = X/n =$ sample proportion

For example, if $N = 200$ million, $M = 100$ million, $p = 1/2$, and $n = 100$, we can calculate that there is a 96 percent chance that X will be between 40 and 60.

However, we want the poll to be even more accurate than that. It makes a big difference to us whether the proportion that supports our candidate is 40 percent or 60 percent. We will try to make the sample more accurate by interviewing more people. We'll now ask 1,000 people.

To make the calculations easier, we'll make two approximations. First, imagine that we select the sample with replacement rather than without replacement. This means that once a person has been selected for the sample, their name is placed back in our big drum so they have a possibility of being selected again. (That person probably will be annoyed, which is one reason why this method is not used in practice. There is even a minuscule chance that we will select the same person 1,000 times.) When selecting with replacement, the probability of success remains constant, so the binomial distribution applies. Next, we can approximate the binomial distribution by the normal distribution, thanks to the Central Limit Theorem.

We found that the expected value of X is $nM/N = np$. Then the expected value of the sample proportion is

$$E(\hat{p}) = E\left(\frac{X}{n}\right) = \frac{E(X)}{n} = \frac{np}{n} = p$$

In words: the expected value of the sample proportion equals the population proportion. This means that the sample proportion $\hat{p}$ is an unbiased estimator of the population proportion p.

The variance of the hypergeometric distribution is

$$n\left(\frac{M}{N}\right)\left(1-\frac{M}{N}\right)\left(\frac{N-n}{N-1}\right)=np(1-p)\left(\frac{N-n}{N-1}\right)$$

Note what happens to the factor in the last parentheses if the population N is very large. For example, if $N = 200$ million and $n = 1,000$, then $(N - n)/(N - 1)$ will be .999995005. This factor is very close to 1, so we can usually ignore it. This means we can approximate X with a normal distribution with mean $np = M/N$ and variance $np(1 - p)$ (which we recognize as the variance of the binomial distribution).

Find the variance of the sample proportion $\hat{p}$:

$$\text{Var}(\hat{p})=\text{Var}\left(\frac{X}{n}\right)=\frac{\text{Var}(X)}{n^2}=\frac{np(1-p)}{n^2}=\frac{p(1-p)}{n}$$

Now to work out the confidence interval:

$$\Pr(\hat{p}-c<p<\hat{p}+c)=.95$$
$$\Pr(-c<p-\hat{p}<c)=.95$$
$$\Pr\left[\frac{-c}{\sqrt{\frac{p(1-p)}{n}}}<\frac{p-\hat{p}}{\sqrt{\frac{p(1-p)}{n}}}<\frac{c}{\sqrt{\frac{p(1-p)}{n}}}\right]=.95$$

Replace the middle term with the standard normal random variable Z.

$$\Pr\left[-c\sqrt{\frac{n}{p(1-p)}}<Z<c\sqrt{\frac{n}{p(1-p)}}\right]=.95$$

From the standard normal table,

$$c\sqrt{\frac{n}{p(1-p)}}=1.96$$
$$c=1.96\sqrt{\frac{p(1-p)}{n}}$$

Therefore, the 95 percent confidence interval for the proportion p is

$$\hat{p}-1.96\sqrt{\frac{\hat{p}(1-\hat{p})}{n}}\quad\text{to}\quad\hat{p}+1.96\sqrt{\frac{\hat{p}(1-\hat{p})}{n}}$$

For example, if $\hat{p} = .62$ and $n = 200$, the confidence interval is from

$$.62-1.96\sqrt{\frac{.62\times .38}{200}}\quad\text{to}\quad .62+1.96\sqrt{\frac{.62\times .38}{200}}$$

or from .553 to .687.

GENERAL PROCEDURE FOR CONFIDENCE INTERVAL FOR PROPORTION

Let n be the number of items in the sample, X be the number of items in the sample with the characteristic you're interested in, $\hat{p} = X/n$ be the proportion of people in the sample with that characteristic, and p be the unknown proportion of people in the population with that characteristic.

1. Determine the confidence level you want.
2. Look up the value of a in Table A3–2. (If the confidence level is 95 percent, then $a = 1.96$.)
3. Calculate $c = a\sqrt{\hat{p}(1 - \hat{p})/n}$.

The confidence interval is from $\hat{p} - c$ to $\hat{p} + c$.

Figure 19–1 shows the density function for the normal distribution corresponding to a population proportion of .25 and a sample size of 50. The function peaks at .25, but there is a lot of spread around this value. In other words, you would not have much confidence that a poll with 50 respondents will accurately reflect the population.

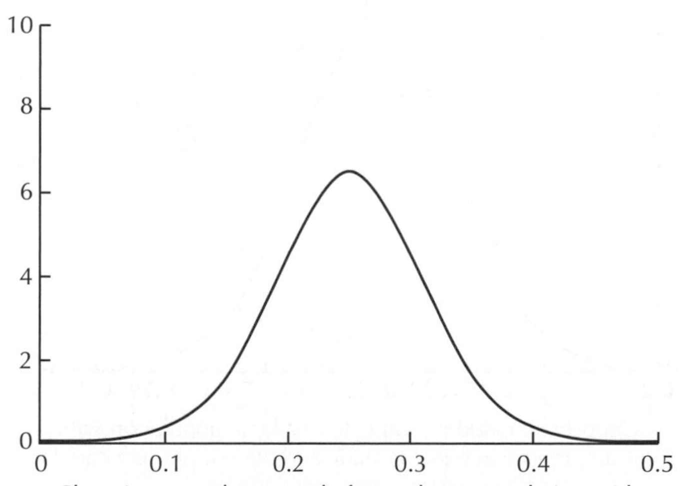

Choosing a random sample from a large population with the population proportion equal to 0.25, and the sample size equal to 50.

FIGURE 19–1

Figure 19–2 shows how the density function becomes steeper and narrower as the sample size increases to 100, allowing you to have a more accurate poll. Figure 19–3 shows the situation with a sample size of 1,000. Be sure to note that this diagram has a different scale than the previous two. With 1,000 respondents, we can see that most likely the sample proportion will be between .22 and .28 when the population proportion is .25.

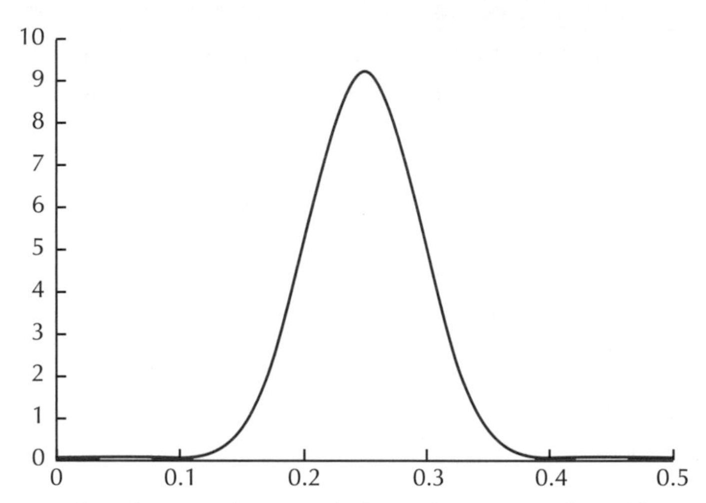

Choosing a random sample from a large population with the population proportion equal to 0.25, and the sample size equal to 100.

FIGURE 19–2

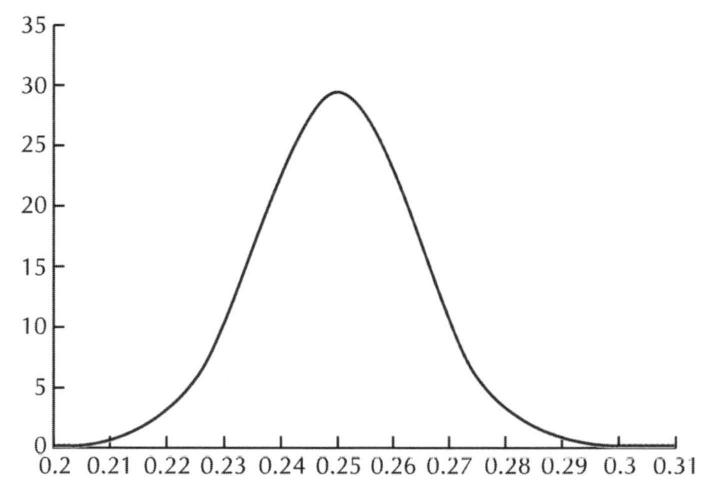

Choosing a random sample from a large population with the population proportion equal to 0.25, and the sample size equal to 1,000.

FIGURE 19–3

Figure 19–4 shows all three cases on one graph so you can see how the density function becomes steeper and narrower as the sample size increases.

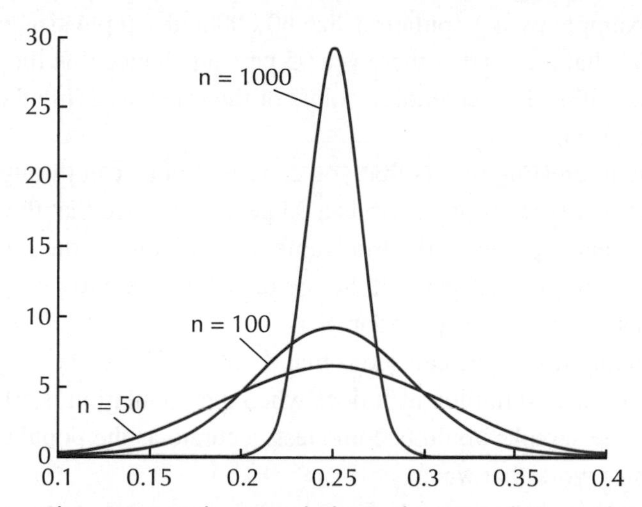

Choosing a random sample from a large population with the population proportion equal to 0.25, for three different sample sizes.

FIGURE 19–4

If you want a slightly more accurate result for the confidence interval, use the formula

$$c = 1.96\sqrt{\frac{\hat{p}(1-\hat{p})}{n}}\left(\sqrt{\frac{N-n}{N-1}}\right)$$

You can recognize the factor $\sqrt{(N-n)/(N-1)}$ as coming from the formula for the variance of the hypergeometric distribution. The factor $\sqrt{(N-n)/(N-1)}$ is called the *finite population correction factor*. If N is much larger than n, the value of the correction factor is close to 1, meaning that we can ignore it.

Table 19–1 shows the margin of error that results from this formula for different values of the population size N and the sample size n.

TABLE 19–1: PERCENT ERROR FOR SAMPLE

N	*n* = 100	*n* = 500	*n* = 1,000	*n* = 5,000	*n* = 10,000	*n* = 50,000
10,000	9.8	4.3	2.9	1.0	0	—
50,000	9.8	4.4	3.1	1.3	0.9	0
100,000	9.8	4.4	3.1	1.4	0.9	0.3
500,000	9.8	4.4	3.1	1.4	1.0	0.4
50,000,000	9.8	4.4	3.1	1.4	1.0	0.4
200,000,000	9.8	4.4	3.1	1.4	1.0	0.4

The table assumes that $p = .5$. If the true value of p is different from .5, then the errors will be less than the values listed in the table. The table lists the percentage-point error. For example, with population size 50,000 and sample size 5,000, the table lists the value 1.3. That means that there is a 95 percent chance that the sample proportion will be within 1.3 percentage points of the true value (in other words, between .487 and .513).

There are some interesting results that you can see if you scan through the table. If you take a sample of 1,000 people, there is a 95 percent chance that the poll result will be within 3.1 percentage points of the true value. If you increase the sample size to 5,000, the error falls to only 1.4 percent. So the table does tend to reinforce one's faith in polls (as long as the poll is truly random).

Another interesting result you can see is that a sample of 1,000 does just as well when the population is 200 million as it does when the population is 50,000. You might expect that the sample would become less accurate as the population becomes larger, but it doesn't work that way.

However, the reverse is also true. If you want to get an accurate sample of a population of 50,000, you need just as large a sample as you would if you had a population of 200 million. Making the population smaller does not reduce the number of people you need in the sample in order to get a representative sample. You can get a good cross section of 200 million people by interviewing 1 person in every 40,000, but if you try to interview 1 person in 40,000 when the population is 50,000, you will end up with a very unrepresentative sample. What this means is that if you want an accurate view of the opinions of people in every state, you will need a much larger sample than you would if you only needed to know the opinions of the entire country.

Another important fact to note is that, although the error does go down as the sample size becomes larger, it reaches a point where large increases in the sample size lead to only small decreases in the error. When you decide what size sample to use, you need to take two factors into consideration. Adding more people will make the sample a bit more accurate, but the cost of taking the sample becomes larger if more people are included.

You can see that a sample of about 1,000 will lead to a margin of error of about 3 percent, which is usually adequate for a national poll. However, such a poll won't be able to predict the election if the difference between the two candidates is less than 3 percent.

Another example of a national sample is the Current Population Survey, which is conducted each month by the Census Bureau for the Bureau of Labor Statistics. Among other things, the survey is used to calculate the unemployment rate every month. However, it would not be very helpful if the unemployment rate is reported with a 3 percent margin of error. Imagine someone saying, "The unemployment rate is somewhere between 2 percent and 8 percent." In order to get a margin of error close to 0.1 percentage points, the Current Population Survey needs to use a much larger sample: more than 100,000 people.

What Can Go Wrong

Polls have a reasonably accurate track record in predicting elections, which tends to increase one's confidence in their accuracy when they are used to survey other kinds of opinions. However, for most polls, we don't have a way of checking the entire population, so we can't be sure about the results. It is important to keep in mind the things that can go wrong.

Nonrandom samples. It is very important to remember that all of these results work only when the sample is a pure random sample. If the sample is not a pure random sample, all bets are off. A good example of the use of a nonrandom sample was the 1936 *Literary Digest* presidential election poll. The *Literary Digest* had 2 million people respond to its poll, which is a much larger number than would have been needed to get an accurate result if the sample had been selected randomly. However, the poll predicted that Alf Landon would be an easy winner, whereas in fact, Franklin Roosevelt won by a landslide. The problem came about because the sample was not a random sample. The magazine mailed out cards to people whose names were obtained from telephone lists and other sources, but at that time the people who had telephones were not representative of the population as a whole. If the sample is not selected randomly, there is no way to estimate how far off it might be.

In practice, modern opinion polls are not able to select pure random samples. You can't put the name of everyone in the country into a hat. For one thing, there is no such thing as a list of names for the whole country. Even if there were, it would be very expensive to interview 1,000 randomly selected people scattered all over the country. Instead, the pollsters select some regions at random, then they select some subregions, and finally they select some households to interview. This procedure guarantees that the people in the sample live in clusters, making it possible for one interviewer to interview quite a few people. However, this procedure does have the effect of making the predicted margin of errors larger, as you will notice if you look at the reported errors for opinion polls.

People might change their minds. A poll taken on Friday will be wrong if people change their mind over the weekend about who they will vote for on Tuesday.

People might lie to the pollster. If one answer is perceived to be socially preferred, people might tell the pollster that answer even if they believe something different.

The result may be affected by the way the question is worded. An advocacy group conducting a poll may begin with some questions designed to encourage the respondents to think one way. Neutral pollsters will do their best to ask questions as neutrally as possible to avoid swaying the views of the respondents, but they may not be totally able to avoid this. Remember that a neutral pollster has no objective other than to be accurate, whereas an advocacy group hopes the poll will support its view. Therefore, the advocacy group poll should be treated with a little bit more skepticism.

Results for subgroups will be less accurate. In addition to reporting the result for the overall sample, polls sometimes report results for different goups. Suppose the poll wishes to learn the opinions of college-educated women over age 40 living in the Northeast. Even if the overall sample shows a respectably small margin of error, the margin of error for a subgroup of the sample will be larger than it will be for the entire sample. The number of people in the sample in that category may be too small to provide accurate representation.

Nonrespondents. For a variety of reasons some people may not be reachable, or they may not want to respond. People are under no obligation to respond to a poll (unless it is the U.S. Census Bureau asking the questions). If the views of nonrespondents are systematically different from the views of respondents, then the poll will be biased. It is very difficult to check this—if you could survey the nonrespondents, then they wouldn't be nonrespondents.

Nonvoters. Elections are decided by voters, so an election poll may accurately predict the views of the entire population but still mispredict the election because the people who actually vote differ from the people who don't vote. Therefore, an election poll needs to avoid counting nonvoters, but it is difficult to determine how to do this. Simply asking people if they intend to vote is not likely to work because they may perceive that there is a stigma attached to not voting, so they may say they plan to vote even if they don't. The pollster can ask other questions, such as, "How informed are you about this election?" in an attempt to collect information that might be useful in predicting who will vote.

TECHNICAL Note: Sampling With or Without Replacement

Suppose that in a population of 200 million, half of the people are on our side. The probability that the first person we select will be on our side is

$$\frac{100{,}000{,}000}{200{,}000{,}000} = .5$$

If the first person we select is on our side and we sample with replacement, the probability doesn't change for the next person we select: there is still a probability of .5 that the second person we select will be on our side. However, if we sample without replacement, then the probability changes slightly. The probability that the second person will be on our side is

$$\frac{99{,}999{,}999}{199{,}999{,}999} = .4999999975$$

This illustrates that with a large population the difference between sampling with or without replacement is extremely insignificant. So, we can make our lives easier by pretending that the sample was chosen with replacement (so we can use the binomial distribution) even if it was chosen without replacement. Since the binomial distribution can be found by summing n independent, identically distributed random variables, the Central Limit Theorem applies; so we can therefore use the normal distribution as an approximation to the binomial distribution.

Experimental Design

In the preceding chapters, we've discussed various inferential statistics, i.e., various calculations that can be used to evaluate hypotheses. We can calculate numbers that will tell us if certain statements are likely to be true.

Now that we have these tools in our toolkit, it is reasonable to seek the best way to get the data for our calculations. Experiments don't just happen; people, perhaps you, need to design and perform them.

If you are in the manufacturing industry and you have to decide between two types of packaging machines, you might want to test both types of machines.

A very simple test would be to take one machine of type A, one of type B, run them both for an hour, and see which one packages more items and/or has less errors.

What if the trial machine of type A had a defect or was damaged in transit? It would perform worse than the machine of type B, while the machine of type A might be faster in general.

Clearly, it would be prudent to test more than one machine of each type, to avoid flukes of this sort. If this were our only consideration, we might want to try out tens, hundreds, even thousands of machines. There are limitations, however, on how many machines that we can test.

Some sort of middle ground would be best. We'd like to test enough machines to feel confident that we will have made the right decision based on our results. We don't want to test any more than we'd have to.

How many is enough? What would be used to determine such a number?

One factor would be the precision that we seek. If we only package 20 items a day, then a difference of one or two items per hour would be significant. It would be worth our while to test a large number of machines to make sure which type was faster.

If we packaged 100,000 items per day, a difference of one or two items per hour might not be significant. It might not be worth our while to test a large number of machines to detect such a difference. The number of tests performed is partly determined by the precision that we seek: the greater the precision, the more tests will be needed.

Other factors might be determined by the testing process itself.

What do we seek in an experiment? Usually we'd like to confirm or disprove some hypothesis.

In our discussion of hypothesis testing, we've talked about setting up a null hypothesis and an alternative hypothesis and have come up with ways to calculate their associated probabilities. These formulas usually have taken a given n, and yielded a given probability. These processes can be reversed.

EXAMPLE Suppose that we have an undetermined number of independent observations of a normally distributed random variable, with known standard deviation 3 and unknown mean. How many observations would we need to be 95 percent sure that our sample mean is within 2 units of the distribution mean? How many times do we have to repeat the experiment before we're satisfied that our sample mean is close enough to the population mean?

SOLUTION

We know that $Z = \sqrt{n}(\bar{x} - \mu)/\sigma$ is a standard normal random variable. We want to cause the following to have a 95 percent likelihood:

$$\begin{aligned} |\bar{x} - \mu| &\leq 2 \\ |\bar{x} - \mu|/\sigma &\leq 2 / 3 \\ \sqrt{n}\,|\bar{x} - \mu|/\sigma &\leq 2\sqrt{n}/3 \\ |Z| &\leq 2\sqrt{n}/3 \end{aligned}$$

We want to find n such that

$$\begin{aligned} \Pr(|Z| \leq 2\sqrt{n}/3) = 2\ \Pr(0 \leq |Z| 2\sqrt{n}/3) &\geq \ 95 \text{ percent } = 0.95 \\ 2(\Pr(Z \leq 2\sqrt{n}/3) - 0.5) &\geq 0.95 \\ 2\Pr(Z \leq 2\sqrt{n}/3) &\geq 1.95 \\ \Pr(Z \leq 2\sqrt{n}/3) &\geq 0.975 \end{aligned}$$

By Table A3–1, we see that

$$\begin{aligned} 2\sqrt{n}/3 &\geq 1.96 \\ \sqrt{n} &\geq 2.94 \\ n &\geq 8.6436 \\ n &\geq 9 \end{aligned}$$

We need at least 9 observations to be 95 percent sure of having so precise a sample mean.

It's not enough to know how many times a test should be performed. All of our statistical work so far has made a very big assumption: that everything about our experiment has been random. We want to make sure that no biases are accidentally introduced into our work.

For example, when we are testing our packaging machines, it would be a poor idea to test the type A machines during the day, and the type B machines during the night, or to test the type A machines on small items and type B machines on large items. Clearly this would bias our results, by possibly favoring one type over the other.

Tests of different procedures should be identically designed wherever possible to avoid this problem.

EXERCISES

1. Estimate the frequency of occurrence of the letters of the alphabet in your favorite book by examining a sample of a few pages. How accurate do you think your results are?

2. Estimate the frequency of the different first names in your city by looking at a random sample of names in the phone book.

3. Suppose you have 60 red marbles and 40 blue marbles in a box. If you pick out 10 marbles at random with replacement, what is the probability that you will pick 6 red marbles? If you select 10 marbles without replacement, what is the probability that you will select 6 red marbles?

4. Suppose that 55 percent of the people in a population of 500,000 support your candidate. If you conduct a poll of 1,000 people, what is the interval that has a 95 percent chance of containing the results of your poll?

In each of the following exercises, you are given the number of people (N) *in the group and the number of people* (m) *who support your candidate. If you ask a sample of* n *people whom they support, what is the probability that the number of people in the sample on your side will have the values listed?*

	N	m	n	People in Sample on Your Side
5.	15	8	5	1, 2, 3, 4
6.	15	3	5	1, 2, 3, 4
7.	15	10	5	1, 2, 3, 4
8.	40	25	5	1, 2, 3, 4
9.	40	25	10	4, 5, 6, 7
10.	40	25	20	9, 10, 11, 12
11.	30	15	20	8, 9, 10, 11, 12
12.	50	25	20	8, 9, 10, 11, 12
13.	100	50	20	8, 9, 10, 11, 12

For each of the following exercises you are given the sample size (n) *and the number of items in the sample that have the characteristic you are interested in* (k). *Calculate 95 and 99 percent confidence intervals for the unknown value of the proportion of items in the population that have this characteristic. Assume that the population size is much larger than the sample size.*

	n	*k*
14.	100	36
15.	200	72
16.	500	180
17.	1,000	360
18.	5,000	1,800
19.	10,000	3,600
20.	824	422
21.	560	328
22.	705	451
23.	234	128
24.	657	503

20

Nonparametric Methods

You may have noticed a common thread running through the last few chapters. When testing hypotheses or generating confidence intervals, we've assumed that the population(s) we were sampling had a given distribution, usually normal. Although we may not have known what the parameters (that is, the mean and variance) were, we could estimate them from the samples and go on from there.

Often, we've assumed also that the data were continuous—that is, someone's height in centimeters could be 180.1, or 180.11, or 180.109, etc. What should we do if the data consist of people's ratings of a given product or service on a scale from 1 to 10? How do we allow for the fact that numerical ratings are subjective, that a product that is given an 8 is not necessarily "twice as good" as a product that is given a 4 by the same person?

In these instances we don't have the usual distributions and parameters to fall back on, and they call for different procedures, called *nonparametric methods*.

Sign Test

One of the simplest nonparametric methods is called the *sign test*. It is useful when evaluating a survey that tests which option the surveyee prefers, but not by how much. For example, suppose Jones Cola Company conducts a blind taste test matching its brand against that of its rival, Smith Cola Company. We will test the null hypothesis that among the population as a whole there is no difference in preferences between the two colas. If each person in the survey provided an accurate estimate of his or her liking for each cola on a scale of 1 to 10, we could use the statistical methods employed before with the t statistic. However, we might not have these rankings available; moreover, we might doubt the accuracy of interpersonal rating comparisons even if they were available. Therefore, we will use the sign test. We will simply ask 20 people in the survey which cola they prefer, using a plus sign to indicate people who prefer Jones Cola and a minus sign to indicate the people who prefer Smith Cola. Suppose the results are as follows:

$$+, -, +, +, +, +, -, +, +, -, +, +, +, -, +, +, -, +, +, +$$

Of the 20 people in the survey, 15 favored Jones Cola and 5 favored Smith Cola. If the null hypothesis is true, then each individual had a 50 percent chance of selecting Jones Cola and a 50 percent chance of selecting Smith Cola. Therefore, the number of plus signs would come from a binomial distribution with $n = 20$ and $p = 5$. We will use n_+ to stand for the number of plus signs. Then

$$E(n_+) = np = 10 \quad \text{and} \quad \text{Var}(n_+) = np(1-p) = \frac{20}{4} = 5$$

We can use the normal distribution as an approximation for the binomial distribution. (For the sign test this approximation will generally be valid when $n > 10$.) Therefore, we'll use a test statistic Z:

$$Z = \frac{n_+ - np}{\sqrt{np(1-p)}} = \frac{n_+ - n/2}{\sqrt{n}/2} = \frac{2n_+ - n}{\sqrt{n}}$$

This statistic will have a standard normal distribution if the null hypothesis is true. In our case $n_+ = 15$ and $n = 20$, so we can calculate $Z = 2.236$. We can reject the null hypothesis at the 5 percent level, since 2.236 is outside 1.96.

One important advantage of the sign test is that you need not make restrictive assumptions about the nature of the population. For example, you do not need to assume that preferences are distributed according to the normal distribution. The disadvantage of the sign test is that it ignores some information that you may have in certain cases. If, as in the cola examples, the only information available tells you which brand was preferred, then you will have to use the sign test. If you also have available information that indicates the strength of each person's preferences, then the sign test is not as effective because it ignores that information.

Friedman F_r Test

Above, where we were testing only two options, we cared only about which one people liked better. Given three or more options, we might also want to know which option people like second best, third best, and so on. Thus, given n options, we would ask people to rank their preferences. This would rule out a null hypothesis assuming a multinomial distribution (a straightforward generalization of the two-option case), which would take into account only which option a person likes best.

We therefore use the Friedman F_r *test* and construct a new test statistic:

$$F_r = \left[\frac{12}{bk(k+1)}\right]\left(\sum_{j=1}^{k} R_j^2\right) - 3b(k+1)$$

Here b is the sample size, k is the number of options, and R_j is the rank sum of the jth option (that is, for the jth option, the sum of the ratings given by those surveyed).

Given the null hypothesis that there is no preference among the k options, and either $k > 5$ or $b > 5$, then F_r has approximately a chi-square distribution with $k - 1$ degrees of freedom, which can be used to test the null hypothesis.

EXAMPLE Suppose 10 salespeople for a company compare their sales (working independently) in four cities. Test to see whether they do better or worse than average in any of the cities. Here is the data table:

Salespeople	City No. 1	Rank No. 1	City No. 2	Rank No. 2	City No. 3	Rank No. 3	City No. 4	Rank No. 4
1	15	3	18	4	9	1	11	2
2	19	3	20	4	13	2	6	1
3	27	4	7	1	15	3	8	2
4	43	4	9	1	19	2	29	3
5	18	4	17	3	16	2	11	1
6	12	4	10	2	11	3	9	1
7	20	4	13	1	16	2	18	3
8	14	1	22	2	23	3	31	4
9	40	4	14	2	25	3	13	1
10	9	1	14	2	26	3	19	4
	$R_1 = 32$		$R_2 = 22$		$R_3 = 24$		$R_4 = 22$	

SOLUTION

Then $b = 10$, $k = 4$, $F_r = 4.08$; and, by checking Table A3–3, we see that we can accept the null hypothesis at the 5 percent level.

Our general procedure, given survey results ranking preferences of choices (either because numerical data are not available, or because they don't fit a known distribution), is to test the null hypothesis that there is no preference among the choices, when either the size of the sample (b) or the number of options (k) is greater than 5.

PROCEDURE FOR THE FRIEDMAN F_r TEST

1. Calculate the rank sums R_j.
2. Calculate

$$F_r = \left[\frac{12}{bk(k+1)}\right]\left(\sum_{j=1}^{k} R_j^2\right) - 3b(k+1)$$

3. F_r has approximately a chi-square distribution with $k - 1$ degrees of freedom. Check Table A3–3 for $k - 1$ degrees of freedom. Accept or reject the null hypothesis at a chosen significance level.

Note: Ties may occur when ranking numerical data. If there are only a few ties, average the rankings among those tied; for example, 10, 20, 10, 15 would be ranked 1.5, 4, 1.5, 3 since there is a tie for first and second rankings. If there are many ties, your data may be unreliable.

Wilcoxon Rank Sum Test

Suppose now that you have samples (not necessarily the same size) from two populations that you think have the same (unknown) distribution. One way of testing whether they really do come from the same distribution is to pool the samples and see how they mix. If all the values of one sample are smaller than all the values of the other, then you should probably reject the null hypothesis (that the two populations have the same distribution). On the other hand, if they are well mixed, then you have no cause to reject the hypothesis.

A way of measuring how well the pooled samples are mixed is the *Wilcoxon rank sum test*. It consists of pooling the samples A and B, ranking them, and adding the ranks given to A to form the rank sum T. If the size of each sample is at least 8, then T has an approximately normal distribution with parameters

$$\mu_T = \frac{n_A(n_A + n_B + 1)}{2}$$

$$\sigma_T^2 = \frac{n_A n_B(n_A + n_B + 1)}{12}$$

Here n_A = the size of sample A and n_B = the size of sample B. Then, by comparing $(T - \mu_T)/\sigma_T$ with Table A3–2, we can determine whether to accept the hypothesis.

EXAMPLE Suppose nine Harvard and eight Yale economics professors make the following predictions for inflation for the coming year:

Harvard	Inflation Prediction (in percent)	Rank	Yale	Inflation Prediction (in percent)	Rank
1	8.6	1	1	15.8	15
2	9.7	3	2	13.4	11
3	11.8	6	3	12.2	7
4	17.1	16	4	17.3	17
5	12.9	10	5	12.6	9
6	11.7	5	6	14.3	14
7	12.4	8	7	14.2	13
8	10.3	4	8	13.5	12
9	8.9	2			

Test the null hypothesis that the professors' predictions have the same distributions.

SOLUTION

Here

$$T = 55 \quad \mu_T = \frac{(18)}{2} = 81$$

$$\sigma_T{}^2 = \frac{(9)(8)(18)}{12} = 108 \quad \frac{T - \mu_T}{\sigma_T} = -2.5$$

By comparing with Table A3–2, we see that we can reject the hypothesis at the 95 percent level.

Thus, we have the procedure shown below to test the null hypothesis that samples come from two populations with the same distribution.

PROCEDURE FOR THE WILCOXON RANK SUM TEST

1. Pool the samples.
2. Rank them (again, if there are relatively few ties, average their ranks).
3. Compute the rank sum T for sample A,

$$\mu_T = \frac{n_A(n_A + n_B + 1)}{2}, \quad \text{and} \quad \sigma_T{}^2 = \frac{n_A n_B(n_A + n_B + 1)}{12}$$

4. Compare the standard normal random variable $(T - \mu_T)/\sigma_T$ with Table A3–2 at a given level to determine whether to accept the hypothesis.

Kruskal-Wallis *H* Test

The *Kruskal-Wallis* H *test* is a generalization of the Wilcoxon rank sum test to more than two populations.

EXAMPLE Suppose that we have samples of k populations, and we want to test the hypothesis that the populations have the same distributions.

SOLUTION

Again we pool the samples and rank them. Our test statistic is a bit more complicated, however:

$$H = \left[\frac{12}{n(n+1)}\right]\left[\sum_{j=1}^{k}\left(\frac{R_j^2}{n_j}\right)\right] - 3(n+1)$$

where

$$n = \sum_{j=1}^{k} n_j,$$

nj = the size of the sample of the jth population, and R_j = the jth rank sum. The statistic H has approximately a chi-square distribution with $k - 1$ degrees of freedom, if each of the nj is larger than 5.

Thus, given a significance level and samples of k different populations (with every sample size larger than 5), calculate H, and check Table A3–3 under $k - 1$ degrees of freedom to decide whether to accept the null hypothesis (that the populations have the same distributions).

Wilcoxon Signed Rank Test

As you might guess, the *Wilcoxon signed rank test* is similar in some ways to the Wilcoxon rank sum test. It is applicable to situations similar to those discussed for the sign test. The null hypothesis states that there is no difference between the two populations.

Suppose we have a survey asking people to evaluate two options A and B. This time, rather than just stating which they prefer, they rate each on a scale (for example, from 1 to 10 or 100). We then rank the absolute values of the differences between the ratings for A and B, and sum the ranks for the positive differences and negative differences, giving T_+ and T_-. If we let T be the smaller of T_+ and T_-, and the sample size N is large, then T is a normal random variable with parameters

$$\mu_T = \frac{N(N+1)}{4} \quad \text{and} \quad \sigma_T = \frac{N(N+1)(2N+1)}{24}$$

if the null hypothesis is correct. We then have a standard normal random variable $(T - \mu_T)\sigma_T$ that we can compare with Table A3–2 to decide whether to accept our hypothesis.

EXAMPLE Suppose we take a random sample of 10 people who have rented cars from both Able Rental Cars and Baker Rental Cars, and ask them to rate both rental services on a scale of 1 to 10. We then have the following table:

Person	*A*	*B*	*A* - *B*	*A* - *B*	Rank	+	-
1	4	7	–3	3	9		9
2	5	6	–1	1	3		3
3	9	6	3	3	9	9	
4	5	7	–2	2	6.5		6.5
5	4	5	–1	1	3		3
6	9	8	1	1	3	3	
7	5	4	1	1	3	3	
8	7	6	1	1	3	3	
9	3	5	–2	2	6.5		6.5
10	3	6	–3	3	9		9
				Totals		**18**	**37**

Test the hypothesis that there is no difference between the two companies.

SOLUTION

Then $T_+ = 18, T_- = 37$. Also,

$$T = T_+ = 18 \quad \mu_T = \frac{10(11)}{4} = 27.5$$

$$\sigma_T{}^2 = \frac{10(11)(21)}{24} = 96.25$$

Thus $(T - \mu_T)/\sigma_T = -0.97$, and we can accept the hypothesis at the 95 percent level.

Again in case of ties, average the ranks involved. If persons give the same rating to *A* and *B* (that is, $A - B = 0$), perform the test twice, first counting 0 as a positive difference, the second time counting 0 as a negative difference. If the results agree, fine. If not, your data may be unreliable.

We then have the procedure shown below, given ratings of two options by *N* people (companies, departments, etc.).

PROCEDURE FOR THE WILCOXON SIGNED RANK TEST

1. Rank the absolute values of the differences of the ratings.
2. Calculate the rank sums T_+ and T_-.
3. Let T be the smaller of T_+ and T_-,

$$\mu_T = \frac{N(N+1)}{4} \quad \text{and} \quad {\sigma_T}^2 = \frac{N(N+1)(2N+1)}{24}$$

4. Compare $(T - \mu_T)/\sigma_T$ with Table A3–2 to determine whether to accept the null hypothesis at a given significance level.

NOTE TO CHAPTER 20

The nonparametric hypothesis tests discussed in this chapter have connections to normal-based parametric tests.

- The Wilcoxon rank sum test is a nonparametric analog of a two-sample T-test on population means (different sample sizes).
- The Kruskal-Wallis H test is a nonparametric analog of an ANOVA.
- The Wilcoxon signed rank test is a nonparametric analog of a paired sample T-test on population means.
- The Friedman F_r test is a nonparametric analog of an ANOVA.

EXERCISES

1. Of 25 rental car users surveyed, 9 preferred Arthur's Rental Cars to Beaumont's Rental Cars, the other 16 preferring Beaumont's. Test the null hypothesis that rental car users have no preference between Arthur's Rental Cars and Beaumont's Rental Cars.

2. On a certain uneven piece of ground, Route 13 splits into two roads that eventually merge, one going over some mountains and the other down through a valley. If a random sampling of 30 drivers showed that 8 drivers took the high road and 22 drivers took the low road, test the null hypothesis that Route 13 drivers have no preference between the two roads.

3. In a taste test, 13 out of 18 people preferred spaghetti sauce made with ground oregano to sauce made with crushed oregano leaves, the other 5 people preferring the crushed leaves. Test the null hypothesis that people are indifferent to which form of oregano is used in their spaghetti sauce.

4. Fifteen frequent fliers are asked to rate in order of preference three airlines: Aircat (A), Bluebird (B), and Condor (C) Airlines. The results are as follows:

A	B	C	A	B	C	A	B	C
1	3	2	3	1	2	1	3	2
1	3	2	3	2	1	3	2	1
1	3	2	2	3	1	2	3	1
1	2	3	1	3	2	3	1	2
2	1	3	1	2	3	1	3	2

Test the null hypothesis that there is no preference among fliers between the three airlines.

5. Four people (A, B, C, D) are asked to rate in order of preference the 12 months of the year. The results are as follows:

	J	F	M	A	M	J	J	A	S	O	N	D
A	12	11	5	3	4	6	7	8	1	2	9	10
B	12	1	2	9	4	3	5	6	7	8	10	11
C	1	2	4	5	6	9	3	7	8	10	11	12
D	1	4	3	12	10	5	9	11	8	7	6	2

Use these data to test the null hypothesis that people have no preference between the months of the year.

6. Six companies show the following profits (in thousands of dollars) from sales in four different cities (A, B, C, D):

A	B	C	D
22	11	16	14
20	19	18	14
19	24	16	13
15	18	17	19
18	17	13	15
17	16	19	12

Use the Friedman F_r statistic to test the null hypothesis that the four cities are equally profitable for the companies.

7. Thirteen voters surveyed showed the following preferences between three presidential candidates (A, B, C):

A	B	C	A	B	C
3	2	1	1	2	3
2	3	1	3	1	2
3	1	2	3	2	1
1	2	3	2	1	3
1	2	3	2	1	3
3	2	1	1	2	3
1	2	3			

Test the null hypothesis that voters have no preference between the candidates.

8. Here are the sales figures (in hundreds) for a company in 10 towns in Arizona and 15 towns in Michigan:

AZ: 16, 8, 7, 14, 22, 27, 19, 23, 18, 30
MI: 13, 28, 17, 26, 12, 11, 33, 35, 5, 40, 39, 10, 6, 32, 31

Test the null hypothesis that the distribution of sales in Arizona is the same as that in Michigan.

9. Here are the defense budgets (in millions of dollars, adjusted for inflation) of two countries (A, B) over a decade:

A: 10, 12, 15, 16, 18, 22, 26, 28, 30, 29
B: 9, 8, 17, 14, 13, 19, 24, 25, 27, 31

Test the null hypothesis that the budgets have the same distribution.

10. Twelve light bulbs of brand X and fifteen of brand Y were tested to see how long they would last, with the following results (in hundreds of hours):

X: 15, 19, 23, 25, 26, 28, 30, 29, 32, 31, 20, 14
Y: 20, 18, 16, 19, 22, 24, 17, 21, 10, 11, 32, 27, 12, 14, 9

Test the null hypothesis that the lifetimes of the two brands of light bulbs have the same distribution.

11. Here are the incomes (in thousands of dollars) of 15 plumbers in California and 13 plumbers in Maine:

    CA: 22, 19, 17, 24, 25, 30, 29, 31, 28, 37, 15, 40, 38, 39, 17
    ME: 21, 20, 18, 17, 23, 26, 28, 29, 32, 35, 19, 27, 16

    Test the null hypothesis that plumbers' incomes are distributed equally in both states.

12. Here are the benefits (in thousands of dollars) paid to employees of three yo-yo manufacturers (A, B, C) over a decade:

A	B	C	A	B	C
10	25	16	23	19	28
26	12	24	30	14	18
29	20	13	31	38	35
21	11	22	39	32	37
17	27	15	33	36	34

    Test the hypothesis that the benefits expenditures of the three companies have the same distribution.

13. Four experimental precision scales (A, B, C, D) are tested on a fixed weight, with the following results:

    A: 103, 121, 106, 120, 114, 128, 116
    B: 112, 105, 132, 136, 109, 138, 135, 126, 124, 117
    C: 131, 104, 130, 108, 123, 119, 113, 133, 127, 134, 125, 115
    D: 129, 111, 122, 137, 107, 110, 139, 118

    Test the null hypothesis that the distributions of values given by the four scales are the same.

14. Fifteen people are asked to rate on a scale of 1–100 two different stereo systems (A, B) with the following results:

A	B	A	B	A	B
85	78	95	70	65	83
60	99	99	65	85	72
70	85	93	62	89	95
75	99	75	80	90	60
88	90	60	78	95	50

    Test the null hypothesis that people have no preference between the two kinds of stereos.

15. Twenty people rate on a scale of 1–10 two political candidates (A, B) with the following results:

A	B	A	B	A	B	A	B
2	8	8	4	6	5	9	7
5	6	9	8	7	4	8	9
3	5	8	9	8	5	4	6
7	8	7	8	8	6	10	7
4	5	5	6	9	6	8	5

Test the null hypothesis that people have no preference between the candidates.

16. Eighteen people rate on a scale of 1–20 two different brands (A, B) of water pistols, with the following results:

A	B	A	B	A	B
18	15	17	13	11	17
14	12	12	15	13	12
17	19	13	17	14	16
16	12	18	15	12	13
18	15	19	16	11	10
11	9	12	16	16	12

Test the null hypothesis that people have no preference between the pistols.

17. Twenty-five people rate on a scale of 1–10 two car models (A, B) with the following results:

A	B	A	B	A	B
4	6	6	4	8	2
8	7	8	7	9	8
5	4	5	8	8	6
4	2	7	6	7	8
3	7	4	8	6	7
6	5	5	9	3	6
9	8	6	3	5	6
7	8	7	3	4	2
2	5				

Test the null hypothesis that people have no preferences between the car models.

21

Regression with One Independent Variable

Often in statistics we want to investigate the question: How much does one quantity affect another quantity? There are many situations in which one quantity (called the *independent variable*) has a big effect on another quantity (called the *dependent variable*). Once we've figured out the relationship between the two variables, we can predict the value of the dependent variable if we know the value of the independent variable.

We will start with a small hypothetical example. In economics, it is assumed that the level of income affects the quantity that will be demanded of a particular good. For most goods, higher income leads to higher demand.

Suppose we have observations of average income and total pizza sales for a 1-month period for eight different towns:

Town	Income (in thousands of dollars)	Pizza Sales (in thousands)
1	5	27
2	10	46
3	20	73
4	8	40
5	4	30
6	6	28
7	12	46
8	15	59

When faced with a real problem it would be better to have more observations, but this sample of eight observations will work well to illustrate the calculations.

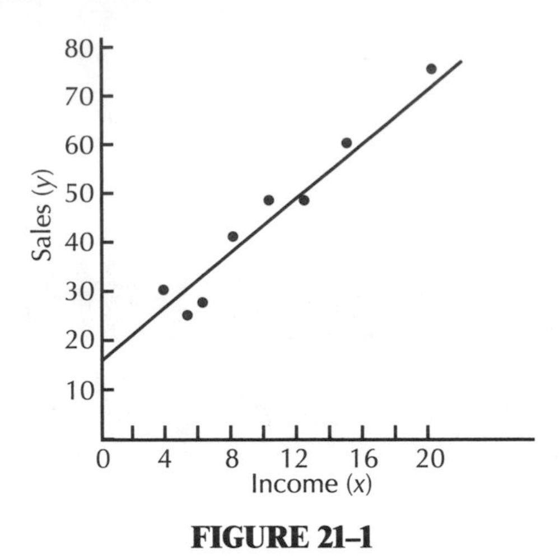

FIGURE 21–1

Figure 21–1 illustrates the relationship between income and pizza sales. Each dot represents a town. The line that best fits these dots is shown on the graph.

We can see clearly that there is a relationship between income and pizza sales, and we can see also that an increase in income leads to an increase in pizza sales.

Calculating a Regression Line

Figure 21–2 illustrates the general situation. We will use x to represent the independent variable, which will be measured along the horizontal axis. We will use y to represent the dependent variable—that is, the variable that depends on x. In the pizza example, income is the independent variable and pizza sales are the dependent variable. We will measure the dependent variable along the vertical axis. Suppose we have four observations. Then the scatter diagram contains four points, which we will call (x_1, y_1), (x_2, y_2), (x_3, y_3), and (x_4, y_4).

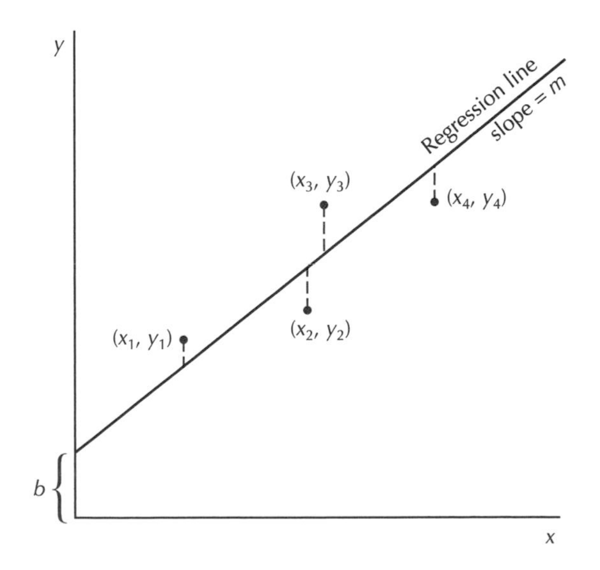

FIGURE 21–2

Any line can be described by specifying two numbers: the *slope* and the *vertical intercept*. We will use $\hat{m}$ to represent the slope and $\hat{b}$ to represent the intercept. (We will explain later why we put the hats on the letters.) The equation of the line can be written as

$$y = \hat{m}x + \hat{b}$$

Suppose we guess that the best regression line is the line shown in Figure 21–3. This line looks like a good choice, but it doesn't fit the data points perfectly. For each point there is a certain amount of vertical distance between the point and the line. We'll call that distance the *error* or *residual* of the line relative to that point. A larger value for the error means that the line does a worse job of representing the points. Each point has its own error. (We'll call these error_1, error_2, error_3, and error_4.) We'd like to choose the line so that the total error is as small as possible. The normal procedure in statistics is to minimize the sum of the squares of all the errors. The square of the error for the point (x_i, y_i) is

$$(\text{error}_i)^2 = [y_i - (\hat{m}x_i + \hat{b})]^2$$

We will call the sum of the squares of all of the errors SE_{line} (short for "squared error about the line"):

$$\begin{aligned} SE_{\text{line}} &= [y_1 - (\hat{m}x_1 + \hat{b})]^2 + [y_2 - (\hat{m}x_2 + \hat{b})]^2 \\ &\quad + [y_3 - (\hat{m}x_3 + \hat{b})]^2 + [y_4 - (\hat{m}x_4 + \hat{b})]^2 \end{aligned}$$

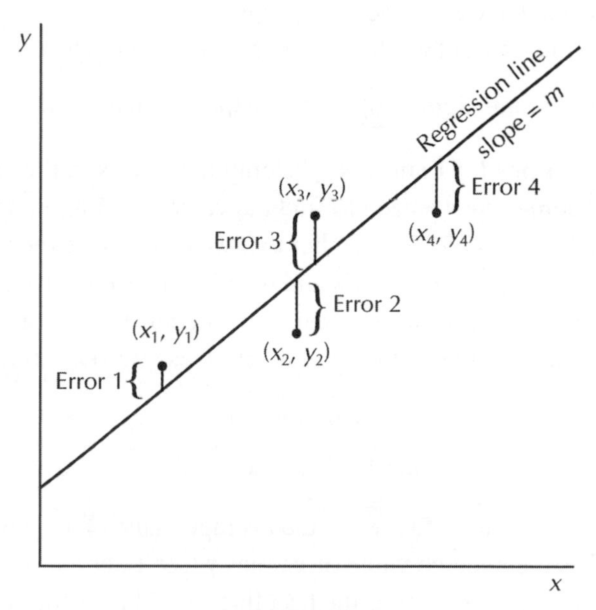

FIGURE 21–3

We originally looked at the question of whether there is a relation between x and y. Here is another way of looking at this question: Does knowing the value of x help you predict the value of y? Suppose you have discovered that there is a very clear relation between x and y that can be represented by this line:

$$y = 2x + 15$$

Then, if you know that next year's value of x will be 10, you can predict that next year's value of y will be $2 \times 10 + 15 = 35$. If next year's value of x will be 30, you can predict that next year's value of y will be 75.

In general, if the variable x has the value x_i, we will use the symbol $\hat{y}_{xi}$ to represent the predicted value of y for that given value of x. In general,

$$\hat{y}_{xi} = \hat{m}x_i + \hat{b}$$

Note that there will be a different value of $\hat{y}_{xi}$ for each different value of x_i. (The predicted value of y is also called the *fitted value*.)

We can write our expression for the first error like this:

$$\text{error}_1 = y_1 - \hat{y}_{x1}$$

In general, we will use n to represent the number of data points, so we can write the expression for SE_{line} with summation notation:

$$\begin{aligned}\text{SE}_{\text{line}} &= \sum [y_i - (\hat{m}x_i + \hat{b})]^2 \\ &= \sum (y_i - \hat{y}_{xi})^2\end{aligned}$$

All through this chapter we will be taking summations from $i = 1$ to n. For convenience we will leave out the little numbers above and below the sigma.

Whenever you see a capital sigma (Σ) in this chapter remember that it means $\sum_{i=1}^{n}$.

We have n observations for x and y, so the only unknowns in the expression for SE_{line} are $\hat{m}$ and $\hat{b}$. Somehow we need to find the values of $\hat{m}$ and $\hat{b}$ that will result in SE_{line} being as small as possible. Remember that we can choose the values of $\hat{m}$ and $\hat{b}$, but we can't change the values of the x's and the y's because we found these values when we conducted our observations. It requires calculus to find the optimum values of $\hat{m}$ and $\hat{b}$. The result is that $\hat{m}$ and $\hat{b}$ must satisfy these two equations:

$$\begin{aligned}\hat{m}\bar{x} + \hat{b} - \bar{y} &= 0 \\ \hat{m}\overline{x^2} + \hat{b}\bar{x} - \overline{xy} &= 0\end{aligned}$$

Here $\bar{x}$ is the average value of x, $\overline{x^2}$ is the average value of x^2, $\bar{y}$ is the average value of y, and $\overline{xy}$ is the average value of x times y (see Exercise 1). The first equation states a fact that makes a lot of sense: the best line should pass through the point $(\bar{x}, \bar{y})$—that is, the point located at the average values of x and y.

Now we can use the two equations given previously to find the formula for the slope:

$$\hat{m} = \frac{\overline{xy} - \overline{x}\,\overline{y}}{\overline{x^2} - \overline{x}^2}$$

Once we know $\hat{m}$, we can calculate $\hat{b}$:

$$\hat{b} = \overline{y} - \hat{m}\overline{x}$$

Now we can calculate the slope and the intercept for the pizza example. We will assume that pizza sales depend on income, so we will call income the independent variable (x) and pizza sales the dependent variable (y). We need to calculate $\overline{x}$, $\overline{y}$, $\overline{x^2}$, and $\overline{xy}$.

	x	*y*	*x*²	*xy*	*y*²
	5	27	25	135	729
	10	46	100	460	2,116
	20	73	400	1,460	5,329
	8	40	64	320	1,600
	4	30	16	120	900
	6	28	36	168	784
	12	46	144	552	2,116
	15	59	225	885	3,481
Total	**80**	**349**	**1,010**	**4,100**	**17,055**
Average	**10.000**	**43.625**	**126.250**	**512.500**	**2,131.875**

Therefore, $\overline{x} = 10$, $\overline{y} = 43.625$, $\overline{x^2} = 126.25$, $\overline{xy} = 512.5$. Note that we also calculated $\overline{y^2} = 2131.88$ because it will be useful later.

We can use the formulas for $\hat{m}$ and $\hat{b}$:

$$\hat{m} = \frac{512.5 - (10 \times 43.625)}{126.25 - 10^2} = 2.905$$

$$\hat{b} = \overline{y} - \hat{m}\overline{x} = 43.625 - 2.905 \times 10 = 14.577$$

Therefore, the equation of the regression line can be written as

$$y = 2.905x + 14.577$$

If you have a scatter diagram with values of x along the horizontal axis and values of y along the vertical axis, then the slope and vertical intercept of the line that best fits these points can be found from these formulas:

$$\text{slope} = \hat{m} = \frac{\overline{xy} - \overline{x}\,\overline{y}}{\overline{x^2} - \overline{x}^2}$$

$$\text{intercept} = \hat{b} = \overline{y} - \hat{m}\overline{x}$$

The bars over the letters represent average values.

Accuracy of the Regression Line

As we pointed out earlier, knowing the slope and the intercept of the regression line doesn't tell us anything about how well the line fits the data. Therefore, we need to develop another measure to tell how well the line fits. Our first inclination is just to use SE_{line}, since that formula measures how much discrepancy there is between the points on the line and the actual data points:

$$SE_{line} = \sum (y_i - \hat{y}_{xi})^2$$

If SE_{line} is zero, then the line fits the data points perfectly. However, if the numerical value of SE_{line} is greater than zero, then we need something to compare this number with so that we can tell whether the fit of the line is any good.

We can compare the predictions of our regression line with those of a very simple-minded prediction plan: We could always predict that the value of y will be $\bar{y}$. For example, suppose you need to predict the total rainfall in your city next year. If you know nothing about what the weather conditions will be next year, but you do know the average rainfall in your city over the last several years, then your best prediction will be to guess that the rainfall next year will be the same as the average.

Let's have a contest between ourselves (using the regression line) and the simple-minded person who always will predict that the value of y will be equal to $\bar{y}$. We can calculate the total squared error of the simple-minded method (call it SE_{av}, since it is the total squared error of y about its average:)

$$SE_{av} = \sum (y_i - \bar{y})^2$$

[We could also write: $SE_{av} = n\text{Var}(y)$.]

If y really does depend on x, and our regression line describes that relation accurately, then the contest will not even be close. We will do a much better job predicting the value of y using our regression line than our simple-minded opponent will be able to do without using the line. In that case our squared error (SE_{line}) will be much less than the simple-minded squared error (SE_{av}). However, suppose that y really does not depend on x. In that case knowing the regression line will not help us. The simple-minded prediction plan will work almost as well, and SE_{line} will be almost as big as SE_{av}. Therefore, we'll define our measure of the accuracy of the regression line as follows:

$$r^2 = 1 - \frac{SE_{line}}{SE_{av}}$$

The quantity r^2 is called the *coefficient of determination*. This measure has two features that our fitness measure should have

1. If $SE_{line} = 0$, then $r^2 = 1$, and the line fits perfectly.
2. If $SE_{line} = SE_{av}$, then $r^2 = 0$, and the line fits very poorly.

The value of r^2 will always be between 0 and 1. The higher the value of r^2 the better the line fits. Here is another interpretation: the value of r^2 is the fraction of the variation in y that can be explained by variations in x. For example, an r^2 value of .75 means that 75 percent of the variations in y can be explained by variations in x. (The symbol r^2 is used because it is the square of the sample correlation coefficient between these two variables. See Exercise 2.)

We can calculate the r^2 value for the pizza example by finding the sum of the squares of all of the residuals. For each value of x we can calculate the predicted value of y from the formula

$$\hat{y}_{xi} = \hat{m}x_i + \hat{b} = 2.905x_i + 14.577$$

Then we can calculate the residual by subtracting the actual value of y from the predicted value. Here is a table of the results:

	x	Actual Value y	Predicted Value $\hat{y}_{xi} = \hat{m}x_i + \hat{b}$	Residual $y_i - \hat{y}_{xi}$	Squared Residual $(y_i - \hat{y}_{xi})^2$
	5	27	29.102	–2.102	4.418
	10	46	43.627	2.373	5.631
	20	73	72.677	0.323	0.104
	8	40	37.817	2.183	4.765
	4	30	26.197	3.803	14.463
	6	28	32.007	–4.007	16.056
	12	46	49.437	–3.437	11.813
	15	59	58.152	0.848	0.719
Total	**80**	**349**	**349**	**0**	**57.970**

This table illustrates two interesting properties of a regression line: (1) the sum of the residuals is always zero, and (2) the sum of the predicted values of y is always the same as the sum of the actual values of y.

We can calculate the variance of the y values from the formula

$$\text{Var}(y) = \overline{y^2} - \overline{y}^2 = 2131.875 - 43.625^2 = 228.734$$

Since $\text{Var}(y) = \text{SE}_{\text{av}}/n$, we can calculate

$$\text{SE}_{\text{av}} = n\text{Var}(y) = 8 \times 228.734 = 1829.875$$

Therefore,

$$r^2 = 1 - \frac{57.970}{1829.875} = .968$$

Just as we suspected, the regression line fits the data points very well; 96.8 percent of the variation in pizza sales can be explained by variations in income.

The value of r^2 can also be calculated from either of these formulas:

$$r^2 = \frac{(\overline{xy} - \overline{x}\,\overline{y})^2}{(\overline{x^2} - \overline{x}^2)(\overline{y^2} - \overline{y}^2)}$$

or

$$r^2 = \frac{(nT_{xy} - T_xT_y)^2}{(nT_{x^2} - T_x^2)(nT_{y^2} - T_y^2)}$$

Here, $T_x = \sum x$, $T_y = \sum y$, $T_{xy} = \sum xy$, $T_{x^2} = \sum x^2$, and $T_{y^2} = \sum y^2$.

Statistical Analysis of Regressions

Now we would like to perform statistical tests on our regression results. To do this we need to make some assumptions about the process relating x and y. In the standard regression model we assume that the true nature of the relation between x and y can be described by this equation:

$$y_i = mx_i + b + e_i$$

Here x_i is the ith observation for the variable x, y_i, is the ith observation for the variable y, and e_i is known as the *random error term*. We are assuming that x is the dominant influence on y, and the relationship can be represented by a straight line with slope m and y-intercept b. Unfortunately the true values of m and b are unknown; but, as you have probably guessed, we will use statistical procedures to estimate their values.

If the true equation were $y = mx + b$, then x would be the only factor that affected the value of y. Every single increase or decrease in y could be explained by an increase or a decrease in x. However, there are almost always some other factors that affect the value of the dependent variable. If the regression line represents the relationship well, these other factors will not be very important. Because all these other factors are mysterious and unknown, we will call them random error. In the equation given above e is a random variable that represents all of these other factors. (When we discuss multiple regression, we will see how it is possible to include some of these other factors in the regression model; but even so there will always be some remaining unexplained factors that make up the random error term.)

If we have n observations, then there will be n different random errors. We will let e_1 represent the random error effect on the first observation. In general, e_1 is a random variable representing the random error effect on the ith observation. We know that the expected value of each random error is zero [$E(e_i) = 0$] because of the fact that we have included the intercept term b in the equation. We will let Var (e_i) be equal to σ^2, but unfortunately we don't know the true value of σ^2. If x is a very good predictor of y, then σ^2 will tend to be small. If σ^2 is large, then there are other important factors that affect y and we should try to include them in our model. We will also assume that each random error term has a normal distribution, that the variance σ^2 is the same for each random error, and that each random error is independent of all of the other random errors.

Even though the true values of m and b are unknown, we can calculate the slope and the intercept of the regression line:

$$\hat{m} = \frac{\overline{xy} - \overline{x}\,\overline{y}}{\overline{x^2} - \overline{x}^2}$$

$$\hat{b} = \overline{y} - \hat{m}\overline{x}$$

Now you see why we included the hats. In statistical inference we often put a hat over a calculated statistic that is being used to estimate the value of an unknown parameter. The regression line is called the *least-squares* line, so $\hat{m}$ and $\hat{b}$ can be called the *least-squares estimators* of the parameters m and b.

Fortunately, it turns out that $\hat{m}$ and $\hat{b}$ have some very desirable statistical properties. For one thing, $\hat{m}$ and $\hat{b}$ are the maximum likelihood estimators of m and b. It can be shown that $\hat{m}$ has a normal distribution with $E(m) = m$ and

$$\text{Var}(\hat{m}) = \frac{\sigma^2}{\sum (x_i - \overline{x})^2}$$

We can write the expression for $\hat{m}$ like this:

$$\hat{m} = \frac{\sum (x_i - \overline{x})(y_i - \overline{y})}{\sum (x_i - \overline{x})^2}$$

This can also be written as

$$\hat{m} = \frac{1}{\sum (x_i - \overline{x})^2} \sum (x_i - \overline{x}) y_i$$

Remember that the xs are constants and the ys are random variables with a normal distribution. From the expression given above we can see that $\hat{m}$ is found by adding together a bunch of normal random variables multiplied by different constants, so we know from the properties of normal distributions that $\hat{m}$ will also have a normal distribution.

The fact that $E(\hat{m}) = m$ is important. It means that $\hat{m}$ is an unbiased estimator of the slope. It can also be shown that $\hat{m}$ is the best estimator (because it has the smallest variance) among a certain general class of all unbiased estimators.

The expression for the variance of $\hat{m}$ tells us that the variance of $\hat{m}$ is larger if σ^2 is larger. This seems reasonable, since a larger value of σ^2 means that we can expect more scatter about the true regression line and the increased scatter will make it more difficult to pin down the true value of m. The expression $\sum (x_i - \overline{x})^2$ represents the squared error about the average of the x's, or we could write

$$\sum (x_i - \overline{x})^2 = n\text{Var}(x)$$

We can see that when there is a greater spread among the x values, then the variance of $\hat{m}$ will be less and it will be easier to pin down the true value of m. Figure 21–4 illustrates two different situations. In each case the number of observations is the same. However, in example (a) there is not very much spread among the x values. In that case there will be much more uncertainty about the true value of the slope than there will be in example (b).

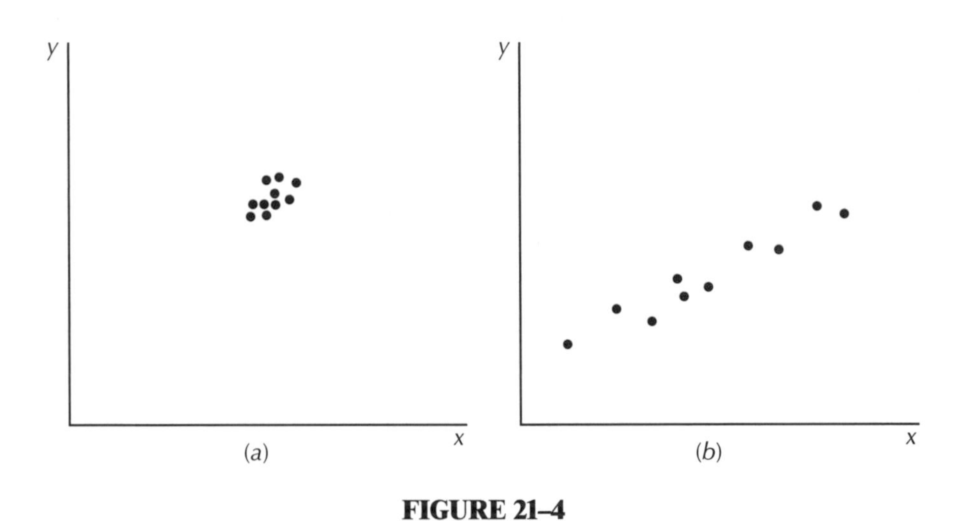

FIGURE 21–4

Since $\hat{m}$ has a normal distribution with mean m and variance $\sigma^2/\sum(x_i - \bar{x})^2$, we know that this random variable:

$$\frac{\hat{m} - m}{\sqrt{\dfrac{\sigma^2}{\sum(x_i - \bar{x})^2}}}$$

will have a standard normal distribution. Therefore, we can determine a 95 percent confidence interval for m:

$$\hat{m} \pm \frac{1.96\sigma}{\sqrt{\sum(x_i - \bar{x})^2}}$$

(See Chapter 17.) However, there is one obvious problem with calculating the confidence interval in this way: we don't know the value of σ. We need a way to estimate it. If we knew the true values of m and b, then we would have n observations of the random variable e:

$$e_i = y_i - (mx_i + b)$$

and so on. Since $E(e) = 0$, we could write

$$\sigma^2 = \text{Var}(e) = E(e^2) - [E(e)]^2 = E(e^2)$$

and we could estimate σ^2 by the average of the squares of the e's. Since we don't know m and b, we can use the residuals from our calculated regression line:

$$[y_1 = (\hat{m}x_1 + \hat{b})], \quad [y_2 - (\hat{m}x_1 + \hat{b})], \quad \text{and so on.}$$

The sum of the squares of all these residuals is what we called SE_{line}. Here we run into a conflict between the different criteria for estimators. The maximum likelihood estimator of σ^2 is equal to $\text{SE}_{\text{line}}/n$. However, in order to have an unbiased estimator of σ^2, we must use $\text{SE}_{\text{line}}/(n - 2)$. (Here $n - 2$ is the degrees of freedom of the squared error about the line, since we started with n points but lost two degrees of freedom when we had to use our observations to estimate the values of the slope and the intercept.) The quantity $\text{SE}_{\text{line}}/(n - 2)$ is called the *mean square error* (*MSE* for short):

$$\text{MSE} = \frac{\text{SE}_{\text{line}}}{n-2} = \frac{\sum(y_i - \hat{y}_{xi})^2}{n-2} = \frac{\sum[y_i - (\hat{m}x_i + \hat{b})]^2}{n-2}$$

The expected value of MSE is σ^2, which makes it an unbiased estimator of σ^2. As you recall, we had established that this random variable:

$$\frac{\hat{m} - m}{\sqrt{\dfrac{\sigma^2}{\sum(x_i - \bar{x})^2}}}$$

has a standard normal distribution. In similar situations in Chapter 17, we substituted an estimator for σ^2 in place of the unknown value of σ^2, and we ended up with something that had a t distribution. In this case, the random variable

$$\frac{\hat{m} - m}{\sqrt{\dfrac{\text{MSE}}{\sum(x_i - \bar{x})^2}}}$$

has a t distribution with $n - 2$ degrees of freedom. Then the confidence interval is

$$\hat{m} \pm a\sqrt{\frac{\text{MSE}}{\sum(x_i - \bar{x})^2}}$$

Here a is a number found from Table A3–5 such that

$$\Pr(-a < t < a) = \text{CL}$$

where t is a random variable having a t distribution with $n - 2$ degrees of freedom and CL is the confidence level.

For the pizza sales example we have MSE $= 57.97/6 = 9.662$ and $\sum(x_i - \overline{x})^2 = 210$. Since $n = 8$, we will have a t distribution with $8 - 2 = 6$ degrees of freedom. If we choose a 95 percent confidence interval, we can see from Table A3–5 that the value of a is 2.447. We already found that $\hat{m} = 2.905$, so the confidence interval is

$$2.905 \pm 2.447\sqrt{\frac{9.662}{210}}$$

which is from 2.38 to 3.43.

We can also perform hypothesis tests on our model. Here is one important null hypothesis: "The values of x do not have any relationship to the values of y." Obviously, if we think our regression is any good, we would like to collect enough statistical evidence to be able to prove this hypothesis wrong. From the equation $y = mx + b + e$, we can see that there will be no relation between x and y if the true value of the slope is zero. Since

$$\frac{\hat{m} - m}{\sqrt{\dfrac{\text{MSE}}{\sum(x_i - \overline{x})^2}}}$$

has a t distribution with $n - 2$ degrees of freedom, it follows that if $m = 0$ then the statistic

$$\frac{\hat{m}}{\sqrt{\dfrac{\text{MSE}}{\sum(x_i - \overline{x})^2}}}$$

will have a t distribution with $n - 2$ degrees of freedom. We can calculate the value of that statistic. If it seems plausible that the calculated value could have come from the t distribution, then we will accept the null hypothesis; otherwise we will reject it. For the pizza sales the calculated t statistic is

$$2.905\sqrt{\frac{210}{9.662}} = 13.542$$

For a two-tailed test at the 5 percent significance level the critical value for a t distribution with 6 degrees of freedom is 2.447. Since 13.542 is in the rejection region, well above the critical value, we can safely reject the null hypothesis that the true value of the slope is zero.

STATISTICAL ANALYSIS OF REGRESSIONS

1. Assume that the true relation between x and y is given by this formula:

$$y = mx + b + e$$

Here e is a normal random variable with mean 0 and unknown variance σ^2.

2. Then the least-squares estimators $\hat{m}$ and $\hat{b}$ are the maximum likelihood estimators of m and b, and they are also unbiased estimators.

3. The mean square error:

$$\text{MSE} = \frac{\sum [y_i - (\hat{m}x_i + \hat{b})]^2}{n - 2}$$

is an unbiased estimator of the unknown value of σ^2.

4. To test the hypothesis that the true value of the slope is zero, calculate this statistic:

$$\frac{\hat{m}}{\sqrt{\frac{\text{MSE}}{\sum (x_i - \bar{x})^2}}}$$

If the true value of the slope is zero, then this statistic will have a t distribution with $n - 2$ degrees of freedom.

Predicting Values of the Dependent Variable

Now we will discuss how to use our regression model to predict values of the dependent variable. Before we get carried away making predictions with a regression model, however, we must heed several important warnings.

- Any prediction based on a regression model is a conditional prediction, since the prediction for the dependent variable depends on the value of the independent variable. Suppose you have found a regression relationship that describes perfectly the relation between y and x. In that case you can predict future values of y if (but only if) you know the future value of x. If pizza sales do depend on income as predicted by our regression relation, then we can predict next year's pizza sales if we know next year's income. If we cannot predict next year's income, then we cannot predict the actual value of next year's pizza sales (although it still may be very useful to have a regression model that tells us how much income affects demand for pizza).

- The regression line has been estimated using past data. This line will not be able to predict the future if the relationship between x and y changes. A sudden change in people's pizza preferences will ruin the ability of the regression line to predict future values of pizza sales.
- Many regression predictions try to predict values of y in situations where the value of x falls outside the range of values previously observed for x. These predictions, known as *extrapolations,* are much less reliable than predictions based on values of the independent variable that fall within the range of previously observed values.

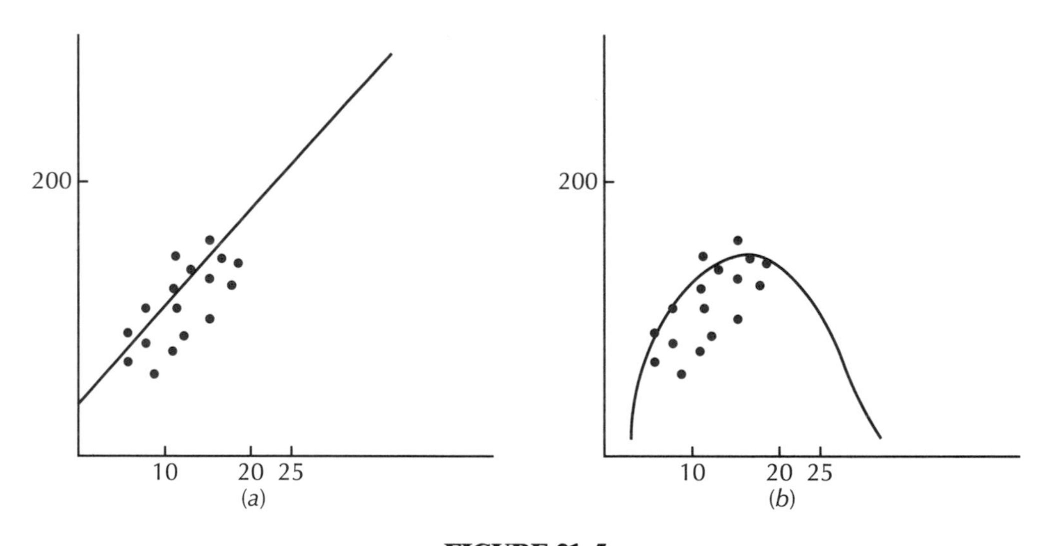

FIGURE 21–5

For example, part (a) of Figure 21–5 shows a number of observations for pizza sales as a function of income in a new set of towns. It seems reasonable to represent the data with a straight line. Suppose that next year we expect income will be 25. Based on the regression line, it seems logical to predict that pizza sales next year will be 200. However, this prediction is an extrapolation. The regression analysis gives us good evidence that the relation between income and sales can be well represented by a straight line when income is in the range from 10 to 20, but we have no way of knowing for sure whether this linear relationship holds for other levels of income. It is quite possible that, unbeknownst to us, the true relation between income and pizza sales is given by the curve shown in part (b). This curve suggests that pizza sales do not continue to increase as income increases. Instead, it appears that people start going to fancier restaurants and consume less pizza once income rises past a certain point. This curve fits the original data points as well as the regression line does, but it predicts a far different value for pizza sales if income is 25. As long as our observations cover only values of income within the narrow range shown, we have no way of distinguishing between these two situations, and extrapolations based on the regression line may be quite erroneous.

- The mere fact that there is a strong association between two variables does not indicate there is a cause and effect relationship between them. If you find a regression line that fits the relationship between y and x very well, then there are four possibilities:
  1. The values of y may really depend on the values of x, as we have assumed so far.
  2. The observed relation may have occurred completely by chance. If we have many observations, this is extremely unlikely, but we have seen that in statistics we cannot rule out the possibility that an event that looks significant may have occurred randomly.
  3. There may be a third variable that affects both x and y. This is the most likely explanation for situations where two variables are closely correlated but there does not seem to be a causal relationship between them. For example, once some researchers discovered a relation between ear wrinkles and certain diseases. Further investigation revealed this relation to be pointless, since it simply reflected the fact that both wrinkles and the diseases increase with age. Also, many unrelated variables tend to increase with time.
  4. There may indeed be a causal link between x and y, but it may be that y is causing x. We may have incorrectly determined which was the dependent variable and which was the independent variable. For example, we assumed that higher incomes cause people to buy more pizza. However, it may work this way: when people buy more pizzas, they create more income for all the people who work in pizza places, and this leads to a multiplier effect that increases the income of the whole community. Therefore, it could be that higher pizza sales cause higher incomes. Here is another example. You may have found a regression result that seems to indicate that higher advertising levels cause higher sales. However, suppose that firms allocate their advertising budgets on the basis of sales. Then it is quite likely that higher sales cause higher advertising budgets.

Now that we have heeded all of these warnings, let's see how we can use our regression model to predict values of y. Assume that we have determined that x really does cause y, that this relationship will still apply in the future, and that it can be accurately described by the regression line $y = 2.905x + 14.577$. If we know that the value of income next year will be 16, then naturally we will predict that next year's value of y will be $2.905 \times 16 + 14.577 = 61.06$.

The next question is how precise this prediction is. We would like to construct an interval such that there is a 95 percent chance that the value of y will be contained in that interval (given that $x = 16$). This type of interval is called a *prediction interval*. Note that it is similar to the confidence intervals we developed for unknown parameters. Suppose for the moment that we know the true values of m, b, and σ. Then, if $x = x_{new}$, we know that y will have a normal distribution with mean $\hat{y}_{xnew} = mx_{new} + b$ and variance σ^2. Therefore, there is a 95 percent chance that the value of y will be between $[(mx_{new} + b) - 1.96\sigma]$ and $[(mx_{new} + b) + 1.96\sigma]$.

Unfortunately, matters become much worse because we don't know the true values of m, b, and σ. Now there are two sources of uncertainty involved with predicting values of y: we don't know the true regression line, and the predicted value of y will deviate randomly about the line. The formula for the estimated variance of y for a given value of x is

$$\mathrm{Var}(y) = \mathrm{MSE}\left[1 + \frac{1}{n} + \frac{(x_{new} - \overline{x})^2}{\sum (x_i - \overline{x})^2}\right]$$

Note that the variance becomes larger when the value of x_{new} becomes farther from $\overline{x}$. When x_{new} is closer to $\overline{x}$, we have greater confidence that our estimated regression line is close to the true regression line. However, if our estimate for the slope of the line is slightly different from the true value, then this difference will cause our estimated regression line to depart farther and farther from the true line when we move farther from $\overline{x}$.

When x has the value x_{new} and we have calculated $\hat{y}_{xnew} = \hat{m}x_{new} + \hat{b}$ and Var (y) using the formula above, then the prediction interval for y is

$$\hat{y}_{xnew} \pm a\sqrt{\mathrm{Var}(y)}$$

where

$\Pr(-a < t < a) = \mathrm{CL}$

t = random variable having a t distribution with $n - 2$ degrees of freedom

CL = confidence level (such as .95)

Here are some sample calculations:

Value of x	Predicted Value of $y = \hat{m}x + \hat{b}$	95 Percent Prediction Interval for y
2	20.39	11.29–29.48
6	32.01	23.67–40.34
10	43.63	35.56–51.69
14	55.25	46.91–63.58
18	66.87	57.77–75.96
20	72.68	63.05–82.30

Figure 21–6 illustrates the prediction intervals compared to the regression line. You can see how the intervals become wider as the value of x is farther from $\bar{x}$.

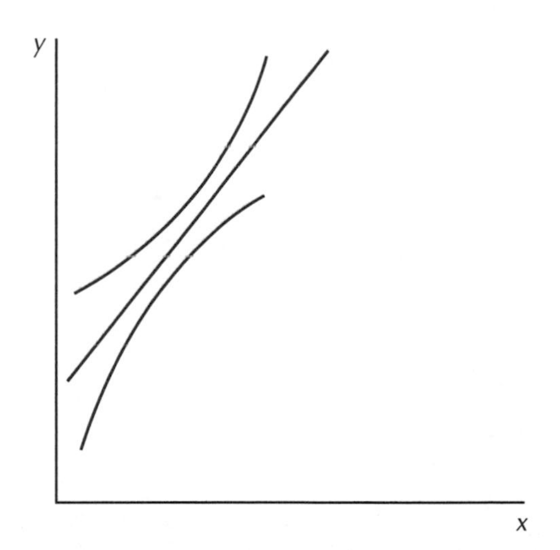

FIGURE 21–6

Analysis of Residuals

Another way to gain valuable information about a regression model is to make a graph of the residuals. For each data point (x_i, y_i) we can calculate the residual:

$$i\text{th residual} = y_i - \hat{y}_{xi} = y_i \; (\hat{m}x_i + \hat{b})$$

Let's make a scatter diagram that measures values of x along the horizontal axis and the residual along the vertical axis. Many computer regression programs will automatically prepare such a diagram if you wish. In the pizza example we have these values:

x	Residual
5	−2.102
10	2.373
20	0.323
8	2.183
4	3.803
6	−4.007
12	−3.437
15	0.848

Figure 21–7 shows the scatter diagram.

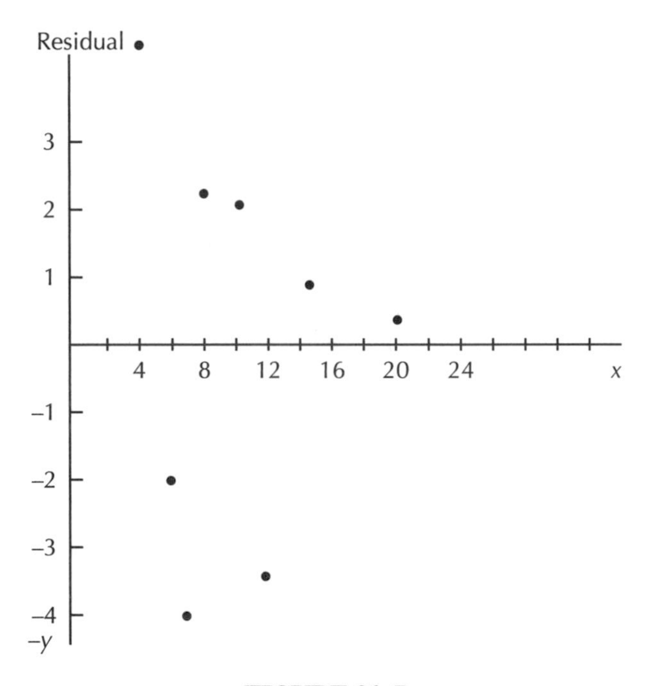

FIGURE 21–7

If the assumptions of the regression model are met, the plot of the residuals should look like a random array of dots. There should be no apparent pattern. Because the error term in the model is assumed to have a normal distribution, there should be more values near zero than far from zero. In particular, here are some things to look for in a residual plot:

- Outliers. An *outlier* is a residual much larger (or much more negative) than the others. On the original scatter diagram an outlier will show up as a point that is far from the estimated regression line. When you have identified which observation corresponds to the outlier point, then you should check to make sure that the observation is correct. An outlier may have occurred because you made a typographical error in entering the data for that observation, in which case you can correct the error and run the regression again. If you are convinced that the outlier observation is indeed correct, then you should investigate to see whether there is some special circumstance that caused that observation to depart so far from the others. Perhaps you can identify a rare natural disaster that corresponds with the observation involved. If you are convinced that the circumstance that caused the outlier will not recur, you may drop that observation and then perform the regression again with the remaining points. If, on the other hand, you can identify the cause of the outlier with a variable that should be included in the model, you should set up a multiple regression model (see the next chapter). If you cannot identify any cause for the outlier, then you must leave the outlier in the regression and treat it as a random error.

- Nonnormal errors. The original regression model assumed that the error terms had normal distributions, but the plot of the residuals may indicate that this is not the case. The least-squares estimators $\hat{m}$ and $\hat{b}$ are unbiased estimators of the true values of m and b whether or not the errors are normally distributed, but the statistical tests we performed were all based on the assumption of normal errors.
- Nonconstant variances. We assumed that the variances of the error terms associated with the observations were all the same. If the residual diagram shows that the residuals are systematically bigger in one area of the diagram, however, this condition may not be met. The pizza regression residual plot indicates that the residuals tend to be greater for smaller values of x, so perhaps there is a situation of nonconstant variance here. (More observations would be needed to make this conclusion definite.) The impressive technical term for the situation with nonconstant variances among the error terms is *heteroscedasticity*. In situations of nonconstant variance it is sometimes possible to transform the model into an equivalent model with constant variances.
- Omitted variables. It may be worthwhile to make a plot of the residuals compared with another independent variable that seems significant but has not been included in the model. If there is an association between the residuals and the new variable, you should set up a multiple regression model including that variable.
- Nonlinearity. If the true relationship is not a straight line, the residual plot will usually reveal the situation immediately. For example, here are observations for two variables x and y:

x	y	Residual
1	10.000	0.786
2	10.800	0.386
3	11.664	0.050
4	12.597	–0.217
5	13.605	–0.409
6	14.693	–0.521
7	15.869	–0.546
8	17.138	–0.476
9	18.509	–0.305
10	19.990	–0.024
11	21.589	0.375
12	23.316	0.902

The simple linear regression calculation for these observations gives a slope of 1.2, an intercept of 8.014, and an r^2 value of .986. The scatter diagram is shown in Figure 21–8.

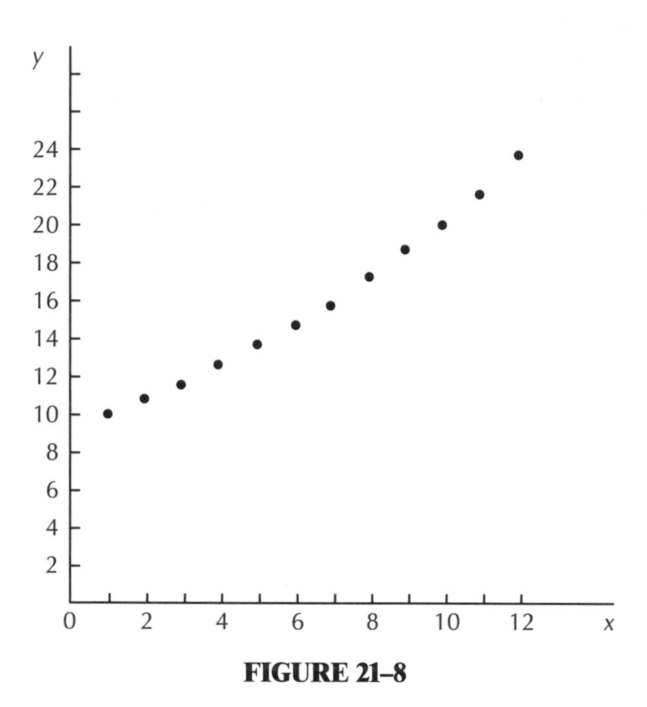

FIGURE 21–8

However, the residual plot shown in Figure 21–9 clearly is not random. When the residuals follow a definite curve like this one, there is strong evidence either that the underlying model is not a linear model, or one of the other problems mentioned above has arisen. In this case we can see that for small or large values of x the residual is always positive, but for medium values of x the residual is always negative. This pattern indicates that the relationship between x and y can be better represented by a curve than by a line.

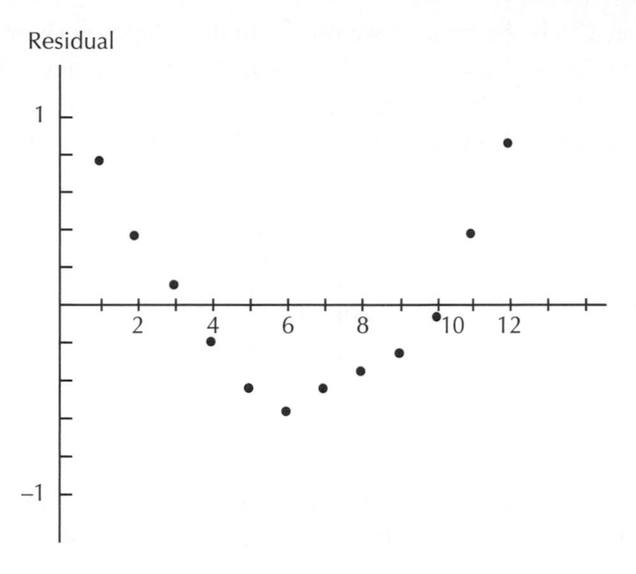

FIGURE 21–9

We will now turn our attention to transformations that can turn a model with a curve into an equivalent model with a straight line.

Transformations with Logarithms

Suppose that the true relationship between x and y can be given by this equation:

$$y = ca^x$$

Here c and a are two unknown constants. We cannot use simple linear regression to find the values for c and a. To handle this situation we must dig back into memory to find the concept called *logarithm*. You should have studied logarithms sometime previously, but we will review the basic ideas here.

Consider this question: What number will we get if we raise 2 to the 7th power? (Remember that raising a number to the power n means writing down that number n different times and then multiplying all of those n numbers together.) Here is a table of powers of 2:

$2^0 = 1$	$2^5 = 32$
$2^1 = 2$	$2^6 = 64$
$2^2 = 4$	$2^7 = 128$
$2^3 = 8$	$2^8 = 256$
$2^4 = 16$	$2^9 = 512$

We can see that 128 is the result if we raise 2 to the 7th power. Now let's ask the opposite question: To what power should we raise 2 in order to get 128 as the result? The result is called the logarithm to the base 2 of 128. In this case we happen to know already that the answer is 7. In logarithm notation this is written as $\log_2 128 = 7$. We could also write

$$\begin{aligned} \log_2 1 &= 0 \\ \log_2 2 &= 1 \\ \log_2 4 &= 2 \\ \log_2 8 &= 3 \\ \log_2 16 &= 4 \\ \log_2 32 &= 5 \text{ and so on} \end{aligned}$$

There are many different logarithm functions. Any positive number except 1 can act as the base for a logarithm function. In general, if a is the base of the logarithm function, then

$$\log_a x = n \text{ means } a^n = x$$

The two most common bases for logarithms are 10 and a special number called e, which is about 2.71828. The expression $\log x$, where no base is specified, usually refers to logarithms to the base 10, which are called *common logarithms*. For example, $\log 10 = 1$, $\log 100 = 2$, $\log 1{,}000 = 3$, and so on. The expression $\ln x$ usually refers to logarithms to the base e, which are called *natural logarithms*. Natural logarithms are especially important in calculus. Calculating logarithms is easy if the number you are interested in happens to be a whole-number power of the base. Otherwise the best thing to do is use a calculator or a computer with a built-in logarithm function to calculate the values.

Logarithms are very useful for our purposes because they satisfy these properties: Multiplications turn into additions:

$$\log (ab) = \log a + \log b$$

Divisions turn into subtractions:

$$\log \left(\frac{a}{b}\right) = \log a - \log b$$

Exponents turn into multiplication:

$$\log a^n = n \log a$$

These properties are true for logarithms of any base. For example, if the true relation between y and x is given by this formula:

$$y = ca^x$$

we can take the logarithm of both sides:

$$\log y = \log(ca^x)$$

Using the properties of logarithms, we can rewrite as follows:

$$\begin{aligned}\log y &= \log c + \log a^x \\ &= \log c + x \log a\end{aligned}$$

Let's make these definitions:

$$\text{let } b = \log c, \quad \text{let } m = \log a$$

Then

$$\log y = b + mx$$

That looks very familiar: we have transformed our model into a situation where the linear regression model is appropriate. All we have to do is use the values of $\log y$ as the dependent variable instead of the original values of y.

y	log *y*	*y*	log *y*
10.000	1.000	15.869	1.201
10.800	1.033	17.138	1.234
11.664	1.067	18.509	1.267
12.597	1.100	19.990	1.301
13.605	1.134	21.589	1.334
14.693	1.167	23.316	1.368

Now we can perform the simple linear regression calculations, using x as the independent variable and $\log y$ as the dependent variable. The result is

$$\text{slope} = m = 0.0334 \quad \text{intercept} = b = 0.966$$

Since the original model is $y = ca^x$ with $b = \log c$ and $m = \log a$, we can calculate that

$$c = 10^{0.966} = 9.3 \quad \text{and} \quad a = 10^{0.0334} = 1.08$$

Therefore, our estimate for the true model is

$$y = 9.3 \times (1.08)^x$$

Note that we used common logarithms for our calculations, but you may use natural logarithms if you wish. This function is an example of a situation called *exponential growth* because the independent variable x appears as an exponent.

Here is another situation where a logarithmic transformation is useful. Suppose we have determined that the quantity demanded of a particular good is given by a formula like this:

$$Q = Q_0 P^{-a}$$

Here P is the price of the good and Q_0 and a are unknown parameters. (In this formula a is the *elasticity* of demand for this product. When the formula is written like this, it is assumed that the elasticity is constant.) Once again we need to take the logarithm of both sides:

$$\begin{aligned} \log Q &= \log(Q_0 P^{-a}) \\ &= \log Q_0 - a \log P \end{aligned}$$

Now let $b = \log Q_0$ and $m = -a$. Then

$$\log Q = b + m \log P$$

Now we can estimate b and m by performing a simple linear regression calculation using $\log P$ as the independent variable and $\log Q$ as the dependent variable.

If the true relation between x and y is $y = ca^x$, then the transformed equation is

$$\log y = \log c + x \log a$$

If the true relation between x and y is $y = cx^n$, then the transformed equation is

$$\log y = \log c + n \log x$$

Examples of Regression Calculations

In Excel it is easy to perform regression calculations. The add-in Data Analysis provides an option for regression calculation. All you need to do is identify the independent variable (X) range and the dependent variable (Y) range.

Here are some examples of different kinds of relationships. (The spreadsheet with these calculations is available at *http://myhome.spu.edu/ddowning/ezstat.html*.)

Figure 21–10 illustrates the relation between height and weight for a group of football players. Most likely you would expect some relationship but not a perfect relationship. The r^2 value for this regression is .558, close to halfway between 0 and 1. In this diagram, each mark represents one player.

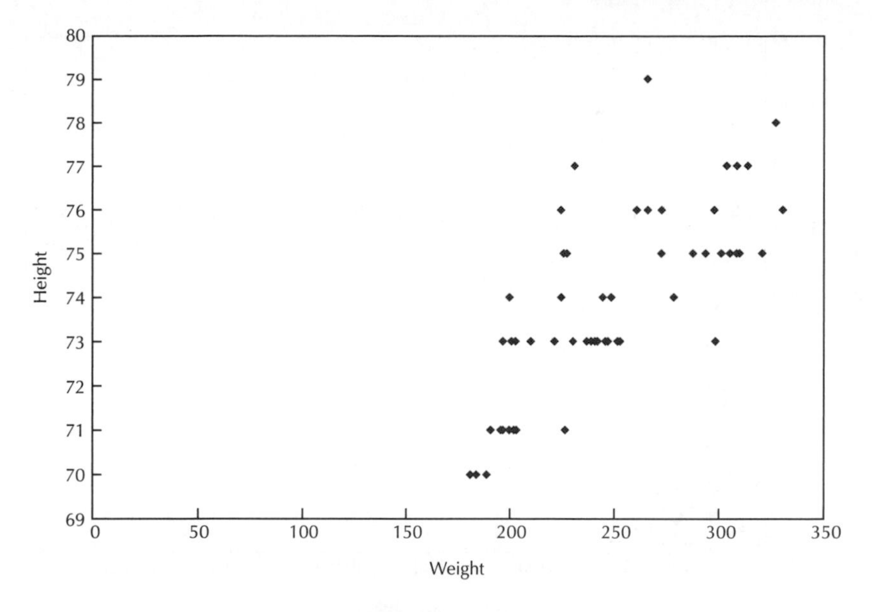

FIGURE 21–10

Figure 21–11 illustrates the relation between the population of a U.S. state and the number of representatives sent by that state to the House of Representatives. This is nearly a perfect linear relationship since representatives are determined by the population of a state. The r^2 value is .9992. The only reason it is not exactly 1 is because the number of representatives is always rounded to a whole number.

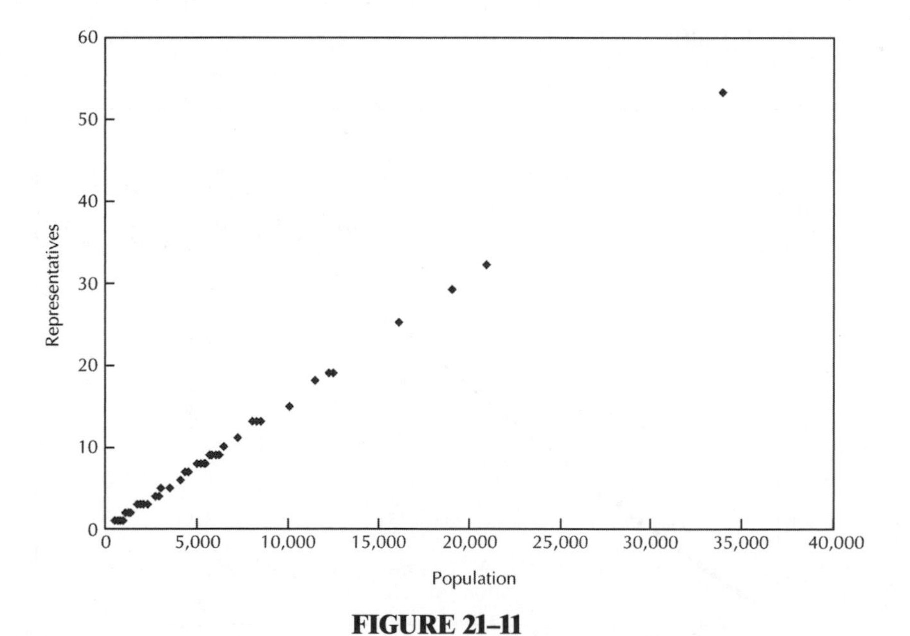

FIGURE 21–11

Figure 21–12 shows there is little relationship between the area of a state and its population. The r^2 value is only .011. (In both Figures 21–11 and 21–12, each mark represents one state.)

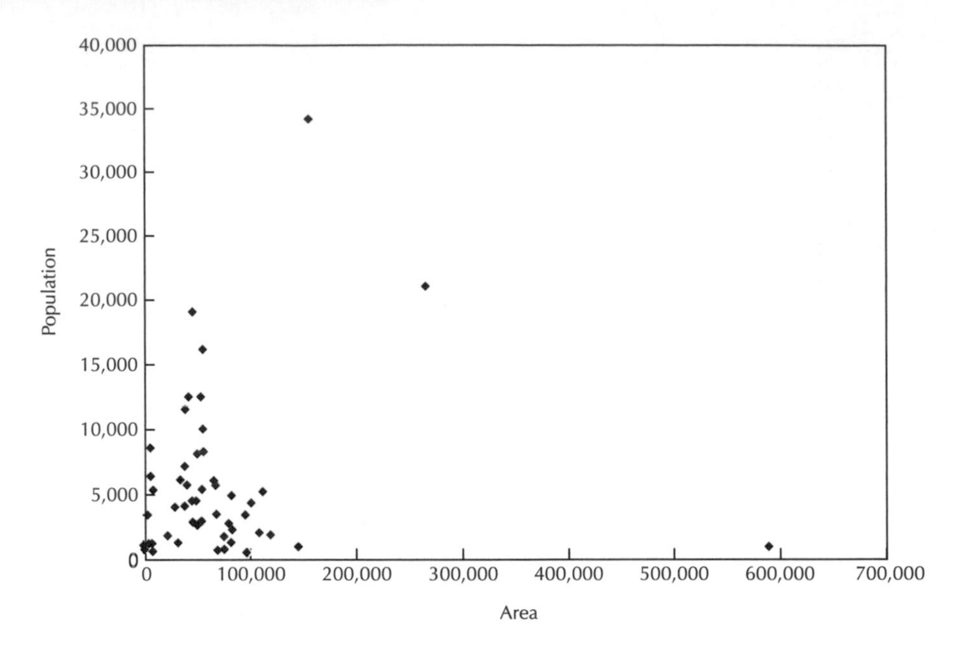

FIGURE 21–12

Sometimes regression analysis is performed on time series data, where each observation represents a time period. Figure 21–13 shows the relation between disposable income and personal consumption spending in the United States from 1959 to 2006. There is a very close relation, indicating that it is possible to forecast consumption spending if you know disposable income. The r^2 value is .9989. Each mark represents a year.

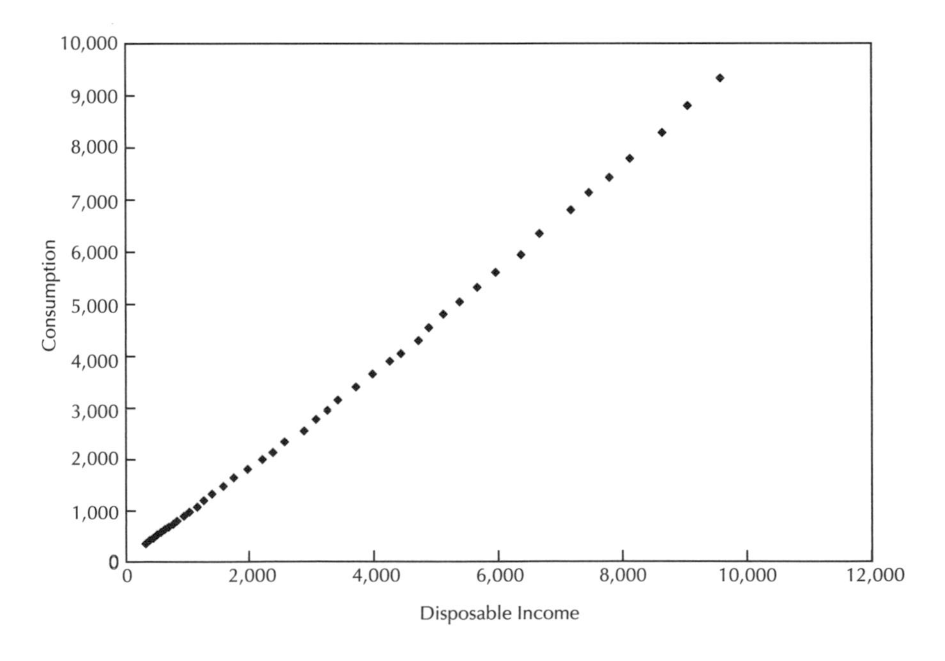

FIGURE 21–13

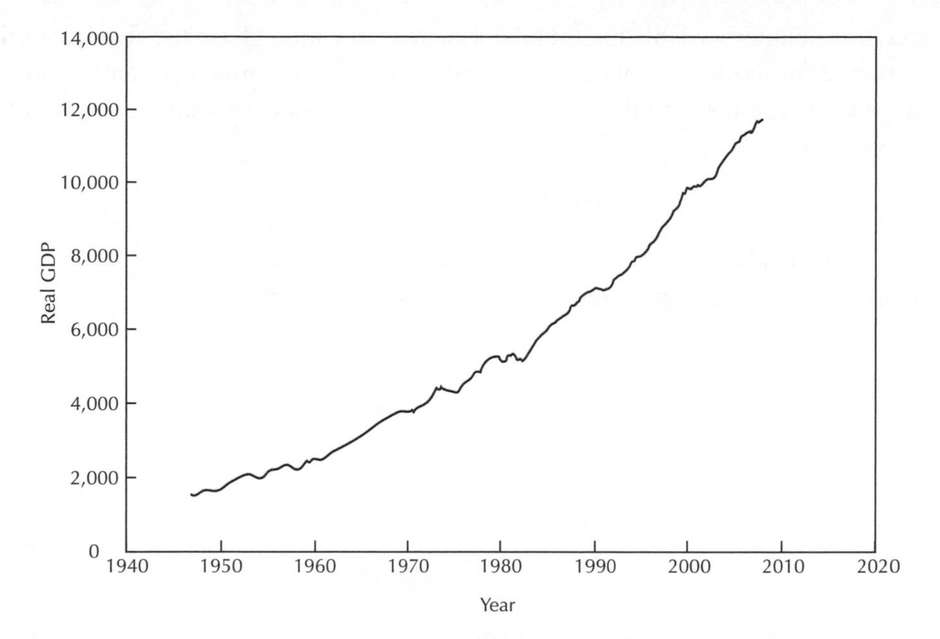

FIGURE 21–14

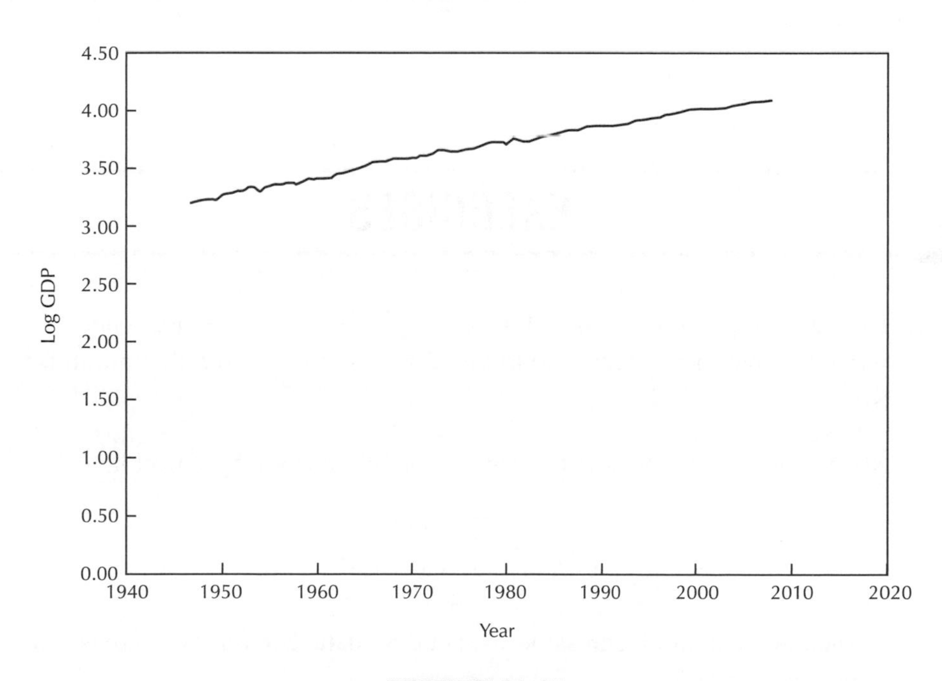

FIGURE 21–15

When analyzing trends, time is used as the independent variable. Figure 21–14 shows the trend in U.S. real (inflation-adjusted) Gross Domestic Product (GDP) from 1947 to 2008. GDP observations are reported each quarter. Because there are so many observations on the graph, a smooth curve is used to illustrate the trend (rather than individual markers). The trend cannot be well represented by a line, but the curved nature of the graph suggests using logarithms. Figure 21–15 shows the trend in the common (base-10) logarithm of real GDP (RGDP), which is nearly a straight line.

Create a time counter variable that numbers each observation (1 for the first quarterly observation, 2 for the second observation, and so on). Use the time counter variable as the independent variable and the log of RGDP as the dependent variable. The resulting regression equation is

$$\log \text{RGDP} = 3.2197 + .003578t$$

The r^2 value is .9948.

Take each side of the regression equation and raise 10 to that power:

$$10^{\log\text{RGDP}} = 10^{3.2197 + .003578t}$$

Use the properties of logarithms to rewrite the equation:

$$\text{RGDP} = 10^{3.2197} \times 10^{.003578t}$$

$$= 1{,}658.44 \times (10^{.003578})\ t$$

$$= 1{,}658.44 \times 1.0083t$$

The final equation indicates that, on average, whenever t increases by 1, the value of RGDP is multiplied by 1.0083, which corresponds to an average quarterly growth rate of 0.83 percent.

EXERCISES

☆ 1. Derive the equations for $\hat{m}$ and $\hat{b}$ for a regression with one independent variable. That means you need to find formulas for $\hat{m}$ and $\hat{b}$ that minimize SE_{line}.

☆ 2. Show that $r^2 = 1 - \text{SE}_{\text{line}}/\text{SE}_{\text{av}}$ can also be found from this formula:

$$r^2 = \frac{(\overline{xy} - \overline{x}\,\overline{y})^2}{(\overline{x^2} - \overline{x}^2)(\overline{y^2} - \overline{y}^2)}$$

3. Update the real GDP regression with current data. See *http://myhome.spu.edu/ddowning/ezstat.html*.

4. Perform a regression using the weather forecasts from the previous day as the independent variable and the actual temperatures as the dependent variable. Are the forecasts very accurate? In other words, is knowing the value of the predicted temperature very much help in knowing the value of the actual temperature the next day? (Note that this regression does not imply a causal connection between the two items.)

5. Perform a regression to investigate the connection between interest rates and housing starts in your area.

In each of the following lists, x represents an independent variable and y represents a dependent variable. Perform a regression calculation to see if there is a relation between x *and* y.

6.

x:	52	12	96	28	22	45	16
y:	16	13	11	17	11	14	17

7.

x:	1	2	3	4	5	6	7	8
y:	14	18	32	38	49	60	79	80

8.

x:	19	11	11	48	16	18	15	12
y:	15	19	34	59	19	46	15	44

9.

x:	3.04	1.98	6.02	3.15	2.78	4.15	2.98
y:	9.24	3.92	36.52	9.92	7.73	17.22	8.88

x:	10.40	25.65	30.29	19.85
y:	108.16	657.92	917.48	380.00

10.

x:	10	11	9	9	8	9	8	7	7	9
y:	5	11	5	8	10	5	4	5	4	7

x:	4	9	6	4	7	6	9	7	4	6
y:	6	11	4	9	3	9	11	2	3	8

11.

x:	15	22	14	17	18	14	12	12	11	16
y:	9	10	7	8	10	8	9	5	4	6

x:	4	7	10	12	10	15	20	9	7	14
y:	2	5	4	7	3	9	11	6	3	6

12. Calculate a regression line for the following values of an independent variable x and a dependent variable y.

x:	.12	.25	.38	.50	.64	.89	.95
y:	.0002	.0039	.0209	.0625	.1678	.6274	.8145

13. Take the logarithm of each of the values for x and y in Exercise 12 and then calculate a regression line using the values of the logarithms.

14. Perform a regression calculation for these observations:

x:	3	4	5	6	7	8	9	10	11
y:	4	11	16	19	20	19	16	11	4

What do the regression results tell you about the relation between these variables? Then draw a scatter diagram. What does the scatter diagram tell you about the relation?

15. Obtain information on the age, weight, height, and experience of the players on a professional football team. Calculate the correlation between each pair of variables. Which two variables are most strongly correlated? Which two variables have the smallest correlation? How do the results compare with your expectations?

16. Obtain information on the daily stock price of a particular company, along with one other variable that you think might affect that stock. Does knowledge of that other variable provide much help in predicting the value of the stock? (You may want to extend this study to include several independent variables after you have read the next chapter.)

22 Multiple Regression

In the last chapter we investigated the relation between one independent variable and the dependent variable. However, there are many times when it is clear that the relationship between the variables is more complicated, and it cannot be represented adequately if we only look at the effect of one independent variable.

Several Independent Variables

For example, suppose we investigated the relation between income and ice cream sales in a sample of eight hypothetical towns during one week and drew a scatter diagram (see Figure 22–1). It looks as if there is no relation. If we perform a simple regression calculation, we find that $y = 70.95 + 1.35x$, with an r^2 value of only .02.

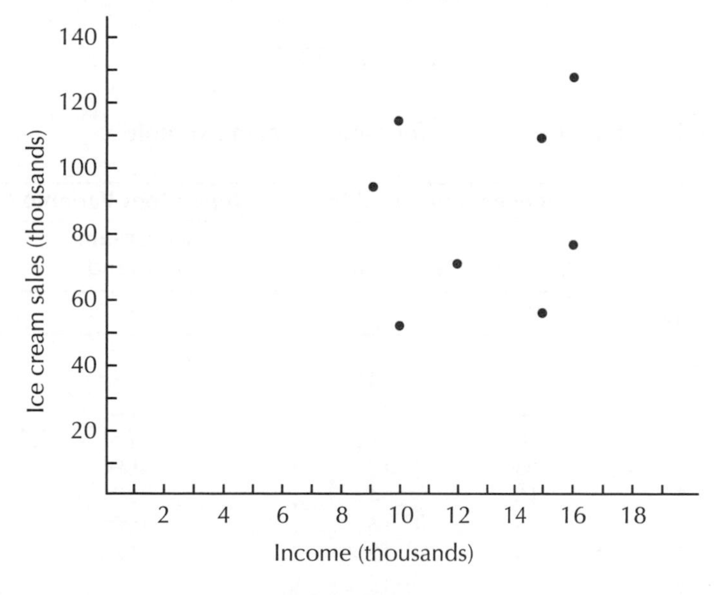

FIGURE 22–1

However, upon further investigation we find that there was considerable variation in temperature between the eight towns. Figure 22–2 shows that two cities had a temperature of 15 (using the Celsius scale) during this week; two had a temperature of 12; two were at 7; and two were at 4. Now it seems that the demand for ice cream is affected by both income and temperature. If everything else is equal, it does seem that higher income leads to higher demand; and it also seems clear that higher temperatures are associated with higher demand. (We are assuming that there are no other relevant variables that vary across the different cities. It may be necessary to conduct further investigation of other possible factors.)

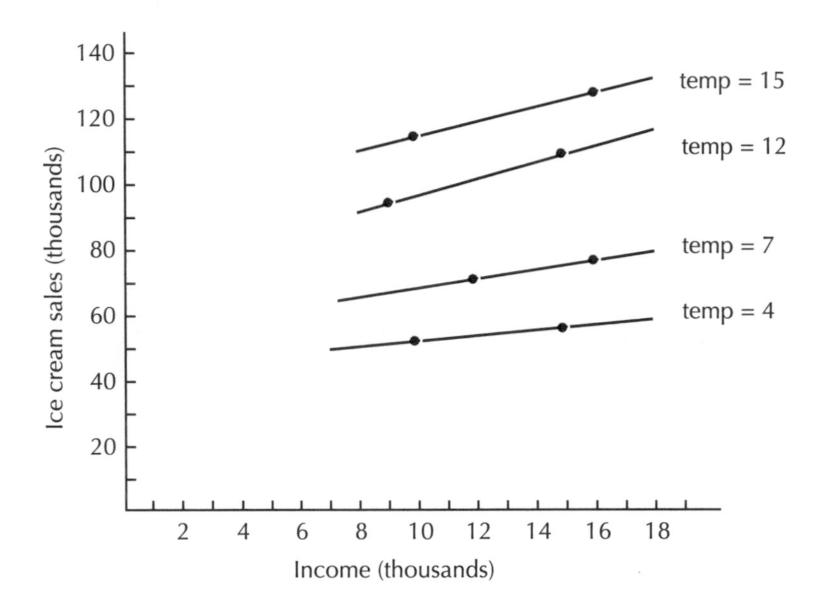

FIGURE 22–2

Here is a table of the observations for the ice cream example:

	Independent Variables		Dependent Variable
	Income	Temperature	Ice Cream Demand
City	X_1	X_2	Y
1	10	4	52
2	12	7	72
3	16	7	78
4	15	12	109
5	15	4	57
6	16	15	128
7	9	12	95
8	10	15	116

We will use y to represent the dependent variable, as we did in the last chapter. We will use x_1 to represent the first independent variable and x_2 to represent the second independent variable.

We will assume that the relation between y, x_1, and x_2 is given by this equation:

$$y_i = B_1 x_{i1} + B_2 x_{i2} + B_3 + e_i$$

Here y_i represents the ith value of the dependent variable, and x_{ij} represents the ith observation of the jth independent variable. We need two subscripts now because we must use one subscript to keep track of the observation number and another subscript to keep track of the variable number. For the observations listed above, x_{11} is 10, x_{21} is 12, x_{12} is 4, x_{22} is 7, and so on.

The true values of B_1, B_2, and B_3 are unknown, but we will try to estimate them. The quantity B_1 represents the effect that x_1 has on y, assuming that x_2 remains constant. Likewise, B_2 represents the effect that x_2 has on y when x_1 remains constant. If x_1 increases by 1 and everything else remains constant, then y will increase by B_1. When we design the model this way, we are assuming that the effects of x_1 and x_2 on y are additive. This means that the amount that x_1 affects y does not depend on the level of x_2, and vice versa. The term B_3 is called the constant term in the model. It is analogous to the y-intercept term in the simple linear regression model.

Again, e is a random variable called the *error term* that represents the effects of all possible factors other than temperature and income that might affect the demand for ice cream. The expectation of e is 0 and the variance of e is σ^2, which is unknown. We will assume that e has a normal distribution.

In principle, we will proceed exactly as we did with simple linear regression, when there was only one independent variable. We will calculate the values of B_1, B_2, and B_3 that minimize the sum of the squares of the errors between the values predicted by the equation and the actual values. There are two main differences between simple regression and multiple regression:

- We can't draw a picture of the relationship. Actually, if there are only two independent variables, we can attempt to draw a three-dimensional perspective drawing with x_1 and x_2 on the horizontal axes, y on the vertical axis, and one dot corresponding to each observation. Then the goal is to identify the plane that minimizes the sum of the squares of the vertical deviations between each observation and the plane. Drawing this type of diagram is very difficult, though. If there are more than two independent variables, it is impossible to draw a diagram. (Mathematicians still think of each observation as defining a point in multidimensional space, and the goal of multiple regression in that case is to find something called a *hyperplane* that best fits all the observations.)
- The calculation process is harder for multiple regression than it is for simple regression. But that won't worry us—we'll have the calculations done by a

computer. We will briefly describe how to perform these calculations in the mathematical section at the end of the chapter. Understanding those calculations requires a knowledge of matrix multiplication and matrix inversion. In a practical situation where you need to perform a multiple regression calculation, you will work with a computer statistical package on a microcomputer or mainframe computer.

Now we will discuss how to interpret the results of a regression analysis.

Multiple Regression Output

Once we have fed the numbers into the computer and ordered it to grind out the calculations, we will be presented with a result something like this:

$$y = 1.8016x_1 + 6.0658x_2 + 7.5546$$

This equation lists the estimated values for the coefficients. These particular numbers are for the ice cream example. The first thing we can learn from this result confirms what we expected: both coefficients are positive, meaning that an increase in income or an increase in temperature is associated with increases in ice cream sales.

The computer will also give an R^2 value for the regression where R^2 is called the *coefficient of multiple determination*. In this case the R^2 value is .9973. This value means that our regression does a good job of accounting for the variation in ice cream sales, now that we have included both of the independent variables that we need. The interpretation of the R^2 value is similar to the interpretation of the r^2 value for a simple regression: R^2 measures the percent of variation in the dependent variable that can be explained by the regression. The value of R^2 will always be between 0 and 1.

The R^2 value is calculated in a manner similar to the way it was in the case with one independent variable:

$$R^2 = 1 - \frac{\sum (y_i - \hat{y}_i)^2}{\sum (y_i - \bar{y})^2}$$

The summation in the denominator represents SE_{av}, the squared error about the average, as it did in the last chapter. The summation in the numerator represents the sum of the squares of the deviation from the actual value to the value predicted by the regression equation. This is analogous to SE_{line} in the previous chapter, except that now the regression does not represent a line. The predicted value for the ith observation is found from this equation if there are two independent variables:

$$\hat{y}_i = \hat{B}_1 x_{i1} + \hat{B}_2 x_{i2} + \hat{B}_3$$

(Here $\hat{B}_1$, $\hat{B}_2$, and $\hat{B}_3$ are the estimated coefficient values that are calculated by the computer program.)

The computer will also present results that indicate whether you have reliable estimates of each individual coefficient in the regression equation. Suppose there are $m - 1$ independent variables. Then there will be m coefficients (one for each of the independent variables, plus one constant term). Then the true regression equation can be written

$$\hat{y}_i = \hat{B}_1 x_{i1} + \hat{B}_2 x_{i2} + \cdots + B_{m-1} x_{i1(m-1)} + B_m$$

The true values of B_1 to B_m are unknown. However, we can estimate those values with the coefficients from the regression calculation, $\hat{B}_1$ to $\hat{B}_m$. It turns out that $\hat{B}_i$ is a random variable with a normal distribution whose mean is equal to B_i. Therefore, $\hat{B}_i$ is an unbiased estimator of the true regression coefficient B_i. The standard deviation of $\hat{B}_i$ is unknown, but the computer will calculate an estimator for it, known as the standard error. It is common to put the value of the calculated standard error in parentheses below the coefficients in the regression equation. For the ice cream example we have

$$\begin{array}{c} y = 1.8016_{x1} + 6.0658_{x2} + 7.5546 \\ (.2202) \quad (.1422) \quad (3.2556) \end{array}$$

Note there are three standard errors, since there are three coefficients (one for each of the two independent variables, plus one for the constant term).

If you divide the coefficient value by the standard error, the result is called the t statistic for that coefficient:

$$\begin{array}{rl} x_1\text{:} & t \text{ statistic} = 1.8016/.2202 = 8.18 \\ x_2\text{:} & t \text{ statistic} = 6.0658/.1422 = 42.66 \\ \text{constant term:} & t \text{ statistic} = 7.5546/3.2556 = 2.32 \end{array}$$

The t statistic is used to test hypotheses about the value of that coefficient. Suppose that temperature really has no effect on the demand for ice cream. Then the true value of the temperature coefficient (B_2) is zero. If this null hypothesis is true, then the t statistic for that coefficient will come from a t distribution with $n - m$ degrees of freedom, where n is the number of observations and m is the number of coefficients that have been estimated. In our case $n = 8$ and $m = 3$, so we need to look in a t distribution table with 5 degrees of freedom to find that the critical value for a test at the 5 percent significance level is 2.571. (See Table A3–5 in the column for 95 percent.) The calculated value of the t statistic for this coefficient is 42.66. Since this value is much larger than 2.571, we will reject the null hypothesis that the true value of B_2 is equal to zero. This means that we believe our regression results show that there is a connection between temperature and ice cream sales. The t statistic for the first coefficient is 8.18, so we can reject the null hypothesis that B_1 is zero (in other words, we believe that income does have an effect on ice cream sales). The t statistic for the constant term is 2.32, so we cannot reject the hypothesis that the true value of the constant term is zero.

Suppose some doubters tell you that they don't believe there is any connection between the dependent variable and the independent variables in the regression. They are making the following null hypothesis:

$$B_1 = B_2 = \ldots = B_{m-1} = 0$$

In words, they think the true value of the coefficients for all $m - 1$ of the independent variables are zero. To test this hypothesis we need to use the F statistic for the regression, which will usually be calculated by the computer program. For our example the calculated value of the F statistic is 928. If the null hypothesis is true, and the true values of all coefficients really are zero, then the F statistic will have come from an F distribution with $m - 1$ degrees of freedom in the numerator and $n - m$ degrees of freedom in the denominator. We can see from Table A3–6 that the critical value for an F distribution with 2 and 5 degrees of freedom is 5.8. Since the calculated value of the F statistic is much larger than the critical value, we can clearly reject the null hypothesis that the independent variables have no effect.

Here is another example of a multiple regression calculation, with residential investment as the dependent variable and unemployment, the federal funds interest rate, and a time trend as the three independent variables. The unemployment rate is an indicator of the state of the economy. Since we would expect residential investment to be higher when the economy is doing well, we expect the unemployment rate to have a negative coefficient. Higher interest rates likely deter residential investment, so we would expect this variable to have a negative coefficient. Finally, we include a time trend variable since we would expect residential investment to increase over time as the economy grows.

Here is a table of the coefficients, standard errors, and t statistics.

Independent Variable	Coefficients	Standard Error	*t* Stat
Intercept	68.221	47.203	1.445
Unemployment	–16.505	7.600	–2.172
FedFunds	–10.844	3.389	–3.200
Trend	12.990	0.750	17.330

The coefficients have the expected signs. The r^2 value is .8868. There are 47 observations (one per year from 1960 to 2006). The number of degrees of freedom for our t statistics is $47 - 4 = 43$. From Table A3–5 we see that the critical value of the t statistic is slightly less than 2.02, so our observed t statistic values will cause us to reject the null hypothesis for all three independent variables. The calculated F-statistic value is 112.3, which is much bigger than the critical value for 3 and 43 degrees of freedom (which is about 2.8). Therefore, we can easily reject the null hypothesis that all of the independent variables have zero coefficients. The data used for this regression are available at *http://myhome.spu.edu/ddowning/ezstat.html*.

Further Analysis of Regression Models

We will mention a few more topics that apply to the analysis of regression.

- **Residuals.** Once you have calculated the regression residuals, you may perform visual analysis just as we did in the last chapter. You may make several scatter diagrams comparing the residuals to the dependent variable and to each of the independent variables. In each case there should be no apparent pattern. If you have time series data, it also helps to make a plot of the residuals with time. It may also help to plot the residuals against another independent variable that you have not included in the model. If the residuals seem to be related to that variable, then you should include it in the model.

- **Transformations.** Again, a model that is not linear to begin with can often be transformed into a linear model by using logarithms or some other type of transformation. For example, if the true relation is

  $$y = b_0 x_1^{b_1} x_2^{b_2} x_3^{b_3}$$

  then you may take the logarithm of both sides:

  $$\log y = \log b_0 + b_1 \log x_1 + b_2 \log x_2 + b_3 \log x_3$$

  Now you may estimate the values of b_0, b_1, b_2, and b_3 by ordinary linear regression.

- **Serial correlation.** In the regression model we have assumed that all of the errors are independent. Suppose we have a set of time series observations where a positive value for the error for one period is more likely to be followed by a positive value for the next period. In that case the errors are not independent. This situation is called *serial correlation* or *autocorrelation*. The least squares estimators are less reliable in this situation. A computer regression program will normally calculate the value of a statistic called the *Durbin-Watson statistic*. A small value of this statistic indicates the presence of one specific type of serial correlation. (How small is small? If the computer does not tell you, you will need to look in a Durbin-Watson table.) In the case of serial correlation the regression results can be made more reliable by trying to find another independent variable to add to the model or by performing a transformation involving the differences between successive values of the variables.

- **Multicollinearity.** When two or more of the independent variables are closely correlated, then the problem of *multicollinearity* arises. If all of the independent variables are uncorrelated, then your regression model will still be able to estimate the coefficients for the variables in the model accurately even if some of the independent variables have been left out. However, the coefficient estimates become less reliable if some of the independent variables are highly correlated. In the extreme case where two of the independent variables are perfectly correlated, it is not even possible to calculate the least squares estimators. Also, the t statistics for the individual coefficients are not reliable when there is multicollinearity.

When two independent variables are highly correlated, it is not possible to separate out their independent effects accurately. For example, suppose you are trying to investigate the effects of income and education on demand for your product. You have collected information from a large sample of households. However, you are likely to find that the people with more income tend to be the people that have more education, so these two variables are highly correlated. A practical rule states that the problem of multicollinearity arises if the correlation coefficient between two variables is greater than .7. If you observe that people with higher income tend to buy your product more often, you do not know whether they do so because they have higher incomes or because they have more education.

Here is a similar type of example. Suppose you have observed that a particular chemical reaction occurs much more rapidly in warm, light environments than it does in cold, dark environments. However, you cannot tell whether it is the warmth or the light that speeds the reaction, since the two independent variables (temperature and amount of light) are correlated in your observations. The best way to solve the problem would be to obtain observations of the reaction in hot dark situations and cold light situations.

By analogy, the best solution in the income/education situation would be to obtain observations of people with high incomes/low educations and high educations/low incomes. However, it may be difficult to find these people. You may have no better alternative than removing either the education or the income variable from the regression and then remembering that the coefficient of the remaining variable represents the combined effect of the two variables.

- **Dummy variables.** Often some of the factors that affect the dependent variable are not quantitative factors that can be given by numbers. For example, suppose you are investigating consumption behavior with time series data for the period 1930 to 1950. You would expect that consumption behavior would have been significantly different during the years of World War II than it was before and after the war. To take this effect into account, you can create an artificial variable that will take the value 1 during each of the war years and the value 0 during each of the other years. This type of variable is called a *dummy variable* or an *indicator variable*. The coefficient of the World War II dummy variable indicates how much effect the war had on the constant term in the regression. Dummy variables can be used in many different situations with regressions.

Consult an advanced book for more information on these and other topics important for the use of regression analysis.

Multiple Regression Calculation

WARNING: We are now coming to the last, and hardest, topic in the book. If you've made it this far, but the rest of the chapter looks too intimidating mathematically, then feel free to bail out now. You've already learned a lot about probability and statistics.

Let us consider the general case where y is determined by the equation

$$y = B_1 x_1 + B_2 x_2 + \cdots + B_{m-1} x_{m-1} + B_m + e$$

[We're assuming that we have $m - 1$ independent variables $(x_1, x_2, \ldots, x_{m-1}.)$ Note that B_m is the coefficient of a constant term, analogous to the y intercept term in the simple linear case. The term e is a normal random variable with mean 0 and unknown variance.] We'll let n be the number of observations—that is, the number of data points.

To save us a horrendous amount of work, we're going to use matrix notation. We will use boldface letters such as **y** or **X** to stand for matrices. We'll define **y** and **e** to be matrices with n rows and 1 column:

$$\mathbf{y} = \begin{pmatrix} y_1 \\ y_2 \\ y_3 \\ y_4 \end{pmatrix} \qquad \mathbf{e} = \begin{pmatrix} e_1 \\ e_2 \\ e_3 \\ e_4 \end{pmatrix} \qquad \text{(shown here for } n = 4\text{)}$$

Here y_i is the ith observation of the dependent variable y. The matrix **y** has n rows because there are n observations.

Define **B** as a matrix with m rows and 1 column:

$$\mathbf{B} = \begin{pmatrix} B_1 \\ B_2 \\ B_3 \end{pmatrix}$$

In this example $m = 3$, which means there are 3 coefficients (one for each of the $m - 1 = 2$ independent variables, and one for the constant term).

We'll write the independent variables in an n by m matrix called **X**:

$$\mathbf{X} = \begin{pmatrix} x_{11} & x_{12} & 1 \\ x_{21} & x_{22} & 1 \\ x_{31} & x_{32} & 1 \\ x_{41} & x_{42} & 1 \end{pmatrix}$$

The element x_{ij} is the ith observation of variable j.

Then we can write the whole equation like this:

$$\mathbf{y} = \mathbf{XB} + \mathbf{e}$$

If we wrote out all of the matrices explicitly, it would look like this (so you can see how much writing the matrix notation saves):

$$\begin{pmatrix} y_1 \\ y_2 \\ y_3 \\ y_4 \end{pmatrix} = \begin{pmatrix} x_{11} & x_{12} & 1 \\ x_{21} & x_{22} & 1 \\ x_{31} & x_{32} & 1 \\ x_{41} & x_{42} & 1 \end{pmatrix} \begin{pmatrix} B_1 \\ B_2 \\ B_3 \end{pmatrix} + \begin{pmatrix} e_1 \\ e_2 \\ e_3 \\ e_4 \end{pmatrix}$$

Here x_{12}, for example, is the observation of the variable x_2 in period 1.

It would be even worse if we wrote all of the equations out the long way:

$$\begin{aligned} y_1 &= B_1x_{11} + B_2x_{12} + B_3 + e_1 \\ y_2 &= B_1x_{21} + B_2x_{22} + B_3 + e_2 \\ y_3 &= B_1x_{31} + B_2x_{32} + B_3 + e_3 \\ y_4 &= B_1x_{41} + B_2x_{42} + B_3 + e_4 \end{aligned}$$

To use matrix notation, we need to know three things: the *transpose* of a matrix $\mathbf{X}$ (we'll symbolize the transpose by $\mathbf{X}'$), the product of two matrices, and the inverse of a matrix $\mathbf{A}$ (we'll symbolize the inverse of $\mathbf{A}$ by $\mathbf{A}^{-1}$).

We will use $\hat{\mathbf{B}}$ to stand for the matrix consisting of our estimates for the coefficients for $\mathbf{B}$. Then $\hat{\mathbf{y}}$, our predictions for $\mathbf{y}$, can be found from the equation

$$\hat{\mathbf{y}} = \mathbf{X}\hat{\mathbf{B}}$$

The vector of the errors is $(\hat{\mathbf{y}} - \mathbf{y})$, so the total squared error (TSE) is

$$\text{TSE} = \sum_{i=1}^{n} (\hat{y}_i - y_i)^2$$

Rewrite the expression for TSE in matrix notation:

$$\text{TSE} = (\mathbf{XB} - \mathbf{y})'(\mathbf{XB} - \mathbf{y})$$

Note that $(\mathbf{XB} - \mathbf{y})'$ is an array with 1 row and n columns and $(\mathbf{XB} - \mathbf{y})$ is an array with n rows and 1 column, so the result of multiplying them will be an array with 1 row and 1 column (in other words, a regular number, or scalar, as it is sometimes called.)

We can rewrite the equation, using some tricks of matrices:

$$\text{TSE} = (\mathbf{B}'\mathbf{X}' - \mathbf{y}')(\mathbf{XB} - \mathbf{y})$$

Matrix $\mathbf{X}$ is an $n \times m$ matrix; $\mathbf{B}$ is an $m \times 1$ matrix; so $\mathbf{XB}$ is an $n \times 1$ matrix (the same as $\mathbf{y}$). The term $(\mathbf{B}'\mathbf{X}' - \mathbf{y}')$ is a $1 \times n$ matrix. Multiplying a $1 \times n$ matrix by an $n \times 1$ matrix gives a 1×1 matrix—that is, a number.

$$\begin{aligned}\text{TSE} &= \mathbf{B'X'(XB - y) - y'(XB - y)}\\ &= \mathbf{B'X'XB - B'X'y - y'XB + y'y}\\ &= \mathbf{B'X'XB - 2B'X'y + y'y}\end{aligned}$$

Now we have to find the m by 1 matrix **B** that gives us the smallest possible value for TSE. If we perform some tricky matrix calculus calculations, it turns out that **B** must satisfy the equation

$$\mathbf{X'XB = X'y}$$

If we multiply both sides by $\mathbf{(X'X)^{-1}}$ we can get the solution for $\hat{\mathbf{B}}$:

$$\hat{\mathbf{B}} = \mathbf{(X'X)^{-1}\, X'y}$$

This formula tells us how we can calculate the optimum value of $\hat{\mathbf{B}}$ (that is, of B_1, $B_2, \ldots, B_m$). Although that equation is nice theoretically, the actual calculations are arduous.

The estimator $\hat{\mathbf{B}}$ calculated from the equation is called the *ordinary least squares estimator,* and this method is called the method of ordinary least squares.

In addition to minimizing the total squared error, $\hat{\mathbf{B}}$ also has three other properties we like estimators to have

1. $\hat{\mathbf{B}}$ is the maximum likelihood of the true values for **B**.
2. $\hat{\mathbf{B}}$ is consistent.
3. $\hat{\mathbf{B}}$ is an unbiased estimator of **B**—in other words, $E(\hat{B}_1) = B_1$.

Here are the matrices for the ice cream example. The matrix of independent variables **X** is this:

$$\begin{pmatrix} 10 & 4 & 1\\ 12 & 7 & 1\\ 16 & 7 & 1\\ 15 & 12 & 1\\ 15 & 4 & 1\\ 16 & 15 & 1\\ 9 & 12 & 1\\ 10 & 15 & 1 \end{pmatrix}$$

Note the column containing all 1's. This column is included because the constant term in the regression can be thought of as the coefficient of an independent variable that always has the value of 1. The matrix of dependent variables **y** is this:

$$\begin{pmatrix} 52 \\ 72 \\ 78 \\ 109 \\ 57 \\ 128 \\ 95 \\ 116 \end{pmatrix}$$

This is the matrix $\mathbf{X}'$:

$$\begin{pmatrix} 10 & 12 & 16 & 15 & 15 & 16 & 9 & 10 \\ 4 & 7 & 7 & 12 & 4 & 15 & 12 & 15 \\ 1 & 1 & 1 & 1 & 1 & 1 & 1 & 1 \end{pmatrix}$$

Now compute the product matrix $\mathbf{X}'\mathbf{X}$:

$$\begin{pmatrix} 1387 & 974 & 103 \\ 974 & 868 & 76 \\ 103 & 76 & 8 \end{pmatrix}$$

A book on matrix algebra will tell how to calculate the inverse of a matrix. Here is $(\mathbf{X}'\mathbf{X})^{-1}$:

$$\begin{pmatrix} 0.0165 & 0.0005 & -0.2168 \\ 0.0005 & 0.0069 & -0.0718 \\ -0.2168 & -0.0718 & 3.5980 \end{pmatrix}$$

The matrix of coefficients is found from the matrix product $(\mathbf{X}'\mathbf{X})^{-1}\mathbf{X}'\mathbf{y}$:

$$\begin{pmatrix} 1.8016 \\ 6.0658 \\ 7.5546 \end{pmatrix}$$

Next, we need to figure out the distribution of $\hat{B}_i$. From the equations,

$$\hat{\mathbf{B}} = (\mathbf{X}'\mathbf{X})^{-1}\mathbf{X}'\mathbf{y}$$

and

$$\mathbf{y} = \mathbf{X}\mathbf{B} + \mathbf{e}$$

we can derive

$$\begin{aligned} \hat{\mathbf{B}} &= (\mathbf{X}'\mathbf{X})^{-1}\,\mathbf{X}'\mathbf{X}\mathbf{B} + (\mathbf{X}'\mathbf{X})^{-1}\,\mathbf{X}'\mathbf{e} \\ &= \mathbf{B} + (\mathbf{X}'\mathbf{X})^{-1}\,\mathbf{X}'\mathbf{e} \\ \hat{\mathbf{B}} - \mathbf{B} &= (\mathbf{X}'\mathbf{X})^{-1}\,\mathbf{X}'\mathbf{e} \end{aligned}$$

Remember that each element of **e** has a normal distribution with variance σ^2. We can use this information to establish that the variance of B_i is

$$\text{Var}(\hat{B}_i) = \sigma^2 q_{ii}$$

In this case q_{ii} stands for the element in row i and column i of the matrix $(\mathbf{X'X})^{-1}$. Note that we can calculate q_{ii}, but we don't know σ^2. However, this problem is similar to the estimation problems we did earlier: Remember, we didn't know the value of σ^2 then, either.

Suppose that the null hypothesis is true and that the true value of B_i is zero. Then

$$\frac{\hat{B}_i}{\sigma\sqrt{q_{ii}}}$$

will have a standard normal distribution. We don't know σ^2, but we can calculate the sample standard deviation s_2:

$$s_2{}^2 = \frac{e_1{}^2 + e_2{}^2 + \cdots + e_n{}^2}{n-1}$$

We know that $(n-1)s_2{}^2/\sigma^2$ has a χ^2 distribution with $n-1$ degrees of freedom. Therefore,

$$\frac{\dfrac{\hat{B}_i}{\sigma\sqrt{q_{ii}}}}{\sqrt{\dfrac{(n-1)s_2{}^2}{(n-1)\sigma^2}}}$$

fits the definition of the t distribution, so the statistic

$$T = \frac{\hat{B}_i}{s_2\sqrt{q_{ii}}}$$

can be used to test the hypothesis that $B_i = 0$. Calculate the value of T. Then compare it with the value of a t distribution shown in Table A3–5 for the appropriate number of degrees of freedom. If the calculated value of T is too large, you can safely reject the hypothesis that X_i has no effect.

We have now covered the last topic that we will include in this book. If you are interested in further work in this area there are many books that cover advanced topics in statistics. You have the foundation that you will need to understand how statistics works. However, there are a few common mistakes that you should be careful to avoid; these are covered in the next chapter.

EXERCISES

1. Here is the output from a multiple regression calculation based on observations from 25 different cities with four independent variables: X1 = temperature; X2 = miles of beaches; X3 = price of ice cream sundaes; X4 = price of meals at first-class restaurants; and one dependent variable: Y = quantity of ice cream sold (in thousands of gallons per month.) The researchers conducting the study believe that the equation has been properly specified and represents the demand for ice cream in these cities. The R^2 value obtained was .9808.

Variable	Coefficient	Standard Error
X1	2.0076	0.1238
X2	0.0481	0.1298
X3	–5.0631	1.6747
X4	11.7193	0.5154
Constant	37.8693	15.1181

   (a) For each coefficient, test the null hypothesis that the coefficient is 0.

   (b) Calculate the value of y that is predicted by the regression equation when the independent variables have these values:

$$X1 = 57.7; \quad X2 = 52.4; \quad X3 = 4; \quad X4 = 17.4$$

2. Here is the output from a multiple regression where X1 = age; X2 = weight; X3 = amount of a particular drug taken in an experiment; Y = tested reaction time. The R^2 value obtained was .2249. There were 24 observations.

Variable	Coefficient	Standard Error
X1	0.0887	0.1062
X2	–0.0973	0.1528
X3	–0.1521	0.0646
Constant	58.5260	11.5365
F statistic:	1.9344	

   For each coefficient, test the null hypothesis that the coefficient is 0. Then test the hypothesis that all of the coefficients are zero.

☆ 3. Create an Excel spreadsheet that uses matrix functions to calculate the regression coefficients.

In the following exercises the X's represent independent variables and Y represents a dependent variable. Perform a multiple regression calculation to find the relationship between Y and the X's. Use a computer statistical package.

4.

Y	X_1	X_2	X_3
100	6.3	2	0
110	10.1	0	1
94	14.2	3	2
96	2.5	4	3
112	6.8	0	4
98	7.3	5	5
94	12.5	2	6
118	16.1	1	7

5.

Y	X_1	X_2
19	2	10
25	0	20
45	3	30
60	4	40
76	6	50
115	10	60
130	12	70
99	4	80

6.

Y	X_1	X_2
17	56	4
9	27	30
8	85	8
10	13	7
1	46	4
5	66	3
17	99	15
9	10	22

7.

Y	X_1	X_2
47	139	15
46	100	50
76	73	47
74	98	53
73	410	45
98	50	54
80	48	74
50	53	46

8.

Y	X_1	X_2
174	27	11
148	13	16
480	56	43
254	13	13
298	46	19
444	99	8
420	66	26
154	10	19

9. Try a multiple regression calculation for the following observations. What happens?

x_1	x_2	y
2	4	100
3	6	68
4	8	87
5	10	200
6	12	45

10. Try a multiple regression calculation for the following observations. What happens?

x_1	x_2	x_3	x_4	y
3	2	6	5	30
12	1	18	8	53
2	9	9	13	1
8	12	1	18	36
2	11	3	3	26

11. Consider these observations (from Chapter 21, Exercise 14):

x:	3	4	5	6	7	8	9	10	11
y:	4	11	16	19	20	19	16	11	4

Perform a multiple regression calculation, using x and x^2 as the two independent variables.

23

Fallacies and Traps

Statistics can be a powerful force for good, as we have seen, but it can also create mischief if it is not applied correctly. This chapter will discuss several different issues where, if you are not careful, you could incorrectly apply a statistical concept. Reading this chapter will also help you gain the ability to expose sinister charlatans who deliberately use statistics to mislead people.

EXAMPLE 1. The XYZ Corporation wishes to emphasize that they have cut the price of their widgets, as shown in Figure 23–1. The fall in price looks dramatic, until you look closely. Note the vertical scale of the graph: it starts just below the lowest data point, and the graph stops just above the highest data point. Any data series will appear to show a tremendous shift if you construct the graph in this fashion. The rule is that a graph of a change in a variable with time should always have a vertical scale that starts at zero, as in Figure 23–2. Otherwise it is inherently misleading. (Note: You may include a graph that does not start at zero as a supplement to illustrate a magnified view of the changes, as long as the reader can also see the nondistortional graph.)

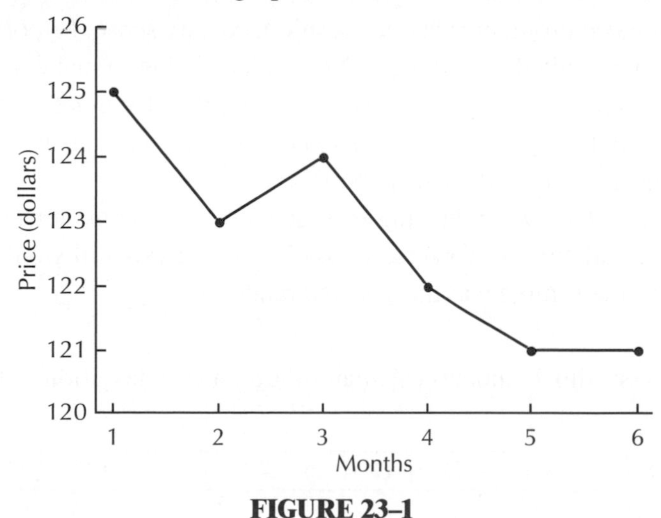

FIGURE 23–1

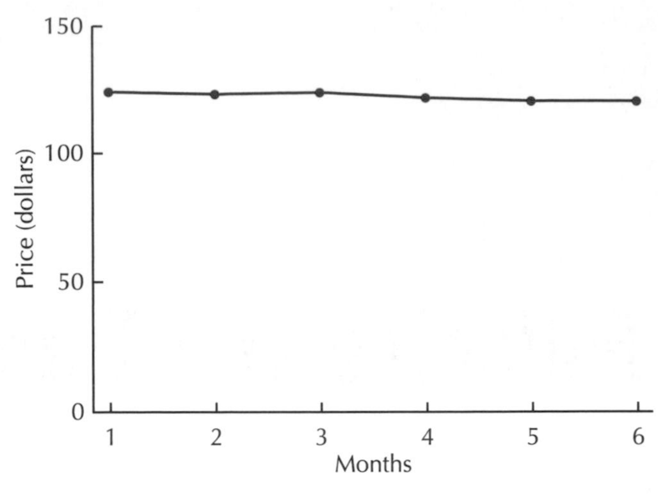

FIGURE 23–2

EXAMPLE 2. A small company wishes to investigate the ages of its workers. It has collected data from all 92 workers and finds an average age of 42 with a standard deviation of 10. It then proceeds to see if a hypothesis testing procedure will show that the age is significantly different from 40, using this test statistic from Chapter 6.

$$\frac{\sqrt{92}(42 - 40)}{10} = 1.92$$

It concludes this is barely within the acceptance region for a standard normal distribution (reasoning that they don't need to use the t distribution for n as large as 92.) Therefore, they will accept the hypothesis that the mean age is not significantly different from 40.

What's wrong with this reasoning?

The basic problem is that it doesn't make any sense. Hypothesis testing is a procedure applied to samples, when you are trying to make inferences about an unknown population parameter. In this case they have data for all 92 workers—the complete population—so they know $\mu = 42$. There is no point in doing any hypothesis testing when the parameter is known.

Perhaps they were thinking that the 92 workers constitute a sample from a larger population of workers, but statistical analysis still would not be applicable because the sample was not selected randomly.

EXAMPLE 3. A teacher with 7 students calculates the standard deviation of their test scores as follows:

$$\sqrt{\frac{(9-7)^2 + (8-7)^2 + (8-7)^2 + (7-7)^2 + (4-7)^2 + (6-7)^2 + (7-7)^2}{7-1}}$$
$$= 1.63299$$

What is wrong with this calculation? The teacher is using the formula for standard deviation of a sample—but these data do not come from a sample. If you are interested in the performance of this class on the exam, then you know all the data for the population, so you will calculate the standard deviation of a population (with n, rather than $n - 1$, in the denominator):

$$\sqrt{\frac{(9-7)^2+(8-7)^2+(8-7)^2+(7-7)^2+(4-7)^2+(6-7)^2+(7-7)^2}{7}}$$
$$= 1.51186$$

Perhaps the teacher was thinking of using this class as a sample of students chosen from the population of all students in the school, but again this is inappropriate because the sample is not selected randomly.

EXAMPLE 4. One day we were visiting New York City, and met a friend named Bill on the street. Out of all the people in New York City, it seems astounding that you could meet someone you know by chance. Suppose we somehow are able to estimate that the chance of meeting Bill is 1 in 100,000. Should we be truly astonished that such an unlikely event happened?

Here is the problem: why are we interested in calculating the probability of meeting Bill? If we really wanted to know this probability, we should have calculated it *before* we met Bill. Then it would have been truly astounding to meet Bill; in fact, we could say, "This is amazing, Bill! Just the other day, we were calculating the probability that we would meet you by chance on the street in New York City, and here you are!" However, this is not the way it happened. Instead, we did not become interested in the probability of meeting Bill until *after* we had already met him.

Now consider this question: what is the probability that we would meet someone—anyone—we know on the streets of New York? If we know 100 people in New York, then we might calculate that the probability of meeting someone we know is 100 times greater than the probability of meeting Bill specifically, so we would call the probability 1 in 1,000. Obviously it is much more likely that we will meet a generic person we know, rather than one specific person, such as Bill. Therefore, the amazing coincidence is not quite as amazing as it seems.

This is the key question you should ask whenever an amazing coincidence seems to have occurred: did you wonder about the probability of this coincidence *before* or *after* it happened? If you truly identified the probability of a particular event as being very small, and then it happens, that is truly amazing. However, you will find life is full of amazing coincidences if you are allowed to consider the likelihood of occurrence after they have happened. For example, suppose you are driving down the street and note with astonishment that the license plate of the car ahead of you reads FLG 427. You realize that the chance of this happening is only 1 in $26^3 10^3 = 17{,}576{,}000$. However, that is only an astonishing coincidence if you had been wondering in advance whether you would be behind car FLG 427; otherwise, it is a totally pointless piece of trivia.

EXAMPLE 5. After taking a sample of four Democrats and four Republicans, you find the following values for broccoli consumption:

Democrats	Broccoli	Republicans	Broccoli
Bill	7	George	1
Jimmy	20	Ron	12
Lyndon	15	Jerry	15
John	10	Dick	11

You will determine if there is a relation between political party and broccoli consumption, so you create the scatterplot shown in Figure 23–3.

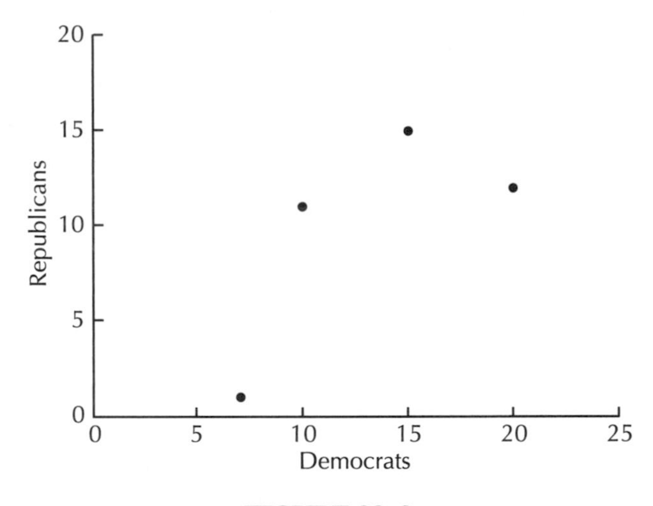

FIGURE 23–3

What is the problem?

This diagram doesn't make any sense, because you don't have matched observations for the variable. Each point in a scatterplot can be labeled because it comes from one person, or one time period, or one something. If you plot the point $(x = 7, y = 1)$, how do you label it? You can't label it, because the 7 comes from Bill, and the 1 comes from George. Also note that the axis labels don't make any sense; the numbers given represent the amount of broccoli, not the number of Democrats or Republicans.

EXAMPLE 6. A researcher is investigating differences between men and women in a characteristic called x-ability. A random sample of $n = 1{,}600$ men and $n = 1{,}600$ women is selected; the sample average for men is $\overline{x_m} = 60.10$ and for women is $\overline{x_w} = 60.85$. The standard deviation of both samples is $\sigma = 10$. The researcher performs a hypothesis test for the difference between two means.

$$Z = \frac{\overline{x_m} - \overline{x_w}}{\sqrt{\dfrac{\sigma^2}{n} + \dfrac{\sigma^2}{n}}} = \frac{60.10 - 60.85}{\sqrt{\dfrac{100}{1{,}600} + \dfrac{100}{1{,}600}}} = \frac{-.75}{\sqrt{.125}} = -2.12$$

The null hypothesis states that there is no difference in mean x-ability between men and women. If this hypothesis is true, then Z will have a standard normal distribution. Since –2.12 is less than 1.96, we can reject the null hypothesis at the 5 percent significance level (see Figure 23–4). Therefore, the researcher can proclaim there is a statistically significant difference in x-ability between men and women.

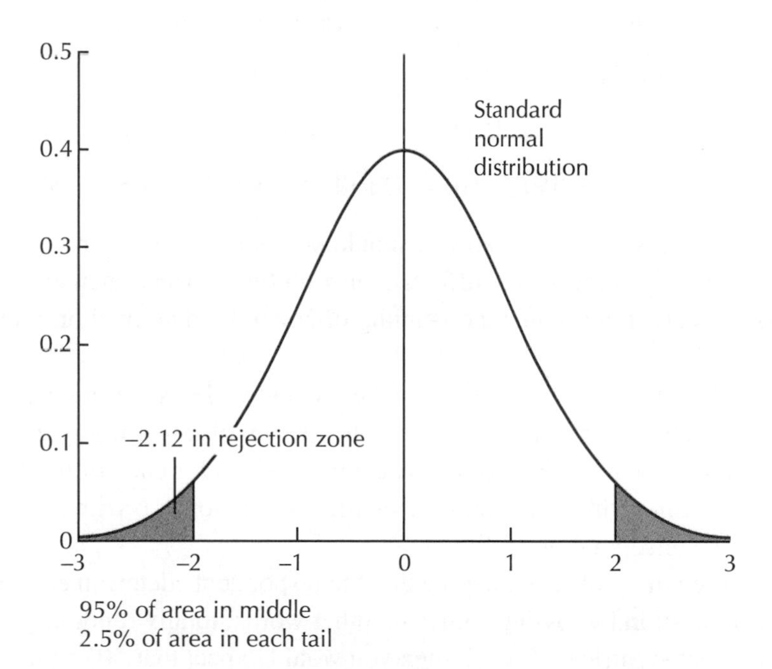

FIGURE 23–4

The question remains: should we be cautious when we discuss this result? Unfortunately, the terminology traditionally used in statistics does obscure what is important. The key word here is "significant." When we say that the result of this study is significant, it means merely that we can reject the hypothesis that the result happened by chance. In other words, we can use this evidence to show that there truly is a difference in mean *x*-ability between men and women.

However, just the fact that there is a difference does not mean it is a big difference. Here it is very important to remember that the word "significant" as used in statistics does *not* mean "important" as we would usually use the word. Someone not schooled in statistics would probably notice the obvious fact right up front: the difference between 60.10 and 60.85 is small, so even if there is truly a difference it is not an important difference. Furthermore, since the standard deviation for both men and women is equal to 10, we realize that there is much greater variability within each sex than there is between the sexes. If we tried to graph both normal curves, they would be practically indistinguishable.

One way to measure this precisely is to ask this question: "What is the probability that a randomly chosen woman will have an *x*-ability score higher than a randomly chosen man?" Assume that our sample mean and standard deviation figures can be used as good estimates of the population figures (a reasonable assumption since we have large random samples). Let X_w be the *x*-ability score for our randomly chosen woman, and X_m be the *x*-ability score for our randomly chosen man. Assume X_w has a normal distribution with mean $\mu_w = 60.85$ and $\sigma = 10$; X_m has a normal distribution with mean $\mu_m = 60.10$ and $\sigma = 10$. Then, $X_w - X_m$ has a normal distribution with mean $\mu_w - \mu_m = .75$ and standard deviation $\sqrt{\sigma^2 + \sigma^2} = \sqrt{200} = 14.1$. Then

$$\Pr(X_w > X_m) = \Pr(X_w - X_m > 0)$$

$$= \Pr(Z > (0 - .75)/14.1) = \Pr(Z > -.05) = .52$$

This result is close to .5, so you should not let the researcher's finding of a "statistically significant" difference in *x*-ability between men and women cause you to start stereotyping the *x*-ability of any individual man or woman.

EXAMPLE 7. A researcher trying to predict whether people will engage in type X behavior administers a 1,000 question survey to a group of randomly selected people. Then the researcher will run separate regressions for each of the 1,000 questions to determine which ones show a significant relationship with type X behavior. Is there a conceptual problem here?

To see the problem, imagine that the respondents determined their answer for each question by tossing coins—in other words, totally randomly. If you use the 5 percent significance level, then you would expect that 50 questions will show a significant connection to type X behavior, solely by chance. You must be very careful about attaching meaning to this result.

The problem here is similar to the example of meeting Bill on the streets of New York, discussed previously. It would have been better to state the hypothesis about which questions matter in advance, then perform the regression analysis to test that hypothesis. If you engage in "data mining" or "data snooping"—that is, testing many regressions before stating any hypotheses—you have a good chance of mistaking coincidences for meaningful results.

Granted, it is hard to resist the temptation to state hypotheses after testing the data. Here is one important question to ask in that case: can the results be replicated with a different sample? Any result in either the natural sciences or the social sciences should not be totally accepted until it can be shown to be replicated by different researchers with independent samples.

EXAMPLE 8. Someone hears this statement: "22 percent of poor people are fatherless children" but then repeats it as "22 percent of fatherless children are poor." The two statements sound similiar, but they say very different things. (In this case, the first one is right, according to the 1994 U.S. *Statistical Abstract*; the second one is incorrect.) To find the percent of poor people who are fatherless children, divide 8 million poor fatherless children by 36.9 million poor people, giving 22 percent. To find the percent of fatherless children that are poor, divide 8 million poor fatherless children by 14.7 million fatherless children, giving 54 percent. Both of these statements are examples of conditional probabilities, and it is important to make sure you state each one correctly.

EXAMPLE 9. A researcher calculates a 95 percent confidence interval for an unknown mean μ, and then states: "There is a 95 percent chance that the value of μ will be between 70 and 78." This is a subtle point, but this is not the correct statement of the meaning of a confidence interval. The mean μ is not a random variable, so it does not make sense to talk about the probability that it takes on certain values. Instead, it is the confidence interval itself that is random, so a correct statement would be: "There is a 95 percent chance that the random interval $\overline{x} - as_2/\sqrt{n}$ to $\overline{x} + as_2/\sqrt{n}$ will contain the unknown true value of μ." This is usually shortened to a saying involving the specific numbers, such as "the 95 percent confidence interval is from 70 to 78," which is fine as long as you remember the correct interpretation of the statement.

EXAMPLE 10. A researcher uses regression to find a strong relationship between x and y, and then claims this proves that x causes y. This is not necessarily true; it may be that y causes x, or it may be that a third variable causes changes in both x and y.

For example, a researcher observes data that the price of cars has increased over the last 4 decades, at the same time that the number of cars sold has increased, thereby concluding that higher prices for cars will cause more people to buy cars. This is very hard to believe; presumably people will buy more cars at lower prices, other things being equal. In reality, other things are not equal; for example, the price increase is caused largely by inflation, and the increase in demand is caused largely by population increases, among other factors. Any simple regression calculation can give misleading results if you have left out variables that should have been included. In that case you need to perform a multiple regression calculation, but even then there is a risk of getting misleading results because some variables have been left out.

EXAMPLE 11. Figure 23–5 shows a regression calculation performed on two variables x and y. There seems to be a good fit, and one is tempted to use this regression line to predict that if x becomes 20, then y will become slightly less than 300. The problem with this type of prediction is that we are going beyond the range of previously observed data. Such a prediction is called an *extrapolation,* and you must be very wary of it. For example, suppose that some more data are collected that follow the pattern shown in Figure 23–6. You can see how far off your prediction would be if it were based on the regression line that came from the limited sample.

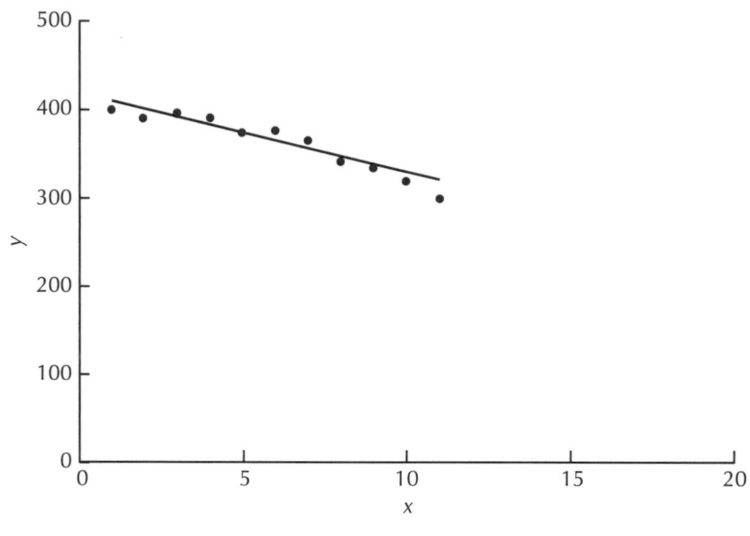

FIGURE 23–5

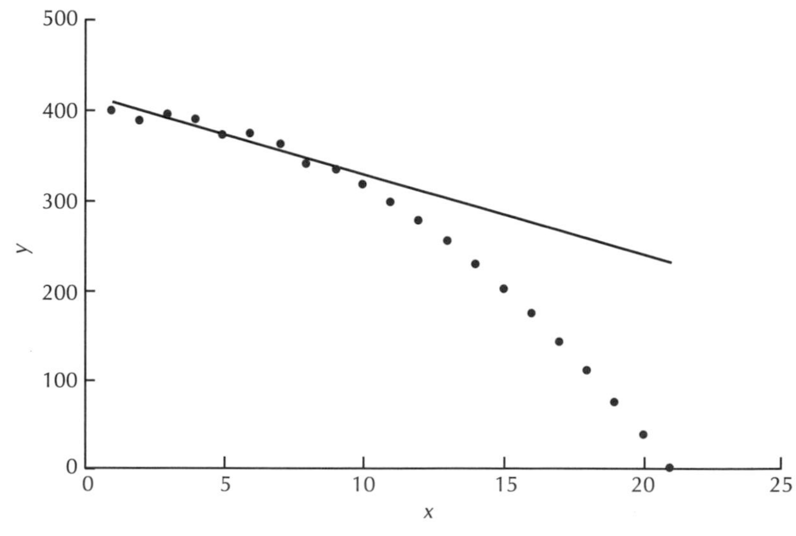

FIGURE 23–6

A sad fact of life is that extrapolation predictions are what we often are very interested in. We would like to know how things would be different if we do something very different than we have ever done before. Since a relationship that is perfectly trustworthy within one range of observations may break down outside that range, we must be very careful of such predictions.

EXAMPLE 12. Suppose you heard this on a news report: "A new study doubts whether increasing the minimum wage will cause a statistically significant increase in unemployment."

What's wrong with the statement?

The use of the phrase "statistically significant" here is wrong for two reasons:

1. The phrase "statistically significant" only applies to the issue of how well a sample represents an unknown population; it does not apply to the actual result of a change in future policy.
2. Presumably the news report intended to say that the study authors doubted that increasing the minimum wage would cause a major increase in unemployment. Perhaps it would cause a minor increase, but there is no objective standard about what separates a minor increase from a major increase. The same point applies as in example 6: "statistically significant" does *not* mean "important."

Here is an example of a proper use of the phrase "statistically significant" (which illustrates what the news report quoted above probably intended to say):

"A new study does not find there has been a statistically significant increase in unemployment from previous minimum wage increases. That means we will accept (at the 5 percent significance level) the null hypothesis that the change in unemployment has been 0." (Note: The actual effects of increasing the minimum wage is a matter of some controversy, which is very hard to settle because of the difficulty of collecting empirical data that show the effect of the minimum wage in isolation from all other factors that affect employment).

EXAMPLE 13. You are given the following results from a multiple regression:

```
r squared:   0.8500           n = 50
Dependent variable:   Y

 Coefficient    1              X1     10.58
                     t statistic:     15.61
                   Standard err:      0.678

 Coefficient    2              X2    424.91
                     t statistic:      4.04
                   Standard err:     105.12

 Coefficient    3        Constant    558.34
                     t statistic:     11.32
                   Standard err:      49.30
```

You observe that the t statistics for both X1 and X2 cause you to reject the null hypothesis that their coefficients are zero, so you could say they both have a statistically significant effect on Y. You notice that the coefficient of X2 is much larger than the coefficient for X1, so you are tempted to say that X2 is the variable that has the dominant effect on X1.

What's wrong with that reasoning?

Remember that "statistically significant" does not mean important. Saying that X2 has a statistically significant effect on Y merely means that we can reject the null hypothesis that it has no effect; it does not mean that it has a large effect. You might think that you can look at the value of the coefficient to tell whether that variable has a large effect, but there is a problem with that reasoning. The coefficient of a particular variable in a regression depends on the units used to measure that variable. If you change the units by a factor of 1,000 (for example, give distances in kilometers instead of meters) you will change the numerical value of the coefficient without changing its meaning at all.

Here is one way to tell whether one independent variable X_i has a large effect on the dependent variable Y. Ask this question: If X_i increases by one percent, what will be the percent change in Y (assuming both variables start at their mean values)? This ratio is called the *elasticity,* and is figured as follows:

$$\text{elasticity for independent variable } X_i = \frac{dY}{dX_i} \times \frac{\overline{x_i}}{\overline{y}}$$

Here dY/dX_i is the regression coefficient for the variable (also the derivative of Y with respect to X_i).

To calculate the elasticities for our example, we first need to know the average values:

$$\overline{y} = 1{,}210; \quad \overline{x_1} = 49.14; \quad \overline{x_2} = 0.311$$

You can see that X2 is measured in units that give it a very small numerical value, which accounts for the fact that its coefficient has such a large numerical value.

The elasticities are as follows:

for X1:

$$10.58 \times \frac{49.14}{1{,}210} = .430$$

for X2:

$$424.91 \times \frac{.311}{1{,}210} = .109$$

The elasticities have the advantage that they are unit-free. We can see that a one percent change in X1 has a larger effect on Y than does a one percent change in X2.

EXAMPLE 14. Suppose you read a news report that says, "A poll shows 60 percent of the people prefer a cat rather than a dog live in the White House. The margin of error is 3 percent, meaning that if we interviewed all people in the population, the proportion favoring cats would be between 57 and 63 percent."
What's wrong with the statement?

There is no way the poll can guarantee with 100 percent accuracy that the true population proportion is within the range 57 percent o 63 percent. Presumably what they mean is that they are giving us a 95 percent confidence interval, so there is a 95 percent chance that the interval contains the true value. This also means there is a 5 percent chance the interval does not contain the true value (a fact not often emphasized in news accounts of polls).

A news report of a poll should contain these items:

1. a statement about the number of people in the poll
2. a statement about how the poll was conducted (by phone? in person? by mail?)
3. a statement about how the sample was selected (if it's not a random sample, don't believe it represents the population)
4. a disclaimer statement that should read something like this: "The margin of error for this poll is plus or minus 3 percent. This means that 95 percent of the time the result found by this poll will be within 3 percentage points of the true population proportion that would be found if everyone in the population had been interviewed. The margin of error for any subgroups reported in this poll will be larger than 3 percent."

Unfortunately many news reports of polls leave out this information. Alas, there is no "truth in polling" laws.

EXAMPLE 15. Finally, we must once again mention one of the most common traps you can fall into: applying statistical analysis to convenience samples. We have discussed the statistical magic that makes it possible to use results from a sample to predict results for a population, but that only works if the sample is randomly selected. That allows us to determine the probabilities that the sample will be like the population, so we can tell if the result could have happened by chance or not. You can't make that kind of generalization if the sample was not selected randomly. A notable example of this was the *Literary Digest* poll of the 1936 presidential election. There are many other cases where statistical analysis is done on convenience samples but the population results are never checked, so the truth is not known. You can make a list of lots of ways not to select a sample: call your friends; ask people at a particular shopping mall; reply to a TV station call-in poll; respond to a magazine poll of its subscribers; or ask a class of college freshmen.

You have no way of knowing how the respondents might be different from the population, and you cannot rely on random selection to cancel out the bias.

Now you're ready to head off on your own. You've learned all you need to know in order to start doing our own statistical calculations. Your knowledge of statistics will be helpful if you intend to study fields such as experimental science, psychology, economics, sociology, and many others. When you are confronted with a pile of raw data, you are now able to determine whether there is a meaningful story that can be learned from those data.

APPENDIX 1

Answers to Selected Exercises

CHAPTER 2

3. 1/2

4. 5/32

5. 0.112. 0.080.

CHAPTER 3

1. 0.039.

2. 1.4×10^{-17}

3. 100 tosses: Accept the hypothesis if there are 40 to 60 heads.

   5 tosses: There is a 6 percent chance that you will get either 0 heads or 5 heads in 5 flips, so you cannot reject the hypothesis in any case.

   10 flips: Accept the hypothesis if the number of heads is between 2 and 8.

4. If the total is 2, 3, 10, 11, or 12, then you will reject the hypothesis; otherwise you will have to accept the hypothesis. In this case there is no probability of a type 1 error.

CHAPTER 4

1. Mean = 18.81; median = 19; mode = 21

2. Mean = 336; median = 256. Each number is a mode, since all of the numbers occur an equal number of times.

3. $\bar{x} = 5.25$. $\overline{x^2} = 47.75$. Standard deviation $= \sqrt{47.75 - 5.25^2} = \sqrt{20.19} = 4.49$; median $= 5.5$

4. The median is not changed: it is still 5.5. The mean and standard deviation become much larger: mean = 16.375; standard deviation = 31.85.

5.

	Mean	Median	σ
English	81.43	84	12.016
Chemistry	89.33	91	6.789

6. When a set of numbers contains a few numbers that are far above the mean, but no numbers that are far below the mean, then the mean will tend to be higher than the median.

7.

	Mean	Median	σ
(a)	4	4	2.58199
(b)	22.667	16	21.4683
(c)	2.6	2	2.41661
(d)	5.2	4	4.83322
(e)	0	0	0.70711
(f)	0	0	0.70711

9. For all 50 states: mean = 72,368; median = 56,222; σ = 87,391

Michigan	58,216	
Illinois	56,400	**East North Central**
Wisconsin	56,154	mean = 49,657
Ohio	41,222	median = 56,154
Indiana	36,291	σ = 9,063
Alabama	51,609	**East South Central**
Mississippi	47,716	mean = 45,491
Tennessee	42,244	median = 44,980
Kentucky	40,395	σ = 4,441
Montana	147,138	
New Mexico	121,666	
Arizona	113,909	
Nevada	110,540	
Colorado	104,247	**Mountain**
Wyoming	97,914	mean = 107,986
Utah	84,916	median = 107,394
Idaho	83,557	σ = 19,384

		Middle Atlantic
New York	49,576	mean = 34,248
Pennsylvania	45,333	median = 45,333
New Jersey	7,836	σ = 18,756
Maine	33,215	
Vermont	9,609	
New Hampshire	9,304	**New England**
Massachusetts	8,257	mean = 11,101
Connecticut	5,009	median = 8,781
Rhode Island	1,214	σ = 10,307
Alaska	589,757	
California	158,693	**Pacific**
Oregon	96,981	mean = 184,015
Washington	68,192	median = 96,981
Hawaii	6,450	σ = 208,704
Georgia	58,876	
Florida	58,560	
North Carolina	52,586	
Virginia	40,817	
South Carolina	31,055	**South Atlantic**
West Virginia	24,181	mean = 34,839
Maryland	10,577	median = 35,936
Delaware	2,057	σ = 20,256
Minnesota	84,068	
Kansas	82,264	
Nebraska	77,227	
South Dakota	77,047	**West North Central**
North Dakota	70,665	mean = 73,892
Missouri	69,686	median = 77,047
Iowa	56,290	σ = 8,723
Texas	267,338	**West South Central**
Oklahoma	69,919	mean = 109,721
Arkansas	53,104	median = 61,512
Louisiana	48,523	σ = 91,348

CHAPTER 5

1. 1/2

2. 2/9

3. 1/38. This is a discrete probability space.

4. This is a continuous probability space. 0. 3/38.

5. $\emptyset$, S

6. $1 = \Pr(S) = \Pr(X \cup X^c) = \Pr(X) + \Pr(X^c)$ (since X and X^c are mutually exclusive). Then $\Pr(X) = 1 - \Pr(X^c)$. Since $\Pr(X) \geq 0$, it follows that $\Pr(X) \leq 1$.

7. See Exercise 6.

9. Let D be the event $B \cup C$. Then A and D are disjoint. $\Pr(A \cup B \cup C) = \Pr(A \cup D) = \Pr(A) + \Pr(D) = \Pr(A) + \Pr(B \cup C) = \Pr(A) + \Pr(B) + \Pr(C)$.

10. $A \cup B = A \cup (B \cap A^c)$; $A \cap (B \cap A^c) = \emptyset$; $\Pr(A \cup B) = \Pr(A) + \Pr(A^c \cap B)$; $B = (A \cap B) \cup (A^c \cap B)$; $(A \cap B) \cap (A^c \cap B) = \emptyset$.

    Then:

    $\Pr(B) = \Pr(A \cap B) + \Pr(A^c \cap B)$; $\Pr(A^c \cap B) = \Pr(B) - \Pr(A \cap B)$;

    So, $\Pr(A \cup B) = \Pr(A) + \Pr(A^c \cap B) = \Pr(A) + \Pr(B) - \Pr(A \cap B)$.

11. 0.000 000 07

12. 3/4

13. 7/8

14. $1 - (5/6)^3 = 0.421$

15. Probability of getting 9 is 0.116, probability of getting 10 is 0.125.

16. If you go for the two points now, you have a 50 percent chance of succeeding, in which case you win the game. If you fail, then you can go for two points next time. If you succeed the next time, you will tie; otherwise you will lose. Therefore, if you go for the two points this time you have a 50 percent chance of winning, a 25 percent chance of tying, and a 25 percent chance of losing. If you kick the extra point this time, then you have a 50 percent chance of winning and a 50 percent chance of losing.

21. $1 - (1/2)^n$

22. 1/21

23. 1/7

24. 1/49

25. 1/4

26. Let A be the event of getting hired by the first firm and B be the event of getting hired by the second firm. Then $\Pr(A \text{ or } B) = \Pr(A \cup B) = 0.40 + 0.40 - 0.16 = 0.64$.

27. 20 percent

CHAPTER 6

1. $\binom{24}{3}\left(\frac{1}{8}\right)^3\left(\frac{7}{8}\right)^{21}$

   $\binom{24}{4}\left(\frac{1}{8}\right)^4\left(\frac{7}{8}\right)^{20}$

2. $\left(\frac{1}{22}\right)\binom{22}{10}$

3. $\frac{32!}{8!8!2!2!2!2!2!2!2!}$

4. $\frac{7!}{4!2!1!} = 105$

5. $\binom{12}{4} = 495$

6. $\frac{52!}{48!} = 6{,}497{,}400$

7. $\frac{\binom{4}{2}+\binom{3}{2}+\binom{2}{2}}{\binom{9}{2}} = \frac{6+3+1}{36} = \frac{5}{18}$

8. $\frac{\binom{4}{2}}{\binom{9}{2}} = 1/6$

9. $9!/6! = 504$

10. $4!\,2!\,2! = 96$

11. $5!\,4!\,2! = 5{,}760$

12. $\binom{15}{10} = 3{,}003$

13. 256/270,725

14. 10

15. 56

17. There is only one way to choose zero objects from a group of n objects, so $\binom{n}{0}$ is 1.

18. There are n ways to choose one object from a group of n objects.

19. There is only one way to choose all n items.

20. There are n ways to select $n - 1$ objects, since you have n choices as to which object you will not select.

21. They are equal. Choosing j objects from the group is the same as *not* choosing $n - j$ objects.

22. $26^5 = 11{,}881{,}376$

23. $1/30^3 = 1/27{,}000$

25. $2^5 \times 6 = 192$

26. 120

27. 4/9

28. $6!/6^6 = 120/7{,}776 = 5/324$

30. 1/36

31. $1/6^4 = 1/1{,}296$

32. $1 - (5/6)^n$

33. 1/5

34. 1/70

35. Discard either the 5D or 5C. Then you can win with either 4H or 8H.

36. 28!

37. $1 - (1{,}460/1{,}461)^n$

38. 2: 0.005; 3: 0.016; 4: 0.031; 5: 0.053; 6: 0.079; 7: 0.111; 8: 0.147; 9: 0.189; 10: 0.237; 15: 0.553; 18: 0.804

40. $\dfrac{8!}{\binom{16}{8}} = \dfrac{(8!)^3}{16!}$

42. When more people are involved in the drawing, the probability that any specified person will select his own name becomes very small. However, this is counteracted by the fact that there are more people who might possibly pick their own names.

43. The number of ways of choosing the first k_1 items is

$$\frac{n!}{k_1!(n-k_1)!}$$

The number of ways of choosing the next k_2 items from the remaining $(n-k_1)$ items is

$$\frac{(n-k_1)!}{k_2!(n-k_1-k_2)!}$$

Keep going like that. When you multiply all of the terms together, then a lot of items cancel out, so the result is

$$\frac{n!}{k_1!k_2!\cdots k_s!}$$

CHAPTER 7

1. Let A be the event that a head appeared and B be the event that the coin was fair.

$$\Pr(B) = 1/2; \quad \Pr(A|B) = 1/2; \quad \Pr(AB^c) = 1;$$

$$\Pr(B|A) = \frac{\Pr(A|B)\Pr(B)}{\Pr(A|B)\Pr(B) + \Pr(A|B^c)\Pr(B^c)}$$

$$= \frac{(1/2)(1/2)}{(1/2)(1/2) + (1)(1/2)} = 1/3$$

2. $\dfrac{4+4+4+3+4+4+4+4}{50} = \dfrac{31}{50}$; yes

3. $\dfrac{2+4+4+3+3+4+4+4}{46} = \dfrac{28}{46} = \dfrac{14}{23}$; yes

4. $4!\left(\frac{1}{51 \times 50 \times 49 \times 48}\right) = \frac{1}{249{,}900}$;

$2! \times \left(\frac{1}{49 \times 48}\right) = \frac{1}{1{,}176}$;

$\frac{1}{48}$

5. Two ways: A, 2, 3, 4, 5; 2, 3, 4, 5, 6

$$2 \times 3!\left(\frac{4}{50} \times \frac{4}{49} \times \frac{4}{48}\right) = \frac{8}{1{,}225}$$

6. Pr (your team wins series in 4 games) $= p^4$
Pr (your team wins series in 5 games) $= 4p^4 (1 - p)$
Pr (your team wins series in 6 games) $= 10p^4(1 - p)^2$
Pr (your team wins series in 7 games) $= 20p^4(1 - p)^3$
Add these together to find the probability your team will win the series.

7. Let A be the event of getting at least one head. B is the event of getting two heads.

$$\Pr(B|A) = \frac{\Pr(A \cap B)}{\Pr(A)} = 1/3$$

8. Let A = ace; $B = AD$;

$$\Pr(B|A) = \frac{\Pr(A \cap B)}{\Pr(A)} = \frac{1/40}{4/40} = 1/4$$

11. $3!\left(\frac{1}{50} \times \frac{1}{49} \times \frac{1}{48}\right) = 1/19{,}600$

12. $\frac{1/13}{3/13} = 1/3$

13. A = one 5; $B = 14$; $\Pr(A) = 25/72$; $\Pr(A \cap B) = 1/36$; $\Pr(B \mid A) = 2/25$

14. (a) The event of getting tails on the first toss.
(b) The event of getting heads on the second toss.
(c) The null set $\emptyset$ is both disjoint and independent of every other event. However, there is no way for two events to be both disjoint and independent if one of them is not the null set.
(d) The event of getting heads on both tosses.

15. $\Pr(J \mid C) = 2/3$; $\Pr(J \mid Co) = 5/9$; $\Pr(C) = \Pr(Co) = 1/2$;
$\Pr(Co \mid J) = 2/5$

16.
$$\Pr(B|A) = \frac{\Pr(A \cap B)}{\Pr(A)}$$
$$\Pr(A|B) = \frac{\Pr(A \cap B)}{\Pr(B)}; \Pr(A \cap B) = \Pr(A|B)\Pr(B)$$
$$\Pr(A) = \Pr(A \cap B) + \Pr(A \cap B^c)$$
$$\Pr(A \cap B^c) = \Pr(A \mid B^c)\Pr(B^c)$$
$$\Pr(B|A) = \frac{\Pr(A|B)\Pr(B)}{\Pr(A|B)\Pr(B) + \Pr(A|B^c)\Pr(B^c)}$$

17. A = left handed; B = O blood type.

$\Pr(A \mid B) = 0.05$; $\Pr(A \mid B^c) = 0.10$; $\Pr(B) = 0.40$;

$$\Pr(B|A) = \frac{(1/20)(2/5)}{(1/20)(2/5) + (1/10)(3/5)} = 1/4$$

18. $\Pr(\text{BrH} \mid \text{BrE}) = 7/10$; $\Pr(\text{BrH} \mid \text{GrE}) = 1/5$;

$\Pr(\text{BrH} \mid \text{B1E}) = 1/20$; $\Pr(\text{BrE}) = 3/4$;

$\Pr(\text{B1E}) = 1/5$; $\Pr(\text{GrE}) = 1/20$;

$$\Pr(\text{GrE}|\text{BrH}) = \frac{(1/5)(1/20)}{(7/10)(3/4) + (1/5)(1/20) + (1/20)(1/5)} = 0.018$$

CHAPTER 8

2. X

3. See table on page 356.

4. $1 - 1/3 - 1/5 = 7/15$

5. 1/12

6. 5

7. $f(k) = 2^{-(k+1)}$

9. $f(k) = 0$ for $k \geq 10$

11. If c is constant, then $E(C) = c$, and $\text{Var}(c) = E[c - E(c)^2] = E[(c - c)^2] = 0$.

12.

i	Outcomes	$\Pr(X=i)$	$i \times \Pr(X=i)$	$i^2 \times \Pr(X=i)$
3	1	0.00463	0.01389	0.04167
4	3	0.01389	0.05556	0.22222
5	6	0.02778	0.13889	0.69444
6	10	0.04630	0.27778	1.66667
7	15	0.06944	0.48611	3.40278
8	21	0.09722	0.77778	6.22222
9	25	0.11574	1.04167	9.37500
10	27	0.12500	1.25000	12.50000
11	27	0.12500	1.37500	15.12500
12	25	0.11574	1.38889	16.66667
13	21	0.09722	1.26389	16.43056
14	15	0.06944	0.97222	13.61111
15	10	0.04630	0.69444	10.41667
16	6	0.02778	0.44444	7.11111
17	3	0.01389	0.23611	4.01389
18	1	0.00463	0.08333	1.50000
Total	**216**	**1**	**10.50000**	**119.00000**

$\text{Var}(X) = 119 - 10.5^2 = 8.75$

13. $\text{Var}(X + Y + Z) = 8.75$ (see previous exercise); $\text{Var}(X) = \text{Var}(Y) = \text{Var}(Z) = 2.916667$ (see page 97); $8.75 = 2.916667 + 2.916667 + 2.916667$

14. $E(X) = (7/2)n$; $\text{Var}(X) = (35/12)n$

15. $1/2(-1) + 1/4(2) + 1/4(-3) = -3/4$;

You will lose 75 cents per game on average.

17. $E(X) = 4$; $E(X^2) = 20$; $\text{Var}(X) = 4$

18.

$$f(k) = \begin{cases} 0.4 & \text{if } k = -2 \\ 0.45 & \text{if } k = 4.5 \\ 0.15 & \text{if } k = 9 \\ 0 & \text{elsewhere} \end{cases}$$

19. The set of all numbers $k/2$, where k is a positive integer.

20. 2

21.

i	*Outcomes*	$\Pr(X = i)$	$i \times \Pr(X = i)$	$i^2 \times \Pr(X = i)$
1.0	1	0.02778	0.02778	0.02778
1.5	2	0.05556	0.08333	0.12500
2.0	3	0.08333	0.16667	0.33333
2.5	4	0.11111	0.27778	0.69444
3.0	5	0.13889	0.41667	1.25000
3.5	6	0.16667	0.58333	2.04167
4.0	5	0.13889	0.55556	2.22222
4.5	4	0.11111	0.50000	2.25000
5.0	3	0.08333	0.41667	2.08333
5.5	2	0.05556	0.30556	1.68056
6.0	1	0.02778	0.16667	1.00000
Total	**36**	**1**	**3.50000**	**13.70833**

$\mathrm{Var}(X) = 13.70833 - 3.5^2 = 1.458333$. If σ^2 is the variance when one die is thrown, then $\sigma^2 = 2.916667$, and the variance of X (the average of two dice) is $\sigma^2/2$.

22. The numbers of the stations broadcasting in your area.

23. $E(X) = (1/13)\left[\sum_{i=1}^{10} i + 30\right] = 85/13 = 6\frac{7}{13}$

24. $E(X) = 0.6$; $E(X^2) = 0.6$; $\mathrm{Var}(X) = 0.24$

25. $f(k) = \Pr(Z = k) = \Pr(X + Y = k)$
$= \Sigma_i \Pr(X = i)\ \Pr(Y = k - i)$

k:	2	3	4	5	6
$f(k)$:	1/36	1/18	1/12	1/9	5/36

k:	7	8	9	10	11	12
$f(k)$:	1/6	5/36	1/9	1/12	1/18	1/36

$E(Z) = 7 = 7/2 + 7/2 = E(X) + E(Y)$

26. $E(cX)\sum_i ca;\quad f(a_i) = c\sum_i a_i f(a_i) = cE(X)$

27.

$$\begin{aligned}\mathrm{Var}(cX) &= E[(cX)^2] - [E(cX)]^2\\ &= E(c^2X^2) - [c\,E(X)]^2\\ &= c^2\,E(X^2) - c^2[E(X)]^2\\ &= c^2\,[E(X^2) - [E(X)]^2]\\ &= c^2\mathrm{Var}(X)\end{aligned}$$

28. Let $W = X + Y$.

$$\begin{aligned} E(X+Y+Z) &= E(W+Z) \\ &= E(W) + E(Z) = E(X) \\ &= E(X) + E(Y) + E(Z) \end{aligned}$$

29. $f(k) = \begin{cases} 0.2 & \text{if } k = 1 \\ 0.2 & \text{if } k = 2 \\ 0.1 & \text{if } k = 4 \\ 0.3 & \text{if } k = 6 \\ 0.2 & \text{if } k = 7 \\ 0 & \text{elsewhere} \end{cases}$

30. (a) Let X_i be the number that appears on dice i. Then:

$$\begin{aligned} T_2 &= X_1 + X_2, \text{ and } T_{50} = X_1 + \ldots + X_{50}. \\ E(T_2) &= E(X_1) + E(X_2) = 3.5 + 3.5 = 7 \\ \text{Var}(T_2) &= \text{Var}(X_1) + \text{Var}(X_2) = 35/12 + 35/12 = 5.8333 \end{aligned}$$

(The last step works because each X is independent.)

$$\begin{aligned} E(T_{50}) &= E(X_1) + \ldots + E(X_{50}) = 175 \\ \text{Var}(T_{50}) &= \text{Var}(X_1) + \ldots + \text{Var}(X_{50}) = 145.833 \end{aligned}$$

(b)

$$\begin{aligned} A_2 &= T_2/2 \\ A_{50} &= T_{50}/50 \\ E(A_2) &= E(T_2)/2 = 3.5 \\ \text{Var}(A_2) &= (1/2)^2\text{Var}(T_2) = 1.4583 \\ E(A_{50}) &= E(T_{50})/50 = 3.5 \\ \text{Var}(A_{50}) &= (1/50)^2\text{Var}(T_{50}) = 0.0583 \end{aligned}$$

Note that the variance of the average number that appears when you roll many dice is much smaller than the variance of the number that appears on any single die. This property is very important in the development of inferential statistics.

33. Note that $[X - E(X)]^2 \geq 0$. Therefore, if we take the expectation of that expression, we get

$$\begin{aligned} &E[(X - E(X))^2] \geq 0 \\ &E[X^2 - 2X\,E(X) + (E(X))^2] \geq 0 \\ &E(X^2) - 2E(X)\,E(X) + [E(X)]^2 \geq 0 \\ &E(X^2) - [E(X)]^2 \geq 0 \\ &E(X^2) \geq [E(X)]^2 \end{aligned}$$

34. $1 = \sum_{k=1}^{\infty} f(k) = \sum_{k=1}^{\infty} c2^{-k} = c\sum_{k=1}^{\infty} 2^{-k} = c;$

Therefore $c = 1$.

CHAPTER 9

1. The number of meteorites that hit Wethersfield will have a binomial distribution with $p = .000\,000\,07$ and $n = 11{,}000$. The probability of two hits is 2.96×10^{-7}. Interestingly enough, the town of Wethersfield *was* hit by two meteorites in an eleven-year period.

2. $\binom{5}{3}\left(\frac{1}{2}\right)^3\left(\frac{1}{2}\right)^2 = \frac{5}{16}$

3. $\binom{5}{4}\left(\frac{1}{2}\right)^4\left(\frac{1}{2}\right) + \binom{5}{5}\left(\frac{1}{2}\right)^5\left(\frac{1}{2}\right)^0 = \frac{3}{16}$

4. $\binom{7}{6}\left(\frac{1}{10}\right)^6\left(\frac{9}{10}\right) + \binom{7}{7}\left(\frac{1}{10}\right)^7\left(\frac{9}{10}\right)^0 - \frac{1}{156{,}250}$

5. $\binom{4}{3}\left(\frac{2}{3}\right)^3\left(\frac{1}{3}\right) + \binom{4}{4}\left(\frac{2}{3}\right)^4\left(\frac{1}{3}\right)^0 = \frac{16}{27}$

6. $\sum_{k=75}^{100}\binom{100}{k} \times \frac{1}{2^{100}}$

8. $1 - (0.9)^3 = 0.271$

9. $\binom{3}{2}\left(\frac{1}{7}\right)^2\left(\frac{6}{7}\right) = \frac{18}{343}$

10. $\binom{6}{4}\left(\frac{2}{5}\right)^4\left(\frac{3}{5}\right)^2 = \frac{432}{3{,}125}$

11. X_1 can be thought of as the number of successes in n_1 trials, with each trial having probability p of success, and X_2 can be thought of as the number of successes in the next n_2 trials; then $X_1 + X_2$ can be interpreted as the number of successes in $n_1 + n_2$ trials, and therefore will have a binomial distribution.

12. $\binom{3}{2}\left(\frac{9}{20}\right)^2\left(\frac{11}{20}\right) + \binom{3}{3}\left(\frac{9}{20}\right)^3\left(\frac{11}{20}\right)^0 = 0.425$

13. If you take 210 reservations, you have a 6.9 percent chance of an overflow. If you take 209 reservations, there is only a 3.8 percent chance of an overflow.

CHAPTER 10

1. Mean = 7.5; variance = 5.12

2. Mean = 75; variance = 150

3. Mean = 64; variance = 192

4. Mean = 10

5. Mean = 180; variance = 3060

6. $\Pr(X > 1) = 1 - (3/2)e^{-1/2}$

7. $\Pr(X = n) = e^{-\lambda}\dfrac{\lambda^n}{n!}$, where $n \le \lambda < n + 1$

8. $e^{-10}(680/3)$

9. $\Pr(X \le 4) = p + p(1-p) + p(1-p)^2 + p(1-p)^3 = 0.802$

10. $n = 2$

11. $448/3^8$

13. $\dfrac{\binom{20}{1}\binom{180}{2}}{\binom{200}{3}} = 0.245$

14. $\dfrac{\binom{2}{2}\binom{6}{1}}{\binom{8}{3}} = 0.107$

15. $e^{-3}(131/8)$

16. $8/81 = 0.099$

17. $\binom{19}{7}\left(\dfrac{1}{3}\right)^8\left(\dfrac{2}{3}\right)^{12} = 4{,}199\left(\dfrac{2^{14}}{3^{19}}\right)$

18. $\dfrac{\binom{30}{2}\binom{270}{3}}{\binom{300}{5}} = 0.072$

19. $\dfrac{11^{19}}{12^{20}}$

20.

$$
\begin{aligned}
f(k) &= e^{-\lambda}\frac{\lambda^k}{k!} \\
\sum_{k=0}^{\infty} f(k) &= \sum_{k=0}^{\infty} e^{-\lambda}\frac{\lambda^k}{k!} \\
&= e^{-\lambda}\sum_{k=0}^{\infty}\frac{\lambda^k}{k!} \\
&= e^{-\lambda}e^{\lambda} \\
&= 1
\end{aligned}
$$

CHAPTER 11

1. Height, weight, temperature

2. F has a maximum at k only if there is no probability that the random variable will be greater than k. F has a minimum at j only if there is no probability that the random variable will be less than j.

4. $r^2/900$

5. $f(r) = r/450$

6. $n + 1$

7. $F(0) = 0 = F(1/2), F(10) = \ln 10$

8. $f(x) = 0$ for $x \leq 0$, $f(x) = e^{-x}$ for $x > 0$

9. (a) g decreases from 2 to 3.

   (b) g is above 1 from 1 to 3.

10. 1/2

11. $F(a) = 0$ for $a \leq -\pi/3$;
    $F(a) = \cos a - 1/2$ for $-\pi/3 < a < 0$;
    $F(a) = 3/2 - \cos a$ for $0 < a < \pi/3$;
    $F(a) = 1$ for $a > \pi/3$

12. $F(x) = 0$ for $x \leq -1$;
    $F(x) = (3/4)(x - x^3/3 + 2/3)$ for $-1 \leq x \leq 1$;
    $F(x) = 1$ for $x > 1$

14. $f(x) = 1/(b - a)$ for $a < x < b$; $f(x) = 0$ otherwise

$$E(X) = \int_a^b x/(b-a)dx = 1/(b-a)x^2/2|_a^b = (a+b)/2$$

$$E(X^2) = \int_a^b x^2 f(x)dx = 1/(b-a)\int_a^b x^2 dx$$
$$= (b^3 - a^3)/[3(b-a)]$$

$\text{Var}(X) = (a^2 + ab + b^2)/3 - (a^2 + 2ab + b^2)/4 = (b - a)^2/12$

15. $E(X) = \int_{-\infty}^{\infty} xf(x)dx$. Let $y = x - c$. Then:

$$\begin{aligned} E(X) &= \int_{-\infty}^{\infty} (c+y)f(c+y)dy \\ &= c\int_{-\infty}^{\infty} f(c+y)dy + \int_{-\infty}^{\infty} y\,f(c+y)dy \\ &= c + \int_{-\infty}^{0} y\,f(c+y)dy + \int_{0}^{\infty} y\,f(c+y)dy \\ &= c + \int_{-\infty}^{0} y\,f(c-y)dy + \int_{0}^{\infty} y\,f(c+y)dy \end{aligned}$$

Let $z = -y$. Then:

$$\begin{aligned} E(X) &= c - \int_0^{\infty} zf(c+z)dz + \int_0^{\infty} yf(c+y)dy \\ &= c \end{aligned}$$

17. $a = \ln 2$

18. 1/30

19. $F(a) = 0$ for $a \le 0$;
$F(a) = a^2/2$ for $0 < a < 1$;
$F(a) = 1/2a^2$ for $a > 1$

20. $f(x) = 0$ for $x < 1$;
$f(x) = (2x + 3)/2$ for $-1 < x < 0$;
$f(x) = 0$ for $0 < x$

21. $0 \le x \le 1$

22. $c = 1/18{,}750$;
$F(a) = 0$ for $a \le 0$;
$F(a) = (30a^2 + 3{,}600a)/18{,}750$ for $0 \le a \le 5$;
$F(a) = 1$ for $a \ge 5$

CHAPTER 12

1. 0.08; 0.66

2. 0.20; 0.63

3. 0.07

4. 0.34

5. 0.31

6. 0.28; 0.58; the normal distribution can only approximately represent this situation, since the number of glasses sold is a discrete random variable.

8. 0.15

9. –2.25

10. 35.43

11. $F(3) = F(\mu) = \Phi(0) = 0.5$, which is greater than $F(4) = 0.4$, thereby contradicting the fact that F must be an increasing function.

12. 0

13. 0

14. $-9.24 < a < -3.16$

15. Mean = 10, variance = 0.0064

17. $1 - \Phi\left[\dfrac{\mu_1 - \mu_2}{\sqrt{{\sigma_1}^2 + {\sigma_2}^2}}\right]$

18. 0.39

19. We know that $aX + b$ is also a normal random variable. Let $a = 1/\sigma$ and $b = -\mu/\sigma$. Then $(X - \mu)/\sigma$ is also a normal random variable. $E[(X - \mu)/\sigma] = (1/\sigma)[E(X) - \mu] = 0$; $\mathrm{Var}[(X - \mu)/\sigma] = 1/\sigma^2\,\mathrm{Var}(X - \mu) = 1/\sigma^2\,\mathrm{Var}(X) = 1$

20. $\Pr(|X - \mu| \leq \sigma) = \Pr[-1 \leq (X - \mu)/\sigma \leq 1] = \Phi(1) - \Phi(-1) = 2\Phi(1) - 1 = 0.68$

22. The mean

23. The mean

24. If X, Y, and Z are normal random variables, let $W = X + Y + Z$, and $U = X + Y$. Then U is a normal random variable, and so is $W = U + Z = X + Y + Z$.

25. Let X be a normal random variable with parameters μ and σ^2.

$$E(X) = \frac{1}{\sigma\sqrt{2\pi}} \int_{-\infty}^{\infty} xe^{-(x-\mu)^2/2\sigma^2} dx$$
$$= \frac{1}{\sigma\sqrt{2\pi}} \int_{-\infty}^{\infty} (x-\mu)e^{-(x-\mu)^2/2\sigma^2} dx + \frac{1}{\sigma\sqrt{2\pi}} \int_{-\infty}^{\infty} \mu e^{-(x-\mu)^2/2\sigma^2} dx$$

Letting $t = x - \mu$,

$$\frac{1}{\sigma\sqrt{2\pi}} \int_{-\infty}^{\infty} (x-\mu)e^{-(x-\mu)^2/2\sigma^2} dx = \frac{1}{\sigma\sqrt{2\pi}} \int_{-\infty}^{\infty} te^{-t^2/2\sigma^2} dt$$
$$= \frac{1}{\sigma\sqrt{2\pi}} (-e^{-t^2/2\sigma^2})\Big|_{t=-\infty}^{t=\infty}$$
$$= 0$$

$$\frac{1}{\sigma\sqrt{2\pi}} \int_{-\infty}^{\infty} \mu e^{-(x-\mu)^2/2\sigma^2} dx = \mu \int_{-\infty}^{\infty} f_X(x) dx$$
$$= \mu$$
$$E(X) = 0 + \mu$$
$$= \mu$$

26. $$\text{Var}(X) = E((X-\mu)^2)$$
$$= \frac{1}{\sigma\sqrt{2\pi}} \int_{-\infty}^{\infty} (x-\mu)^2 e^{-(x-\mu)^2/2\sigma^2} dx$$
$$= \frac{1}{\sigma\sqrt{2\pi}} \int_{-\infty}^{\infty} t^2 e^{-t^2/2\sigma^2} dt \qquad (\text{letting } t = x - \mu)$$
$$= \frac{1}{\sigma\sqrt{2\pi}} (t)(-\sigma^2 e^{-t^2/2\sigma^2})\Big|_{t=-\infty}^{t=\infty} + \frac{\sigma^2}{\sigma\sqrt{2\pi}} \int_{-\infty}^{\infty} e^{-t^2/2\sigma^2} dt$$

Using integration by parts,

$$\text{Var}(X) = 0 + \sigma^2 \int_{-\infty}^{\infty} f_Y(t) dt$$
$$= \sigma^2$$

where Y is a normal random variable with parameters $\mu = 0, \sigma^2$.

27. Let $I = \int_{-\infty}^{\infty} e^{-x^2/2}dx$

$$
\begin{aligned}
I^2 &= \int_{-\infty}^{\infty} e^{-x^2/2}dx \int_{-\infty}^{\infty} e^{-y^2/2}dy \\
&= \int_{-\infty}^{\infty}\int_{-\infty}^{\infty} e^{-(x^2+y^2)/2}dx\,dy \\
&= \int_{0}^{2\pi}\int_{0}^{\infty} e^{-r^2/2}rdr\,d\theta
\end{aligned}
$$

Shifting to polar coordinates,

$$
\begin{aligned}
I^2 &= \int_0^{2\pi} d\theta \\
&= 2\pi \\
I &= \sqrt{2\pi}
\end{aligned}
$$

$$\frac{1}{\sigma\sqrt{2\pi}}\int_{-\infty}^{\infty} e^{-x^2/2}dx = 1$$

= the area under the standard normal density function $\frac{1}{\sqrt{2\pi}}e^{-x^2/2}$

30.

$$
\begin{aligned}
f(x) &= \frac{1}{\sigma\sqrt{2\pi}}e^{-0.5[(x-\mu)/\sigma]^2} \\
f'(x) &= -\left(\frac{x-\mu}{\sigma}\right)\frac{1}{\sigma\sqrt{2\pi}}e^{-0.5[(x-\mu)/\sigma]^2} \\
f''(x) &= -\left(\frac{x-\mu}{\sigma}\right)\left[-\left(\frac{x-\mu}{\sigma}\right)\frac{1}{\sigma\sqrt{2\pi}}e^{-0.5[(x-\mu)/\sigma]^2}\right] \\
&\quad - \frac{1}{\sigma\sqrt{2\pi}}e^{-0.5[(x-\mu)/\sigma]^2} \\
f''(x) &= \left[\left(\frac{x-\mu}{\sigma}\right)^2 - 1\right]\frac{1}{\sigma\sqrt{2\pi}}e^{-0.5[(x-\mu)/\sigma]^2}
\end{aligned}
$$

The expression in the brackets will be zero either when $x = \mu + \sigma$ or $x = \mu - \sigma$, causing the entire second derivative to be zero.

32.

i	$\Pr(X = i)$	e^{tX}
1	1/6	e^{t}
2	1/6	e^{2t}
3	1/6	e^{3t}
4	1/6	e^{4t}
5	1/6	e^{5t}
6	1/6	e^{6t}

The sum of the right-hand column gives the moment generating function, that is, $E(^{tX})$:

$(e^t + e^{2t} + e^{3t} + e^{4t} + e^{5t} + e^{6t})/6$

$\psi'(t) = (e^t + 2e^{2t} + 3e^{3t} + 4e^{4t} + 5e^{5t} + 6e^{6t})/6$

$\psi''(t) = (e^t + 4e^{2t} + 9e^{3t} + 16e^{4t} + 25e^{5t} + 36e^{6t})/6$

$\psi''(0) = (e^0 + 4e^0 + 9e^0 + 16e^0 + 25e^0 + 36e^0)/6$

$= (1 + 4 + 9 + 16 + 25 + 36)/6 = 91/6 = 15.167 = E(X^2)$

CHAPTER 13

1. A bit less than 0.25
2. About 0.10
3. About 0.30
4. About 0.15
5. There is a probability of about 0.2 that this random variable might be as big as 16, so you can believe the person.
6. There is less than a 0.005 probability that this random variable will be less than 5, so you should not believe this claim.
7. There is only a 0.05 probability that this random variable will be smaller than –2.3, so this value is implausible but you cannot reject it with certainty.
8. Yes
9. F distribution with 1 and n degrees of freedom

10.
$$
\begin{aligned}
E(X) &= \int_0^\infty \lambda x e^{-\lambda x} dx \\
&= -xe^{-\lambda x} \Big|_{x=0}^{x=\infty} + \int_0^\infty e^{-\lambda x} dx \qquad \text{(integrating by parts)} \\
&= 0 + \frac{1}{\lambda} \\
&= \frac{1}{\lambda} \\
E(X^2) &= \int_0^\infty \lambda x^2 e^{-\lambda x} dx \\
&= -\lambda^2 e^{-\lambda x} \Big|_{x=0}^{x=\infty} + 2\int_0^\infty xe^{-\lambda x} \qquad \text{(integrating by parts)} \\
&= 0 + \frac{2}{\lambda} E(X) \\
&= \frac{2}{\lambda^2} \\
\text{Var}(X) &= E(X^2) - [E(X)]^2 \\
&= \frac{2}{\lambda^2} - \frac{1}{\lambda^2} \\
&= \frac{1}{\lambda^2}
\end{aligned}
$$

11.
$$
\begin{aligned}
F(a) &= \int_0^\infty \lambda e^{-\lambda x} dx \\
&= -e^{\lambda x} \Big|_{x=0}^{x=a} \\
&= 1 - e^{-\lambda a} \qquad \text{for } a \geq 0
\end{aligned}
$$

12. $\dfrac{\lambda}{\lambda - t}$ for $t < \lambda$.

13.
$$
\begin{aligned}
\Pr(X > s) &= 1 - F(s) \\
&= e^{-\lambda s} \\
\Pr(X > a + b) &= e^{-\lambda(a+b)} \\
&= e^{-\lambda s} e^{-\lambda b} \\
&= \Pr(X > a)\Pr(X > b)
\end{aligned}
$$

14. $$\begin{aligned} F(y) &= \Pr(Y < y) \\ &= \Pr(Z^2 < y) \\ &= \Pr(-\sqrt{y} < Z < \sqrt{y}) \\ &= \Phi(\sqrt{y}) - \Phi(-\sqrt{y}) \end{aligned}$$

$$\begin{aligned} f(y) &= \Phi'(\sqrt{y})\left(\frac{1}{2\sqrt{y}}\right) - \Phi'(-\sqrt{y})\left(\frac{-1}{2\sqrt{y}}\right) \\ &= \frac{1}{2\sqrt{2}}(\Phi'(\sqrt{y}) + \Phi'(\sqrt{-y})) \end{aligned}$$

$$\Phi'(x) = \frac{1}{\sqrt{2\pi}}(e^{-x^2/2})$$

$$\begin{aligned} f(y) &= \frac{1}{\sqrt{2\pi}}\left(\frac{1}{2\sqrt{y}}\right)(2e^{-y/2}) \\ &= \frac{e^{-y/2}}{\sqrt{2\pi y}} \end{aligned}$$

16. Let X be a chi-squared random variable with m degrees of freedom.

$$fx(x) = \frac{x^{(m-2)/2}e^{-x/2}}{2^{m/2}\Gamma\left(\frac{m}{2}\right)} \quad \text{for } x > 0$$

$$\begin{aligned} \psi x(t) &= E(e^{tx}) \\ &= \frac{1}{2^{m/2}\Gamma\left(\frac{m}{2}\right)}\int_0^\infty e^{tx}x^{(m-2)/2}e^{-x/2}dx \end{aligned}$$

Let $u = (1 - 2t)x$ with $t < 1/2$; then

$$\begin{aligned} \psi x(t) &= \frac{1}{2^{m/2}\Gamma\left(\frac{m}{2}\right)}\int_0^\infty \frac{u^{(m-2)/2}e^{-u/2}}{(1-2t)^{m/2}}du \\ &= (1-2t)^{-m/2}\int_0^\infty fx(u)du \\ &= (1-2t)^{-m/2} \end{aligned}$$

17. $$\begin{aligned} \psi_{Y_1} + {}_{Y_2}(t) &= \psi_{Y_1}(t)\psi_{Y_2}(t) \\ &= (1-2t)^{-n_1/2}(1-2t)^{-n_2/2} \\ &= (1-2t)^{-(n_1+n_2)/2} \end{aligned}$$

and $Y_1 + Y_2$ has a chi-squared distribution with $n_1 + n_2$ degrees of freedom.

18. Let X_n be a random variable with a t distribution with n degrees of freedom.

$$fx_n(x) = \frac{c_n}{\left(1 + \frac{x^2}{n}\right)^{(n+1)/2}} \quad \text{where } c_n = \frac{\Gamma\left(\frac{n+1}{2}\right)}{\sqrt{m\pi}\,\Gamma\left(\frac{n}{2}\right)}$$

Assume that $\lim_{n\to\infty} c_n = \frac{1}{\sqrt{2\pi}}$

$$\lim_{n\to x}\left(1 + \frac{x^2}{n}\right)^{-(n+1)/2} = \lim_{n\to\infty} \exp\left[\frac{-(n+1)}{2}\ln\left(1 + \frac{x^2}{n}\right)\right]$$

$$= \exp \lim_{n\to\infty}\left[\frac{-(n+1)}{2}\ln\left(1 + \frac{x^2}{n}\right)\right]$$

$$\lim_{n\to\infty}\left[\frac{-(n+1)}{2}\ln\left(1 + \frac{x^2}{n}\right)\right] = \lim_{n\to\infty}\frac{-\ln\left(1 + \frac{x^2}{n}\right)}{2/(n+1)}$$

$$= \lim_{n\to\infty}\frac{\left(\frac{x^2}{n^2}\right)\left(1 + \frac{x^2}{n}\right)^{-1}}{2/(n+1)}$$

(by l'Hôpital's rule)

$$= \lim_{n\to\infty}\left(\frac{n+1}{n}\right)^2\left[\frac{-x^2}{2\left(1 + \frac{x^2}{n}\right)}\right]$$

$$= -\frac{x^2}{2}$$

$$\lim_{n\to\infty}\left(1 = \frac{x^2}{n}\right)^{-(n+1)/2} = e^{-x^2/2}$$

$$\lim_{n\to\infty} fx_n(x) = \frac{1}{\sqrt{2\pi}}e^{-x^2/2}$$

As $n \to \infty$ the distribution of X_n approaches that of a standard normal random variable.

CHAPTER 14

1.

X	Y = 2	Y = 3	Y = 4	Y = 5	Y = 6	Y = 7	Y = 8	Y = 9	Y = 10	Y = 11	Y = 12
2	1/216	1/216	1/216	1/216	1/216	1/216	0	0	0	0	0
3	1/216	2/216	2/216	2/216	2/216	2/216	1/216	0	0	0	0
4	1/216	2/216	3/216	3/216	3/216	3/216	2/216	1/216	0	0	0
5	1/216	2/216	3/216	4/216	4/216	4/216	3/216	2/216	1/216	0	0
6	1/216	2/216	3/216	4/216	5/216	5/216	4/216	3/216	2/216	1/216	0
7	1/216	2/216	3/216	4/216	5/216	6/216	5/216	4/216	3/216	2/216	1/216
8	0	1/216	2/216	3/216	4/216	5/216	5/216	4/216	3/216	2/216	1/216
9	0	0	1/216	2/216	3/216	4/216	4/216	4/216	3/216	2/216	1/216
10	0	0	0	1/216	2/216	3/216	3/216	3/216	3/216	2/216	1/216
11	0	0	0	0	1/216	2/216	2/216	2/216	2/216	2/216	1/216
12	0	0	0	0	0	1/216	1/216	1/216	1/216	1/216	1/216

2.

	1	2	3	4	5	6
2	1/36	0	0	0	0	0
3	1/36	1/36	0	0	0	0
4	1/36	1/36	1/36	0	0	0
5	1/36	1/36	1/36	1/36	0	0
6	1/36	1/36	1/36	1/36	1/36	0
7	1/36	1/36	1/36	1/36	1/36	1/36
8	0	1/36	1/36	1/36	1/36	1/36
9	0	0	1/36	1/36	1/36	1/36
10	0	0	0	1/36	1/36	1/36
11	0	0	0	0	1/36	1/36
12	0	0	0	0	0	1/36

3. $p_X(x) \geq p_{X,Y}(x,y)$

4. 3

5.

x:	–2	4	5	
$f_X(x)$:	0.4	0.2	0.4	
y:	0	2	4	6
$f_X(y)$:	0.2	0.1	0.2	0.5

The covariance and correlation are 0.

6.

x:	−2	−1	0	1	2
$f_X(x)$:	0.2	0.25	0.1	0.35	0.1

y:	3	5	9
$f_X(y)$:	0.35	0.25	0.4

$\text{Cov}(X, Y) = 0.49$; $r(XY) = 0.139$

7.

x:	0	5	9	11	13
$f_X(x)$:	0.2	0.25	0.05	0.35	0.15

y:	1	4	5	6	9
$f_Y(y)$:	0.25	0.45	0.1	0.05	0.15

$\text{Cov}(X, Y) = -6.28$; $r(X, Y) = -0.920$

8.

x:	−3	2	4	5	7
$f_X(x)$:	0.25	0.05	0.15	0.05	0.50

y:	−9	−4	−3	−1	0
$f_Y(y)$:	0.05	0.35	0.15	0.1	0.35

$\text{Cov}(X, Y) = -0.17$; $r(X, Y) = -0.018$

9.

x:	1	2	3	4
$f_X(x)$:	0.4	0	0.4	0.2

y:	1	2	3	4
$f_Y(y)$:	0.2	0.2	0.3	0.3

11. If y is between 1 and 31, then $f(x, y) = 1/365$ if x is $1, 3, 5, 7, 8, 10$, or 12. If y is between 1 and 30, and x is $4, 6, 9$, or 11, then $f(x, y) = 1/365$. Finally, if y is between 1 and 28, and x is 2, then $f(x, y) = 1/365$. $f(x, y) = 0$ for all other values of x and y.

13. 7

15.

	$y < 7$	$7 \le y < 10$	$10 \le y < 13$	$13 \le y < 14$	$14 \le y$
$x < 1$	0	0	0	0	0
$1 \le x < 2$	0	0.1	0.1	0.2	0.2
$2 \le x < 5$	0	0	0.1	0.2	0.35
$5 \le x < 9$	0	0	0.35	0.45	0.7
$9 \le x$	0	0.2	0.55	0.75	1

16. Because the sum of the values is greater than 1.

17. $\text{Cov}(X, Y) = -2.917; \quad r(X, Y) = -1$

18. $\text{Cov}(X, Z) = 0; \quad r(X, Z) = 0$

27. $$f(x,y) = \frac{1}{2\pi\sqrt{1-\rho^2}\sigma_x\sigma_y} \exp\left[-\frac{1}{2(1-\rho^2)}\left[\left(\frac{x-\mu_x}{\sigma_x}\right)^2 - 2\rho\left(\frac{x-\mu_x}{\sigma_x}\right)\left(\frac{y-\mu_y}{\sigma_y}\right) + \left(\frac{y-\mu_y}{\sigma_y}\right)^2\right]\right]$$

$$f_x(x) = \int_{y=-\infty}^{y=\infty} f(x,y)dy$$

$$= \int_{y=-\infty}^{y=\infty} \frac{1}{2\pi\sqrt{1-\rho^2}\sigma_x\sigma_y} \exp\left[-\frac{1}{2(1-\rho^2)}\left[\left(\frac{x-\mu_x}{\sigma_x}\right)^2 - 2\rho\left(\frac{x-\mu_x}{\sigma_x}\right)\left(\frac{y-\mu_y}{\sigma_y}\right) + \left(\frac{y-\mu_y}{\sigma_y}\right)^2\right]\right]dy$$

$$= \frac{1}{2\pi\sqrt{1-\rho^2}\sigma_x\sigma_y} \int_{y=-\infty}^{y=\infty} \exp\left[-\frac{1}{2(1-\rho^2)}\left[\left(\frac{x-\mu_x}{\sigma_x}\right)^2 - 2\rho\left(\frac{x-\mu_x}{\sigma_x}\right)\left(\frac{y-\mu_y}{\sigma_y}\right) + \left(\frac{y-\mu_y}{\sigma_y}\right)^2\right]\right]dy$$

Let $x' = \dfrac{x-\mu_x}{\sigma_x}$; $y' = \dfrac{y-\mu_y}{\sigma_y}$; $dy' = (1/\sigma_y)dy$; $dy = \sigma_y dy'$

$$f_x(x) = \frac{1}{2\pi\sqrt{1-\rho^2}\sigma_x\sigma_y} \int_{y'=-\infty}^{y'=\infty} \exp\left[-\frac{1}{2(1-\rho^2)}\left[x'^2 - 2px'y' + y'^2\right]\right](\sigma_y dy')$$

$$= \frac{1}{2\pi\sqrt{1-\rho^2}\sigma_x} \int_{y'=-\infty}^{y'=\infty} \exp\left[-\frac{1}{2(1-\rho^2)}\left[x'^2 - 2px'y' + y'^2\right]\right]dy'$$

Let $k = \dfrac{1}{2(1-\rho^2)}$

$$f_x(x) = \frac{1}{2\pi\sqrt{1-\rho^2}\sigma_x}\int_{y'=-\infty}^{y'=\infty} \exp\left[-kx'^2 + 2kpx'y' - ky'^2\right]dy'$$

$$= \frac{1}{2\pi\sqrt{1-\rho^2}\sigma_x}\int_{y'=-\infty}^{y'=\infty} e^{-kx'^2}e^{2kpx'y'-ky'^2}dy'$$

$$= \frac{1}{2\pi\sqrt{1-\rho^2}\sigma_x}e^{-kx'^2}\int_{y'=-\infty}^{y'=\infty} e^{2kpx'y'-ky'^2}dy'$$

Let $y'' = y' - px'$; then $y' = y'' + px'$; $dy' = dy''$

$$f_x(x) = \frac{1}{2\pi\sqrt{1-\rho^2}\sigma_x}e^{-kx'^2}\int_{y''=-\infty}^{y''=\infty} e^{2kpx'(y''+px')-k(y''+px')^2}dy''$$

$$= \frac{1}{2\pi\sqrt{1-\rho^2}\sigma_x}e^{-kx'^2}\int_{y''=-\infty}^{y''=\infty} e^{2kpx'y''+2kp^2x'^2-k(y''^2+2y''px'+\rho^2x'^2)}dy''$$

$$= \frac{1}{2\pi\sqrt{1-\rho^2}\sigma_x}e^{-kx'^2}\int_{y''=-\infty}^{y''=\infty} e^{2kpx'y''+2kp^2x'^2-ky''^2-2kpx'y''-k\rho^2x'^2}dy''$$

$$= \frac{1}{2\pi\sqrt{1-\rho^2}\sigma_x}e^{-kx'^2}\int_{y''=-\infty}^{y''=\infty} e^{-ky''^2+k\rho^2x'^2}dy''$$

$$= \frac{1}{2\pi\sqrt{1-\rho^2}\sigma_x}e^{-kx'^2+kp^2x'^2}\int_{y''=-\infty}^{y''=\infty} e^{-ky''^2}dy''$$

Let $y''' = \sqrt{2k}y''$; $dy''' = \sqrt{2k}dy''$

$$f_x(x) = \frac{1}{2\pi\sqrt{1-\rho^2}\sigma_x}e^{-kx'^2+k\rho^2x'^2}\int_{y'''=-\infty}^{y'''=\infty} e^{-y'''^2/2}(1/\sqrt{2k})dy'''$$

$$= \frac{1}{2\pi\sqrt{1-\rho^2}\sigma_x}e^{-kx'^2+k\rho^2x'^2}\sqrt{2\pi}(1/\sqrt{2k})$$

$$k = \frac{1}{2(1-\rho^2)}$$

$$f_x(x) = \frac{1}{\sqrt{2\pi}\sqrt{1-\rho^2}\sigma_x}e^{-(x'^2-\rho^2x'^2)/[2(1-\rho^2)]}[1/\sqrt{1/(1-\rho)^2}]$$

$$= \frac{1}{\sqrt{2\pi}\sqrt{1-\rho^2}\sigma_x}e^{-x'^2(1-\rho^2)/[2(1-\rho^2)]}(\sqrt{(1-\rho)^2})$$

$$= \frac{1}{\sqrt{2\pi}\sigma_x}e^{-(x'^2)/2}$$

$$f_x(x) = \frac{1}{\sqrt{2\pi}\sigma_x}e^{-\frac{1}{2}[(x-\mu_x)/\sigma_x]^2}$$

The marginal density function for X is a normal distribution. Similar reasoning will show the marginal density function for Y is also a normal density function. The joint density function in this exercise is called a *bivariate normal* distribution. The next exercise shows how X and Y are related in a bivariate normal distribution.

28. $f(x,y) = \frac{1}{2\pi\sqrt{1-\rho^2}} \exp\left[-\frac{1}{2(1-\rho^2)}(x^2 - 2\rho xy + y^2)\right]$

$f_y(y) = \frac{1}{\sqrt{2\pi}} e^{-y^2/2}$

To find the conditional density function, divide the joint density function $f(x, y)$ by the marginal density function $f_x(x)$:

$$\frac{f(x,y)}{f_y(y)} = \frac{\frac{1}{2\pi\sqrt{1-\rho^2}} \exp\left[-\frac{1}{2(1-\rho^2)}(x^2 - 2\rho xy + y^2)\right]}{\frac{1}{\sqrt{2\pi}} e^{-y^2/2}}$$

$$= \frac{1}{\sqrt{(2\pi)(1-\rho^2)}} \exp\left[-\left(\frac{1}{2(1-\rho^2)}\right)(x^2 - 2\rho xy + y^2) + y^2/2\right]$$

$$= \frac{1}{\sqrt{(2\pi)(1-\rho^2)}}$$

$$\exp\left[-\left(\frac{1}{2(1-\rho^2)}\right)(x^2 - 2\rho xy + y^2 - 2(1-\rho^2)y^2/2\right]$$

$$= \frac{1}{\sqrt{(2\pi)(1-\rho^2)}} \exp\left[-\left(\frac{1}{2(1-\rho^2)}\right)(x^2 - 2\rho xy + \rho^2 y^2)\right]$$

If $\rho = 0$, then the conditional density function becomes:

$$\frac{1}{\sqrt{(2\pi)}} e^{-x^2/2}$$

This is the density function for a standard normal random variable, which is also the marginal density function for X (see the previous exercise). Therefore, if $\rho = 0$, the conditional density function for X is the same as its marginal density function, which means that X and Y are independent. It turns out that the parameter ρ in the bivariate joint normal density function is in fact the correlation between the two random variables X and Y.

CHAPTER 15

1. 0

2. 0.68

3. 0.02

4. 0.32

7. 0.94

8. 1/4

9. 1/16

12. It will have a normal distribution.

13. Suppose that X is a discrete random variable. (The proof is very similar for a continuous random variable.) Let $f(x)$ be the density function of X. Then we can write out the definition of expectation:

$$E(X) = x_1 f(x_1) + x_2 f(x_2) + \cdots + x_n f(x_n)$$

Let's say that $x_1, x_2, \ldots x_a\bullet$ are all of the possible values of X that are less than a, and that $x_k, x_{k+1}, \ldots x_n$ are the possible values of X that are greater than a. Then we can break the expectation into two parts, like this:

$$E(X) = [x_1 f(x_1) + x_2 f(x_2) + \cdots + x_a \cdot f(x_a\cdot)]$$
$$+ [x_k f(x_k) + x_{k+1} f(x_{k+1}) + \cdots + x_n f(x_n)]$$

We know that the first term is positive, so we have

$$E(X) = \text{(something positive)}$$

$$+ [x_k f(x_k) + x_{k+1} f(x_{k+1}) + \cdots + x_n f(x_n)]$$

We're just going to ignore the first term, and change the equal sign into an inequality:

$$E(X) \geq x_k f(x_k) + x_{k+1} f(x_{k+1}) + \cdots + x_n f(x_n)$$

(There's no reason why we can't do this, although if the term we're ignoring is very large, $E(X)$ will be much bigger than the right hand side.)

Every value of x that appears in the right hand side is bigger than a, so we also have this inequality:

$$x_k f(x_k) + x_{k+1} f(x_{k+1}) + \cdots + x_n f(x_n)$$
$$\geq af(x_k) + af(x_{k+1}) + \cdots + af(x_n)$$

Putting these two inequalities together,

$$E(X) \geq \sum_{i=k}^{n} x_i f(x_i) \geq \sum_{i=k}^{n} af(x_i)$$

Now we're going to ignore the middle expression:

$$E(X) \geq af(x_k) + af(x_{k+1}) + \cdots + af(x_n)$$
$$E(X) \geq a[f(x_k) + f(x_{k+1}) + \cdots + f(x_n)]$$

The term in the brackets is equal to $\Pr(X \geq a)$. Therefore,

$$E(X) \geq a \Pr(X \geq a)$$

We can rewrite that as

$$\Pr(X \geq a) \leq \frac{E(X)}{a}$$

And Markov's inequality has been proved.

CHAPTER 16

1. $\Pr(6 < \bar{x} < 8) = 0.7062$
   $\Pr(8 < \bar{x} < 9) = 0.1295$
   $\Pr(9 < \bar{x} < 10) = 0.0166$
2. Mean 19.53 Variance 13.85
3. Mean 10.33 Variance 6.89
4. Mean 14.53 Variance 12.52

CHAPTER 17

2. 17.39 to 21.66
3. 8.82 to 11.83
4. 12.50 to 16.56
8. 8.14 to 16.57
9. 9.80 to 16.19
10. 11.58 to 16.77
11. 10.80 to 15.55
13. $E(X^2) - [E(X)]^2$

	$\bar{x}$	n	s_2	a	c	Lower Boundary	Upper Boundary
19.	60	10	8	2.262	5.722	54.278	65.722
20.	60	20	8	2.093	3.744	56.256	63.744
21.	60	50	8	2.009	2.273	57.727	62.273
22.	60	100	8	1.984	1.587	58.413	61.587
23.	60	200	8	1.96	1.109	58.891	61.109
24.	60	500	8	1.96	0.701	59.299	60.701
25.	60	1000	8	1.96	0.496	59.504	60.496
26.	60	50	0.5	2.009	0.142	59.858	60.142
27.	60	50	1	2.009	0.284	59.716	60.284
28.	60	50	5	2.009	1.421	58.579	61.421
29.	60	50	10	2.009	2.841	57.159	62.841
30.	60	50	50	2.009	14.206	45.794	74.206

Note: when $n = 50$, we need the value from the t table for 49 degrees of freedom. This value is not included, but it is very close to the value for 50 degrees of freedom. When $n \geq 200$, we use the value 1.96 from the standard normal distribution.

CHAPTER 18

1. Test statistic: 0.383; accept

2. Test statistic: 0.380; accept

3. Test statistic: –3.15; reject

4. Test statistic: –.101; accept

	$\bar{x}$	n	s_2	μ^*	Test Statistic	Critical Value	
22.	60	10	8	62	–0.791	2.262	accept
23.	60	20	8	62	–1.118	2.093	accept
24.	60	50	8	62	–1.768	2.009	accept
25.	60	100	8	62	–2.500	1.984	reject
26.	60	500	8	62	–5.590	1.960	reject
27.	60	50	1	62	–14.142	2.009	reject
28.	60	50	5	62	–2.828	2.009	reject
29.	60	50	10	62	–1.414	2.009	accept
30.	60	50	20	62	–0.707	2.009	accept

CHAPTER 19

	k	Probability That the Number of People in the Sample on Your Side will Equal k
5.	1	0.093
	2	0.326
	3	0.392
	4	0.163
6.	1	0.495
	2	0.220
	3	0.022
	4	0.000
7.	1	0.017
	2	0.149
	3	0.400
	4	0.350
8.	1	0.052
	2	0.207
	3	0.367
	4	0.288
9.	4	0.075
	5	0.188
	6	0.285
	7	0.258
10.	9	0.020
	10	0.071
	11	0.162
	12	0.243
11.	8	0.097
	9	0.227
	10	0.300
	11	0.227
	12	0.097
12.	8	0.119
	9	0.193
	10	0.227
	11	0.193
	12	0.119
13.	8	0.122
	9	0.175
	10	0.197
	11	0.175
	12	0.122

	95% Confidence Interval	99% Confidence Interval
14.	0.266 to 0.454	0.236 to 0.484
15.	0.293 to 0.427	0.272 to 0.448
16.	0.318 to 0.402	0.305 to 0.415
17.	0.330 to 0.390	0.321 to 0.399
18.	0.347 to 0.373	0.342 to 0.378
19.	0.351 to 0.369	0.348 to 0.372
20.	0.478 to 0.546	0.467 to 0.557
21.	0.545 to 0.627	0.532 to 0.639
22.	0.604 to 0.675	0.593 to 0.686
23.	0.483 to 0.611	0.463 to 0.631
24.	0.733 to 0.798	0.723 to 0.808

CHAPTER 20

1. $n = 25, n_+ = 16, Z = 1.4$; accept the null hypothesis.

2. $n = 30, n_+ = 22, Z = 2.56$; reject the null hypothesis.

3. $n = 18, n_+ = 13, Z = 1.89$; accept the null hypothesis.

4. $R_A = 26, R_B = 35, R_C = 29$

$$F_r = \left(\frac{12}{(15)(3)(4)}\right)(26^2 + 35^2 + 29^2) - 3(15)(4) = 2.8$$

and you would accept the null hypothesis at the 90 percent level.

5. $R_1 = 26, R_2 = 18, R_3 = 14, R_4 = 29, R_5 = 24, R_6 = 23, R_7 = 24, R_8 = 32,$
$R_9 = 24, R_{10} = 27, R_{11} = 36, R_{12} = 35,$

$$F_r = \left(\frac{12}{(4)(12)(13)}\right)(26^2 + 18^2 + 14^2 + 29^2 + 24^2 + 23^2 + 24^2 + 32^2 + 24^2 + 27^2 + 36^2 + 35^2) - 3(4)(13) = 8.77$$

and you would accept the null hypothesis at the 90 percent level.

6. $R_A = 19, R_B = 16, R_C = 14, R_D = 11,$

$$F_r = \left(\frac{12}{(6)(4)(5)}\right)(19^2 + 16^2 + 14^2 + 11^2) - 3(6)(5) = 3.4$$

and you would accept the hypothesis at the 90 percent level.

7. $R_A = 26, R_B = 23, R_C = 29$

$$F_r = \left(\frac{12}{(13)(3)(4)}\right)(26^2 + 23^2 + 29^2) - 3(13)(4) = 1.38$$

and you would accept the null hypothesis at the 90 percent level.

8. $T = 116, \mu_T = 10(26)/2 = 130, \sigma_T^2 = (10)(15)(26)/12 = 325, (T - \mu_T)/\sigma_T = -0.776$, and you would accept the null hypothesis at the 90 percent level.

9. $T = 113, \mu_T = 10(21)/2 = 105, \sigma_T^2 = (10)(10)(21)/12 = 175, (T - \mu_T)/\sigma_T = 0.605$, and you would accept the null hypothesis at the 90 percent level.

10. $T = 214, \mu_T = 12(28)/2 = 168, \sigma_T^2 = (12)(15)(28)/12 = 420, (T - \mu_T)/\sigma_T = 2.25$, and you would accept the null hypothesis at the 95 percent level.

11. $T = 240.5, \mu_T = 15(29)/2 = 217.5, \sigma_T^2 = (15)(13)(29)/12 = 471.25$, $(T - \mu_T)/\sigma_T = 1.06$, and you would accept the null hypothesis at the 90 percent level.

12. $R_A = 169, R_B = 144, R_C = 152$,

$$H = \left(\frac{12}{30(31)}\right)\left(\frac{169^2}{10} + \frac{144^2}{10} + \frac{152^2}{10}\right) - 3(31) = 0.421$$

and you would accept the null hypothesis at the 90 percent level.

13. $R_A = 94, R_B = 214, R_C = 238, R_D = 157$

$$H = \left(\frac{12}{37(38)}\right)\left(\frac{94^2}{7} + \frac{214^2}{10} + \frac{238^2}{12} + \frac{157^2}{8}\right) - 3(38) = 2.44$$

and you would accept the hypothesis at the 90 percent level.

14. $T_+ = 70, T_- = 50, T = 50, \mu_T = 15(16)/4 = 60, \sigma_T^2 = (15)(16)(31)/24 = 310$, $(T - \mu_T)/\sigma_T = -0.568$, and you would accept the null hypothesis at the 90 percent level.

15. $T_+ = 132, T_- = 78, T = 78, \mu_T = 20(21)/4 = 105, \sigma_T^2 = (20)(21)(41)/24 = 717.5$, $(T - \mu_T)/\sigma_T = -1.007$, and you would accept the null hypothesis at the 90 percent level. Note: In view of the large number of ties in this sample, the data may not be reliable.

16. $T_+ = 100, T_- = 71, T = 71, \mu_T = 18(19)/4 = 85.5, \sigma_T^2 = (18)(19)(37)/24 = 527.25$, $(T - \mu_T)/\sigma_T = -0.630$, and you would accept the null hypothesis at the 90 percent level.

17. $T_+ = 164, T_- = 161, T = 161, \mu_T = 25(26)/4 = 162.5, \sigma_T^2 = (25)(26)(51)/24$ $= 1{,}381.25, (T - \mu_T)/\sigma_T = -0.040$, and you would accept the null hypothesis at the 90 percent level. Note: In view of the large number of ties in this sample, the data may not be reliable.

CHAPTER 21

1. The expression for SE_{line} is:

$$SE_{line} = \sum_{i=1}^{n}(mx_i + b - y_i)^2$$

We can rewrite the equation for SE_{line} like this:

$$\begin{aligned} SE_{line} &= \sum_{i=1}^{n}[(mx_i + b)^2 - 2(mx_i + b)y_i + y_i^2] \\ &= \sum_{i=1}^{n}[(m^2x_i^2 + 2bmx_i + b^2 - 2mx_iy_i - 2by_i + y_i^2] \\ &= m^2\sum_{i=1}^{n}x_i^2 + 2bm = \sum_{i=1}^{n}x_i + nb^2 \\ &\quad - 2m\sum_{i=1}^{n}x_iy_i - 2b\sum_{i=1}^{n}y_i + \sum_{i=1}^{n}y_i^2 \end{aligned}$$

We can simplify this expression by using $\overline{x}$ to represent the average value of x, $\left[\overline{x} = (1/n) \times \sum_{i=1}^{n} x_i\right]$ $\overline{y}$ to represent the average value of y, $\overline{x^2}$ to represent the average value of x^2, $\overline{y^2}$ to represent the average value of y^2, and $\overline{xy}$ to represent the average value of xy.

$$SE_{line} = m^2n\overline{x^2} + 2bmn\overline{x} + nb^2 - 2mn\overline{xy} - 2bn\overline{y} + n\overline{y^2}$$

Finding the values of b and m that minimize this expression requires a little bit of calculus. Take the derivatives of TSE with respect to m and with respect to b and then set both derivatives equal to zero. Then optimum values of m and b must satisfy these two equations:

$$\begin{aligned} 2mn\overline{x} + 2nb - 2n\overline{y} &= 0 \\ 2mn\overline{x^2} + 2nb\overline{x} - 2n\overline{xy} &= 0 \end{aligned}$$

We can divide both of these equations by $2n$:

$$\begin{aligned} m\overline{x} + b - \overline{y} &= 0 \\ m\overline{x^2} + b\overline{x} - \overline{xy} &= 0 \end{aligned}$$

Now we can use these two equations to find the value for the slope:

$$\hat{m} = \frac{\overline{xy} - \overline{x}\,\overline{y}}{\overline{x^2} - \overline{x}^2}$$

Once we know $\hat{m}$, we can calculate $\hat{b}$:

$$\hat{b} = \overline{y} - \hat{m}\overline{x}$$

2. First, find an expression for SE_{line}:

$$\begin{aligned}
SE_{line} &= \sum[y_i - (\hat{m}x_i + \hat{b})]^2 \\
&= \sum[y_i - (\hat{m}x_i + \overline{y} - \hat{m}\overline{x})]^2 \\
&= \sum[(y_i - \overline{y}) - \hat{m}(x_i - \overline{x})]^2 \\
&= \sum(y_i - \overline{y})^2 - 2\hat{m}\sum(x_i - \overline{x})(y_i - \overline{y}) + \hat{m}^2\sum(x_i - \overline{x})^2 \\
r^2 &= 1 - SE_{line}/\sum(y_i - \overline{y})^2 \\
&= \frac{2\hat{m}\sum(x_i - \overline{x})(y_i - \overline{y}) - \hat{m}^2\sum(x_i - \overline{x})^2}{\sum(y_i - \overline{y})^2}
\end{aligned}$$

Simplify the first summation:

$$\begin{aligned}
\sum(x_i - \overline{x})(y_i - \overline{y}) &= \sum(x_i y_i - x_i\overline{y} - \overline{x}y_i + \overline{x}\,\overline{y}) \\
&= n\overline{xy} - n\overline{x}\,\overline{y} - n\overline{x}\,\overline{y} + n\overline{x}\,\overline{y} \\
&= n(\overline{xy} - \overline{x}\,\overline{y})
\end{aligned}$$

Use these formulas:

$$\begin{aligned}
\hat{m} &= (\overline{xy} - \overline{x}\,\overline{y})/\mathrm{Var}(x) \\
\sum(x_i - \overline{x}) &= n\mathrm{Var}(x) \\
\sum(y_i - \overline{y}) &= n\mathrm{Var}(y)
\end{aligned}$$

to rewrite the expression for r^2.

$$\begin{aligned}
r^2 &= \frac{2n(\overline{xy} - \overline{x}\,\overline{y})^2}{n\mathrm{Var}(y)\,\mathrm{Var}(x)} - \frac{(\overline{xy} - \overline{x}\,\overline{y})^2 n\,\mathrm{Var}(x)}{[\mathrm{Var}(x)]^2 n\,\mathrm{Var}(y)} \\
&= \frac{2(\overline{xy} - \overline{x}\,\overline{y})^2}{\mathrm{Var}(x)\,\mathrm{Var}(y)} - \frac{(\overline{xy} - \overline{x}\,\overline{y})^2}{\mathrm{Var}(x)\,\mathrm{Var}(y)} \\
&= \frac{(\overline{xy} - \overline{x}\,\overline{y})^2}{(\overline{x^2} - \overline{x}^2)(\overline{y^2} - \overline{y}^2)}
\end{aligned}$$

Note that this is the square of the correlation coefficient between x and y.

	Slope	Intercept	r^2
6.	−0.033	15.418	0.136
7.	10.262	0.071	0.979
8.	0.842	15.585	0.373
9.	30.16	−106.3	0.959
10.	0.48	2.87	0.112
11.	0.50	0.19	0.721
12.	0.979	−0.279	0.830
13.	4.02	0.007	0.999

This result suggests that the true relation between y and x is of the form $y = x^4$.

14. The r^2 value is zero. However, the scatter diagram indicates that there is a definite relation between the variables, although it is a curve that cannot be represented by a line (see Figure A1–1). See Chapter 22, Exercise 11.

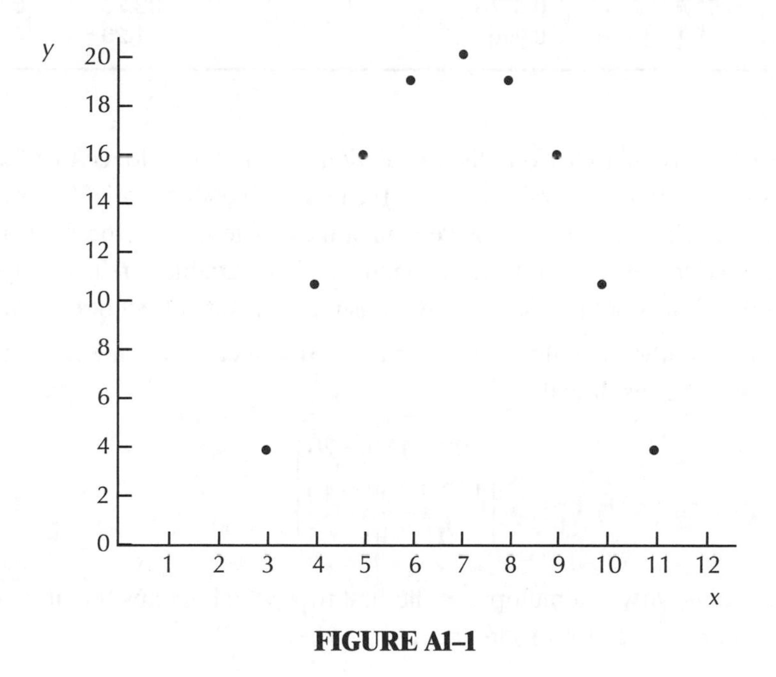

FIGURE A1–1

CHAPTER 22

1. (a) Calculate the t statistic for each coefficient:

$$X1 : 16.22;\quad X2 : 0.37;\quad X3 : -3.02;\quad X4 : 22.74$$

The 95 percent critical value for a t distribution with $25 - 5 = 20$ degrees of freedom is 2.086. It is possible to reject the null hypothesis that the coefficient equals zero for all coefficients except for $X2$.

(b) 339.9

2. The t statistics are as follows:

$$X1 : 0.84; \quad X2 : -0.64; \quad X3 : -2.35$$

The 95 percent critical value for a t distribution with $24 - 5 = 19$ degrees of freedom is 2.093. It is possible to reject the null hypothesis that the coefficient of $X3$ equals zero. Accept the hypothesis for the other coefficients. The critical value for an F distribution with 3 and 20 degrees of freedom is 3.1. Accept the hypothesis that all of the coefficients are zero. It may just be by chance that the coefficient for $X3$ looks significant.

	Coefficient 1	Coefficient 2	Coefficient 3	Constant	r^2
4.	–0.254	–4.051	1.316	109.16	0.620
5.	5.129	0.968		1.26	0.994
6.	0.062	0.160		4.54	0.146
7.	0.017	0.677		33.38	0.326
8.	3.980	5.996		1.20	0.999

9. It is not possible to perform the calculations in this case. Notice that the value of x_2 is always twice the value of x_1, so the two independent variables are perfectly correlated. This is the extreme version of the problem of multicollinearity discussed in the chapter. When two independent variables are perfectly correlated, it is not possible to distinguish the separate effect of each variable.

(Note to mathematically inclined readers: If you calculate the **X′X** matrix in this case the result is this:

$$\begin{pmatrix} 90 & 180 & 20 \\ 180 & 360 & 40 \\ 20 & 40 & 5 \end{pmatrix}$$

The second row is a multiple of the first row, which means that it is impossible to find an inverse for this matrix.)

10. The regression equation is as follows:

$$y = 5.298x_1 - 0.7895x_2 - 1.7155x_3 - 1.6286x_4 + 34.121$$

This equation fits the observations perfectly, so you might think that you have found a clear relationship between the variables. However, the number of observations (5) is equal to the number of coefficients you are estimating. When that happens it is always possible to find an equation that perfectly fits the available data, but you have no evidence that this equation works in any other case. If you try the F statistic you will find that you need to check a distribution with 4 and 0 degrees of freedom; the 0 tips you off that something is wrong. As a general rule, the regression technique is most reliable when the number of observations is much greater than the number of coefficients. It is meaningless if there are as many coefficients as observations.

11. The regression result is $y = -x^2 + 14x - 29$, with an R^2 value of 1. Therefore, the equation fits all observations perfectly. This example shows that a regression calculation can sometimes be used when there is a curved relationship between two variables, provided that both x and x^2 are used as independent variables. (Recall that the simple regression calculation in Chapter 21, Exercise 14, reported an r^2 value of zero.)

10. The regression equation is as follows:

[illegible]

[illegible]

[illegible]

APPENDIX 2

Using Calculators and Computers

Statistical Calculations on a Calculator

Important statistical calculations usually involve large amounts of data, more than anyone would want to try to analyze by hand. Fortunately, calculators and computers are designed to handle repetitive arithmetic calculations. Most scientific calculators available today have a wide range of statistical features. (Software packages for analyzing statistics will be discussed in the next section.)

What are the advantages and disadvantages of using a calculator instead of a computer software package? Calculators are portable and affordable, but there are limits to the amount of data that they can handle.

Since there are many different types of calculators on the market, we will not list the keystrokes for every single one; rather, we will try to summarize the common procedures necessary to perform statistical analyses on a calculator. Once you purchase one yourself, it will be easier then to consult the owner's manual for your calculator.

1. Entering data. With improvements in the way calculators display data, it has become much easier to enter and edit data. Typically you will need to choose a STATISTICS menu on your calculator and then choose either a DATA or EDIT sub-menu. By using the arrow keys, and where appropriate the DELETE key, you can enter your data into one or more lists or columns for analysis. Again, using the arrow keys, you can proofread and if necessary correct your data. (Remember this basic principle of statistical analysis: garbage in, garbage out.)

2. Saving data. One of the basic principles in working with calculators and computers is that you should never have to type in the same set of numbers twice. If there is any chance that you will want to analyze your data set further at a later date, it would be a good idea to save your data. Some calculators will provide you with a limited choice of names for your data set, while others will allow you to choose any name. You can then recall the data set at a later date when you need it.
3. Transferring data. If you are working with someone else who has the same kind of calculator, you may wish to share your data with them. This is possible with many scientific calculators, either through a cable that is plugged in to both calculators or via an infrared data link.
4. Statistical calculations. Going back to the STATISTICS menu on your calculator, you will usually find the following options under a CALC submenu:

- basic descriptive statistics (mean, variance, standard deviation, minimum, maximum)
- various types of regression models
- hypothesis tests on some later model calculators
- statistical displays. If there is a STATISTICS PLOT option on your calculator, you will be able to graph frequency histograms, scatterplots (and simple regression models superimposed over the data points), and box-plots.

Statistical Calculations Using a Computer Software Package

For most practical applications of statistics, it is best to use a computer software package such as SAS, Minitab, SPSS, etc. Why?

- A computer software package can handle much more in the way of arithmetic calculations than you can by hand or with a calculator.
- A computer software package can offer a much wider array of statistical calculations and types of data displays than a calculator can.
- It is possible to insert the results of your work directly into a word-processing document.
- It is easier to transmit your data and/or analysis to someone else via electronic mail.
- Computer displays can show more information in their screens at a given time than a calculator, making data review and display much easier.

Statistical software packages all follow the same sort of basic operations:

1. Entering data. Data can be entered at the keyboard (which can involve a substantial amount of work for large data sets) or loaded from an existing data file. Data can be entered or stored outside of the statistical software package in an ASCII file, but it is usually easier to enter the data from within the package itself. Typically the data will be displayed in a spreadsheet format; you may then use the arrow keys to move around to review or edit data.
2. Saving data. One of the basic principles in working with calculators and computers is that you should never have to type in the same set of numbers twice. If there is any chance that you will want to analyze your data set further at a later date, it would be a good idea to save your data. Software packages will always prompt you before you exit them to save your data first into a data file; even so, it is best to save your data as soon as you have entered it in case of a system crash.
3. Transferring data and results to a word-processing document. Within the Windows and Macintosh operating systems it is possible to save your data and/or results onto a clipboard and then insert them into your document. You can also dynamically link the document to your data file so that if you change your data file at a later date the document will automatically be updated.
4. Statistical calculations. Statistical software packages will perform for you every calculation described in this book. The functions will typically be grouped together into menus. For example, descriptive statistics such as the mean, variance, and standard deviation will typically be contained in a menu item called "One-variable statistics," regression models in a menu item called "Regression," and so on.
5. Graphs. Statistical software packages will create for you every display discussed in this book. The different kinds of graphs will typically be contained in a menu item called "Plot."

If you are working from a Windows or Macintosh environment you will start your program by clicking with your mouse on the program's icon. Once the program is running, you should be able to get specific advice on any topic from the program's "Help" feature (usually listed as a menu item).

Statistical Functions in the Microsoft Excel Spreadsheet

A third possible choice of tools for statistical calculations is a general purpose spreadsheet, such as Microsoft Excel. If you use a spreadsheet for other purposes anyway, then learning the spreadsheet commands would be a good way to do statistical calculations.

The following lists some examples of the statistical functions included with Microsoft Excel.

Description of Function	Example	Result
BINOMDIST (*k*, *n*, *p*,FALSE) calculates the probability that $X = k$, where X has a binomial distribution with parameters *n* and *p*.	=BINOMDIST(6,10,0.75,FALSE)	0.14600
BINOMDIST (*k*, *n*, *p*,TRUE) calculates the cumulative probability that X is less than or equal to *k*; in other words, it sums the probabilities from $X = 0$ up to $X = k$.	=BINOMDIST(6,10,0.75,TRUE)	0.22412
CHIDIST(*a*, *df*) gives the probability that a chi-square random variable with *df* degrees of freedom will be greater than *a*.	=CHIDIST(11.07,5)	0.05001
CHINV(*p*, *df*) gives the value *a* such that $\Pr(\chi^2_{df} > a) = p$, where χ^2_{df} is a chi-square random variable with *df* degrees of freedom. This function is the inverse of the previous one.	=CHINV(0.05,5)	11.07048
COMBIN(*n*, *j*) gives the number of combinations when *j* objects are selected from *n* objects.	=COMBIN(52,5)	2,598,960
FACT(*n*) gives *n*! (*n* factorial).	=FACT(9)	362,880
FDIST(*a*, df_{num}, df_{den}) gives the probability that an *F* random variable with df_{num} and df_{den} degrees of freedom will be greater than *a*.	=FDIST(5.96,10,4)	0.05006
FINV(*p*, df_{num}, df_{den}) gives the value *a* such that $\Pr(F > a) = p$, where *F* is an *F* random variable with df_{num} and df_{den} degrees of freedom. This function is the inverse of the previous one.	=FINV(0.05,10,4)	5.96435
HYPGEOMDIST(*k*, *ns*, *M*, *N*) gives $\Pr(X = k)$, where *X* has a hypergeometric distribution with population size *N*, sample size *ns*, and *M* objects in the population of type *M*, and *k* objects in the sample of type *M*.	=HYPGEOMDIST(3,10,13,52)	0.27806
NORMDIST(*x*, *mu*, *sigma*,TRUE) gives the probability that a normal random variable (with mean *mu* and standard deviation *sigma*) will be less than *x*.	=NORMDIST(12,10,2,TRUE)	0.84134

NORMINV(*p, mu, sigma*) gives the value *a* such that Pr $(X < a) = p$, where *X* has a normal distribution. This function is the inverse of the previous one.	=NORMINV(0.841345,10,2)	12.00000
NORMSDIST(*a*) gives Pr $(Z < a)$, where *Z* has a standard normal distribution.	=NORMSDIST(1.96)	0.97500
NORMSINV(*p*) gives the value *a* such that Pr $(Z < a) = p$, where *Z* has a standard normal distribution. This function is the inverse of the previous one.	=NORMSINV(0.975)	1.95996
TDIST(*a, df,* 1) gives the one-tail probability for a *t* distribution: Pr $(T > a)$, where *T* is a random variable with the *T* distribution with *df* degrees of freedom.	=TDIST(2.365,7,1)	0.02499
TDIST(*a, df,* 2) gives the two-tail probability for a *t* distribution: Pr $(T > a)$ + Pr $(T < -a)$, or 2 Pr $(T > a)$.	=TDIST(2.365,7,2)	0.04997
TINV(*p, df*) give the value *a* such that Pr $(T > a) = p$, where *T* is a random variable with the *T* distribution with *df* degrees of freedom.	=TINV(0.05,7)	2.36462

Excel also provides several functions that calculate descriptive statistics for a range of data:

AVERAGE(*range*)	calculate average
STDEVP(*range*)	standard deviation of a population
STDEV(*range*)	standard deviation of a sample
MEDIAN(*range*)	calculate median
PERCENTRANK(*range, value*)	give the percentile rank of *value* within the list *range*

The Data Analysis add-in performs regression and other statistical calculations. See the HELP menu for more details on these and other statistical functions.

APPENDIX 3

Statistical Tables

TABLE A3–1: ONE-TAILED STANDARD NORMAL (Z) RANDOM VARIABLE TABLE

$\Pr(Z < a) = p$ (page 1)

a	p	a	p	a	p	a	p	a	p	a	p
−2.99	.0014	−2.49	.0064	−1.99	.0233	−1.49	.0681	−0.99	.1611	−0.49	.3121
−2.98	.0014	−2.48	.0066	−1.98	.0239	−1.48	.0694	−0.98	.1635	−0.48	.3156
−2.97	.0015	−2.47	.0068	−1.97	.0244	−1.47	.0708	−0.97	.1660	−0.47	.3192
−2.96	.0015	−2.46	.0069	−1.96	.0250	−1.46	.0721	−0.96	.1685	−0.46	.3228
−2.95	.0016	−2.45	.0071	−1.95	.0256	−1.45	.0735	−0.95	.1711	−0.45	.3264
−2.94	.0016	−2.44	.0073	−1.94	.0262	−1.44	.0749	−0.94	.1736	−0.44	.3300
−2.93	.0017	−2.43	.0075	−1.93	.0268	−1.43	.0764	−0.93	.1762	−0.43	.3336
−2.92	.0018	−2.42	.0078	−1.92	.0274	−1.42	.0778	−0.92	.1788	−0.42	.3372
−2.91	.0018	−2.41	.0080	−1.91	.0281	−1.41	.0793	−0.91	.1814	−0.41	.3409
−2.90	.0019	−2.40	.0082	−1.90	.0287	−1.40	.0808	−0.90	.1841	−0.40	.3446
−2.89	.0019	−2.39	.0084	−1.89	.0294	−1.39	.0823	−0.89	.1867	−0.39	.3483
−2.88	.0020	−2.38	.0087	−1.88	.0301	−1.38	.0838	−0.88	.1894	−0.38	.3520
−2.87	.0021	−2.37	.0089	−1.87	.0307	−1.37	.0853	−0.87	.1922	−0.37	.3557
−2.86	.0021	−2.36	.0091	−1.86	.0314	−1.36	.0869	−0.86	.1949	−0.36	.3594
−2.85	.0022	−2.35	.0094	−1.85	.0322	−1.35	.0885	−0.85	.1977	−0.35	.3632
−2.84	.0023	−2.34	.0096	−1.84	.0329	−1.34	.0901	−0.84	.2005	−0.34	.3669
−2.83	.0023	−2.33	.0099	−1.83	.0336	−1.33	.0918	−0.83	.2033	−0.33	.3707
−2.82	.0024	−2.32	.0102	−1.82	.0344	−1.32	.0934	−0.82	.2061	−0.32	.3745
−2.81	.0025	−2.31	.0104	−1.81	.0351	−1.31	.0951	−0.81	.2090	−0.31	.3783
−2.80	.0026	−2.30	.0107	−1.80	.0359	−1.30	.0968	−0.80	.2119	−0.30	.3821
−2.79	.0026	−2.29	.0110	−1.79	.0367	−1.29	.0985	−0.79	.2148	−0.29	.3859
−2.78	.0027	−2.28	.0113	−1.78	.0375	−1.28	.1003	−0.78	.2177	−0.28	.3897
−2.77	.0028	−2.27	.0116	−1.77	.0384	−1.27	.1020	−0.77	.2206	−0.27	.3936
−2.76	.0029	−2.26	.0119	−1.76	.0392	−1.26	.1038	−0.76	.2236	−0.26	.3974
−2.75	.0030	−2.25	.0122	−1.75	.0401	−1.25	.1056	−0.75	.2266	−0.25	.4013
−2.74	.0031	−2.24	.0125	−1.74	.0409	−1.24	.1075	−0.74	.2296	−0.24	.4052
−2.73	.0032	−2.23	.0129	−1.73	.0418	−1.23	.1093	−0.73	.2327	−0.23	.4090
−2.72	.0033	−2.22	.0132	−1.72	.0427	−1.22	.1112	−0.72	.2358	−0.22	.4129
−2.71	.0034	−2.21	.0136	−1.71	.0436	−1.21	.1131	−0.71	.2389	−0.21	.4168
−2.70	.0035	−2.20	.0139	−1.70	.0446	−1.20	.1151	−0.70	.2420	−0.20	.4207
−2.69	.0036	−2.19	.0143	−1.69	.0455	−1.19	.1170	−0.69	.2451	−0.19	.4247
−2.68	.0037	−2.18	.0146	−1.68	.0465	−1.18	.1190	−0.68	.2483	−0.18	.4286
−2.67	.0038	−2.17	.0150	−1.67	.0475	−1.17	.1210	−0.67	.2514	−0.17	.4325
−2.66	.0039	−2.16	.0154	−1.66	.0485	−1.16	.1230	−0.66	.2546	−0.16	.4364
−2.65	.0040	−2.15	.0158	−1.65	.0495	−1.15	.1251	−0.65	.2578	−0.15	.4404
−2.64	.0041	−2.14	.0162	−1.64	.0505	−1.14	.1271	−0.64	.2611	−0.14	.4443
−2.63	.0043	−2.13	.0166	−1.63	.0516	−1.13	.1292	−0.63	.2643	−0.13	.4483
−2.62	.0044	−2.12	.0170	−1.62	.0526	−1.12	.1314	−0.62	.2676	−0.12	.4522
−2.61	.0045	−2.11	.0174	−1.61	.0537	−1.11	.1335	−0.61	.2709	−0.11	.4562
−2.60	.0047	−2.10	.0179	−1.60	.0548	−1.10	.1357	−0.60	.2743	−0.10	.4602
−2.59	.0048	−2.09	.0183	−1.59	.0559	−1.09	.1379	−0.59	.2776	−0.09	.4641
−2.58	.0049	−2.08	.0188	−1.58	.0570	−1.08	.1401	−0.58	.2810	−0.08	.4681
−2.57	.0051	−2.07	.0192	−1.57	.0582	−1.07	.1423	−0.57	.2843	−0.07	.4721
−2.56	.0052	−2.06	.0197	−1.56	.0594	−1.06	.1446	−0.56	.2877	−0.06	.4761
−2.55	.0054	−2.05	.0202	−1.55	.0605	−1.05	.1469	−0.55	.2912	−0.05	.4801
−2.54	.0055	−2.04	.0207	−1.54	.0618	−1.04	.1492	−0.54	.2946	−0.04	.4840
−2.53	.0057	−2.03	.0212	−1.53	.0630	−1.03	.1515	−0.53	.2981	−0.03	.4880
−2.52	.0059	−2.02	.0217	−1.52	.0642	−1.02	.1539	−0.52	.3015	−0.02	.4920
−2.51	.0060	−2.01	.0222	−1.51	.0655	−1.01	.1562	−0.51	.3050	−0.01	.4960
−2.50	.0062	−2.00	.0228	−1.50	.0668	−1.00	.1587	−0.50	.3085	0.00	.5000

TABLE A3–1: ONE-TAILED STANDARD NORMAL (Z) RANDOM VARIABLE TABLE

Pr ($Z < a$) = p (page 2)

a	p	a	p	a	p	a	p	a	p	a	p
0.01	.5040	0.51	.6950	1.01	.8438	1.51	.9345	2.01	.9778	2.51	.9940
0.02	.5080	0.52	.6985	1.02	.8461	1.52	.9357	2.02	.9783	2.52	.9941
0.03	.5120	0.53	.7019	1.03	.8485	1.53	.9370	2.03	.9788	2.53	.9943
0.04	.5160	0.54	.7054	1.04	.8508	1.54	.9382	2.04	.9793	2.54	.9945
0.05	.5199	0.55	.7088	1.05	.8531	1.55	.9394	2.05	.9798	2.55	.9946
0.06	.5239	0.56	.7123	1.06	.8554	1.56	.9406	2.06	.9803	2.56	.9948
0.07	.5279	0.57	.7157	1.07	.8577	1.57	.9418	2.07	.9808	2.57	.9949
0.08	.5319	0.58	.7190	1.08	.8599	1.58	.9429	2.08	.9812	2.58	.9951
0.09	.5359	0.59	.7224	1.09	.8621	1.59	.9441	2.09	.9817	2.59	.9952
0.10	.5398	0.60	.7257	1.10	.8643	1.60	.9452	2.10	.9821	2.60	.9953
0.11	.5438	0.61	.7291	1.11	.8665	1.61	.9463	2.11	.9826	2.61	.9955
0.12	.5478	0.62	.7324	1.12	.8686	1.62	.9474	2.12	.9830	2.62	.9956
0.13	.5517	0.63	.7357	1.13	.8708	1.63	.9484	2.13	.9834	2.63	.9957
0.14	.5557	0.64	.7389	1.14	.8729	1.64	.9495	2.14	.9838	2.64	.9959
0.15	.5596	0.65	.7422	1.15	.8749	1.65	.9505	2.15	.9842	2.65	.9960
0.16	.5636	0.66	.7454	1.16	.8770	1.66	.9515	2.16	.9846	2.66	.9961
0.17	.5675	0.67	.7486	1.17	.8790	1.67	.9525	2.17	.9850	2.67	.9962
0.18	.5714	0.68	.7517	1.18	.8810	1.68	.9535	2.18	.9854	2.68	.9963
0.19	.5753	0.69	.7549	1.19	.8830	1.69	.9545	2.19	.9857	2.69	.9964
0.20	.5793	0.70	.7580	1.20	.8849	1.70	.9554	2.20	.9861	2.70	.9965
0.21	.5832	0.71	.7611	1.21	.8869	1.71	.9564	2.21	.9864	2.71	.9966
0.22	.5871	0.72	.7642	1.22	.8888	1.72	.9573	2.22	.9868	2.72	.9967
0.23	.5910	0.73	.7673	1.23	.8907	1.73	.9582	2.23	.9871	2.73	.9968
0.24	.5948	0.74	.7704	1.24	.8925	1.74	.9591	2.24	.9875	2.74	.9969
0.25	.5987	0.75	.7734	1.25	.8944	1.75	.9599	2.25	.9878	2.75	.9970
0.26	.6026	0.76	.7764	1.26	.8962	1.76	.9608	2.26	.9881	2.76	.9971
0.27	.6064	0.77	.7794	1.27	.8980	1.77	.9616	2.27	.9884	2.77	.9972
0.28	.6103	0.78	.7823	1.28	.8997	1.78	.9625	2.28	.9887	2.78	.9973
0.29	.6141	0.79	.7852	1.29	.9015	1.79	.9633	2.29	.9890	2.79	.9974
0.30	.6179	0.80	.7881	1.30	.9032	1.80	.9641	2.30	.9893	2.80	.9974
0.31	.6217	0.81	.7910	1.31	.9049	1.81	.9649	2.31	.9896	2.81	.9975
0.32	.6255	0.82	.7939	1.32	.9066	1.82	.9656	2.32	.9898	2.82	.9976
0.33	.6293	0.83	.7967	1.33	.9082	1.83	.9664	2.33	.9901	2.83	.9977
0.34	.6331	0.84	.7995	1.34	.9099	1.84	.9671	2.34	.9904	2.84	.9977
0.35	.6368	0.85	.8023	1.35	.9115	1.85	.9678	2.35	.9906	2.85	.9978
0.36	.6406	0.86	.8051	1.36	.9131	1.86	.9686	2.36	.9909	2.86	.9979
0.37	.6443	0.87	.8078	1.37	.9147	1.87	.9693	2.37	.9911	2.87	.9979
0.38	.6480	0.88	.8106	1.38	.9162	1.88	.9699	2.38	.9913	2.88	.9980
0.39	.6517	0.89	.8133	1.39	.9177	1.89	.9706	2.39	.9916	2.89	.9981
0.40	.6554	0.90	.8159	1.40	.9192	1.90	.9713	2.40	.9918	2.90	.9981
0.41	.6591	0.91	.8186	1.41	.9207	1.91	.9719	2.41	.9920	2.91	.9982
0.42	.6628	0.92	.8212	1.42	.9222	1.92	.9726	2.42	.9922	2.92	.9982
0.43	.6664	0.93	.8238	1.43	.9236	1.93	.9732	2.43	.9925	2.93	.9983
0.44	.6700	0.94	.8264	1.44	.9251	1.94	.9738	2.44	.9927	2.94	.9984
0.45	.6736	0.95	.8289	1.45	.9265	1.95	.9744	2.45	.9929	2.95	.9984
0.46	.6772	0.96	.8315	1.46	.9279	1.96	.9750	2.46	.9931	2.96	.9985
0.47	.6808	0.97	.8340	1.47	.9292	1.97	.9756	2.47	.9932	2.97	.9985
0.48	.6844	0.98	.8365	1.48	.9306	1.98	.9761	2.48	.9934	2.98	.9986
0.49	.6879	0.99	.8389	1.49	.9319	1.99	.9767	2.49	.9936	2.99	.9986
0.50	.6915	1.00	.8413	1.50	.9332	2.00	.9772	2.50	.9938	3.00	.9987

TABLE A3–2: TWO-TAILED STANDARD NORMAL (*Z*) RANDOM VARIABLE TABLE

This table gives the probability p that a standard normal random variable Z (mean 0, standard deviation 1) will be between $-a$ and a.

$$\Pr(-a < Z < a) = p$$

a	p	a	p
0.100	0.0796	1.400	0.8384
0.200	0.1586	**1.439**	**0.8500**
0.300	0.2358	1.500	0.8664
0.400	0.3108	1.600	0.8904
0.500	0.3830	**1.645**	**0.9000**
0.600	0.4514	1.700	0.9108
0.700	0.5160	1.800	0.9282
0.800	0.5762	1.900	0.9426
0.900	0.6318	**1.960**	**0.9500**
1.000	0.6826	2.000	0.9544
1.100	0.7286	2.500	0.9876
1.200	0.7698	**2.576**	**0.9900**
1.282	**0.8000**	3.000	0.9974
1.300	0.8064	3.100	0.9980

The values in boldface are those commonly used for confidence intervals and hypothesis testing.

The table on the previous two pages gives the probability p that a standard normal random variable Z will be less than the specified value a: $p = \Pr(Z < a)$. Also, it gives the area under the standard normal density function to the left of the specified value a.

Here is the connection between the two tables. If $p_1 = \Pr(Z < a)$ (the value from the one-tailed table), and $p_2 = \Pr(-a < Z < a)$ (the value from the two-tailed table), then $p_2 = 2p_1 - 1$.

TABLE A3–3: CHI-SQUARE TABLE

The table gives the value of *a* such that $\Pr(\chi^2_{DF} < a) = p$, where χ^2_{DF} is a chi-square random variable with *DF* degrees of freedom. For example, there is a probability of .95 that a chi-square random variable with 6 degrees of freedom will be less than 12.6.

DF	*p* = .005	.01	.025	.05	.25	.5	.75	.9	.95	.975	.99
1	.000	.000	.001	.004	.10	.45	1.32	2.71	3.84	5.02	6.64
2	.010	.020	.051	.10	.58	1.39	2.77	4.61	5.99	7.38	9.21
3	.072	.11	.22	.35	1.21	2.37	4.11	6.25	7.81	9.35	11.3
4	.21	.30	.48	.71	1.92	3.36	5.39	7.78	9.49	11.1	13.3
5	.41	.55	.83	1.15	2.67	4.35	6.63	9.24	11.1	12.8	15.1
6	.68	.87	1.24	1.64	3.45	5.35	7.84	10.6	12.6	14.4	16.8
7	.99	1.24	1.69	2.17	4.25	6.35	9.04	12.0	14.1	16.0	18.5
8	1.34	1.65	2.18	2.73	5.07	7.34	10.2	13.4	15.5	17.5	20.1
9	1.73	2.09	2.70	3.33	5.90	8.34	11.4	14.7	16.9	19.0	21.7
10	2.16	2.56	3.25	3.94	6.74	9.34	12.5	16.0	18.3	20.5	23.2
11	2.60	3.05	3.82	4.57	7.58	10.3	13.7	17.3	19.7	21.9	24.7
12	3.07	3.57	4.40	5.23	8.44	11.3	14.8	18.5	21.0	23.3	26.2
13	3.56	4.11	5.01	5.89	9.30	12.3	16.0	19.8	22.4	24.7	27.7
14	4.07	4.66	5.63	6.57	10.2	13.3	17.1	21.1	23.7	26.1	29.1
15	4.60	5.23	6.26	7.26	11.0	14.3	18.2	22.3	25.0	27.5	30.6
16	5.14	5.81	6.91	7.96	11.9	15.3	19.4	23.5	26.3	28.8	32.0
17	5.70	6.41	7.56	8.67	12.8	16.3	20.5	24.8	27.6	30.2	33.4
18	6.26	7.01	8.23	9.39	13.7	17.3	21.6	26.0	28.9	31.5	34.8
19	6.84	7.63	8.91	10.1	14.6	18.3	22.7	27.2	30.1	32.9	36.2
20	7.43	8.26	9.59	10.9	15.5	19.3	23.8	28.4	31.4	34.2	37.6
21	8.03	8.90	10.3	11.6	16.3	20.3	24.9	29.6	32.7	35.5	38.9
22	8.64	9.54	11.0	12.3	17.2	21.3	26.0	30.8	33.9	36.8	40.3
23	9.26	10.2	11.7	13.1	18.1	22.3	27.1	32.0	35.2	38.1	41.6
24	9.89	10.9	12.4	13.8	19.0	23.3	28.2	33.2	36.4	39.4	43.0
25	10.5	11.5	13.1	14.6	19.9	24.3	29.3	34.4	37.7	40.6	44.3
26	11.2	12.2	13.8	15.4	20.8	25.3	30.4	35.6	38.9	41.9	45.6
27	11.8	12.9	14.6	16.1	21.7	26.3	31.5	36.7	40.1	43.2	47.0
28	12.5	13.6	15.3	16.9	22.7	27.3	32.6	37.9	41.3	44.5	48.3
29	13.1	14.3	16.0	17.7	23.6	28.3	33.7	39.1	42.6	45.7	49.6
30	13.8	14.9	16.8	18.5	24.5	29.3	34.8	40.3	43.8	47.0	50.9
35	17.2	18.5	20.6	22.5	29.0	34.3	40.2	46.1	49.8	53.2	57.3
40	20.7	22.2	24.4	26.5	33.7	39.3	45.6	51.8	55.8	59.3	63.7
50	28.0	29.7	32.4	34.8	42.9	49.3	56.3	63.2	67.5	71.4	76.2
60	35.5	37.5	40.5	43.2	52.3	59.3	67.0	74.4	79.1	83.3	88.4
70	43.3	45.4	48.8	51.7	61.7	69.3	77.6	85.5	90.5	95.0	100.4
80	51.2	53.5	57.2	60.4	71.1	79.3	88.1	96.6	101.9	106.6	112.3
90	59.2	61.8	65.6	69.1	80.6	89.3	98.6	107.6	113.1	118.1	124.1
100	67.3	70.1	74.2	77.9	90.1	99.3	109.1	118.5	124.3	129.6	135.8

TABLE A3–4: ONE-TAILED *t* DISTRIBUTION TABLE

If T is a random variable with a t distribution with DF degrees of freedom, then the table gives the value of a such that $\Pr(T < a) = p$. For example, there is a .975 probability that a T random variable with 7 degrees of freedom will be less than 2.365.

DF	*p* = .750	*p* = .900	*p* = .950	*p* = .975	*p* = .990	*p* = .995
1	1.000	3.078	6.314	12.706	31.821	63.657
2	0.816	1.886	2.920	4.303	6.965	9.925
3	0.765	1.638	2.353	3.182	4.541	5.841
4	0.741	1.533	2.132	2.776	3.747	4.604
5	0.727	1.476	2.015	2.571	3.365	4.032
6	0.718	1.440	1.943	2.447	3.143	3.707
7	0.711	1.415	1.895	2.365	2.998	3.499
8	0.706	1.397	1.860	2.306	2.896	3.355
9	0.703	1.383	1.833	2.262	2.821	3.250
10	0.700	1.372	1.812	2.228	2.764	3.169
11	0.697	1.363	1.796	2.201	2.718	3.106
12	0.695	1.356	1.782	2.179	2.681	3.055
13	0.694	1.350	1.771	2.160	2.650	3.012
14	0.692	1.345	1.761	2.145	2.624	2.977
15	0.691	1.341	1.753	2.131	2.602	2.947
16	0.690	1.337	1.746	2.120	2.583	2.921
17	0.689	1.333	1.740	2.110	2.567	2.898
18	0.688	1.330	1.734	2.101	2.552	2.878
19	0.688	1.328	1.729	2.093	2.539	2.861
20	0.687	1.325	1.725	2.086	2.528	2.845
21	0.686	1.323	1.721	2.080	2.518	2.831
22	0.686	1.321	1.717	2.074	2.508	2.819
23	0.685	1.319	1.714	2.069	2.500	2.807
24	0.685	1.318	1.711	2.064	2.492	2.797
25	0.684	1.316	1.708	2.060	2.485	2.787
26	0.684	1.315	1.706	2.056	2.479	2.779
27	0.684	1.314	1.703	2.052	2.473	2.771
28	0.683	1.313	1.701	2.048	2.467	2.763
29	0.683	1.311	1.699	2.045	2.462	2.756
30	0.683	1.310	1.697	2.042	2.457	2.750
35	0.682	1.306	1.690	2.030	2.438	2.724
40	0.681	1.303	1.684	2.021	2.423	2.704
45	0.680	1.301	1.679	2.014	2.412	2.690
50	0.679	1.299	1.676	2.009	2.403	2.678
55	0.679	1.297	1.673	2.004	2.396	2.668
60	0.679	1.296	1.671	2.000	2.390	2.660
80	0.678	1.292	1.664	1.990	2.374	2.639
100	0.677	1.290	1.660	1.984	2.364	2.626
120	0.677	1.289	1.658	1.980	2.358	2.617

TABLE A3–5: TWO-TAILED *t* DISTRIBUTION TABLE

If T is a random variable with a t distribution with DF degrees of freedom, then the table gives the value of a such that $\Pr(-a < T < a) = p$. For example, there is a 95 percent probability that a T random variable with 7 degrees of freedom will be between –2.365 and 2.365.

DF	$p = .900$	$p = .950$	$p = .99$
1	6.314	12.706	63.657
2	2.920	4.303	9.925
3	2.353	3.182	5.841
4	2.132	2.776	4.604
5	2.015	2.571	4.032
6	1.943	2.447	3.707
7	1.895	2.365	3.499
8	1.860	2.306	3.355
9	1.833	2.262	3.250
10	1.812	2.228	3.169
11	1.796	2.201	3.106
12	1.782	2.179	3.055
13	1.771	2.160	3.012
14	1.761	2.145	2.977
15	1.753	2.131	2.947
16	1.746	2.120	2.921
17	1.740	2.110	2.898
18	1.734	2.101	2.878
19	1.729	2.093	2.861
20	1.725	2.086	2.845
21	1.721	2.080	2.831
22	1.717	2.074	2.819
23	1.714	2.069	2.807
24	1.711	2.064	2.797
25	1.708	2.060	2.787
26	1.706	2.056	2.779
27	1.703	2.052	2.771
28	1.701	2.048	2.763
29	1.699	2.045	2.756
30	1.697	2.042	2.750
35	1.690	2.030	2.724
40	1.684	2.021	2.704
45	1.679	2.014	2.690
50	1.676	2.009	2.678
55	1.673	2.004	2.668
60	1.671	2.000	2.660
80	1.664	1.990	2.639
100	1.660	1.984	2.626
120	1.658	1.980	2.617

TABLE A3–6: *F* DISTRIBUTION TABLE

In each of the following three tables, the numerator degrees of freedom are read along the top, and the denominator degrees of freedom are read along the left side. The table gives the value of *a* such that $\Pr(F < a) = p$, where *F* is a random variable with an *F* distribution with DF_{num} numerator degrees of freedom and DF_{den} denominator degrees of freedom. Each table has a different value of *p*—first .99; then .95; then .90. These are values commonly used for hypothesis testing. Another way to describe the table is to give the value of $1 - p$, which is the right tail area (that is, the area to the right of the given value of *a*). For example, there is a .99 probability that an *F* random variable with 5 numerator degrees of freedom and 9 denominator degrees of freedom will be less than 6.06.

$\Pr(F < a) = 0.99$; right tail area = 0.01

DF_{den}	$DF_{num} = 2$	3	4	5	10	15	20	30	60	120
2	99.00	99.16	99.25	99.30	99.40	99.43	99.45	99.47	99.48	99.49
3	30.82	29.46	28.71	28.24	27.23	26.87	26.69	26.50	26.32	26.22
4	18.00	16.69	15.98	15.52	14.55	14.20	14.02	13.84	13.65	13.56
5	13.27	12.06	11.39	10.97	10.05	9.72	9.55	9.38	9.20	9.11
6	10.92	9.78	9.15	8.75	7.87	7.56	7.40	7.23	7.06	6.97
7	9.55	8.45	7.85	7.46	6.62	6.31	6.16	5.99	5.82	5.74
8	8.65	7.59	7.01	6.63	5.81	5.52	5.36	5.20	5.03	4.95
9	8.02	6.99	6.42	6.06	5.26	4.96	4.81	4.65	4.48	4.40
10	7.56	6.55	5.99	5.64	4.85	4.56	4.41	4.25	4.08	4.00
15	6.36	5.42	4.89	4.56	3.80	3.52	3.37	3.21	3.05	2.96
20	5.85	4.94	4.43	4.10	3.37	3.09	2.94	2.78	2.61	2.52
30	5.39	4.51	4.02	3.70	2.98	2.70	2.55	2.39	2.21	2.11
60	4.98	4.13	3.65	3.34	2.63	2.35	2.20	2.03	1.84	1.73
120	4.79	3.95	3.48	3.17	2.47	2.19	2.03	1.86	1.66	1.53

Pr($F < a$) = 0.95; right tail area = 0.05

DF_{den}	DF_{num} = 2	3	4	5	10	15	20	30	60	120
2	19.00	19.16	19.25	19.30	19.40	19.43	19.45	19.46	19.48	19.49
3	9.55	9.28	9.12	9.01	8.79	8.70	8.66	8.62	8.57	8.55
4	6.94	6.59	6.39	6.26	5.96	5.86	5.80	5.75	5.69	5.66
5	5.79	5.41	5.19	5.05	4.74	4.62	4.56	4.50	4.43	4.40
6	5.14	4.76	4.53	4.39	4.06	3.94	3.87	3.81	3.74	3.70
7	4.74	4.35	4.12	3.97	3.64	3.51	3.44	3.38	3.30	3.27
8	4.46	4.07	3.84	3.69	3.35	3.22	3.15	3.08	3.01	2.97
9	4.26	3.86	3.63	3.48	3.14	3.01	2.94	2.86	2.79	2.75
10	4.10	3.71	3.48	3.33	2.98	2.85	2.77	2.70	2.62	2.58
15	3.68	3.29	3.06	2.90	2.54	2.40	2.33	2.25	2.16	2.11
20	3.49	3.10	2.87	2.71	2.35	2.20	2.12	2.04	1.95	1.90
30	3.32	2.92	2.69	2.53	2.16	2.01	1.93	1.84	1.74	1.68
60	3.15	2.76	2.53	2.37	1.99	1.84	1.75	1.65	1.53	1.47
120	3.07	2.68	2.45	2.29	1.91	1.75	1.66	1.55	1.43	1.35

Pr($F < a$) = 0.90; right tail area = 0.10

DF_{den}	DF_{num} = 2	3	4	5	10	15	20	30	60	120
2	9.00	9.16	9.24	9.29	9.39	9.42	9.44	9.46	9.47	9.48
3	5.46	5.39	5.34	5.31	5.23	5.20	5.18	5.17	5.15	5.14
4	4.32	4.19	4.11	4.05	3.92	3.87	3.84	3.82	3.79	3.78
5	3.78	3.62	3.52	3.45	3.30	3.24	3.21	3.17	3.14	3.12
6	3.46	3.29	3.18	3.11	2.94	2.87	2.84	2.80	2.76	2.74
7	3.26	3.07	2.96	2.88	2.70	2.63	2.59	2.56	2.51	2.49
8	3.11	2.92	2.81	2.73	2.54	2.46	2.42	2.38	2.34	2.32
9	3.01	2.81	2.69	2.61	2.42	2.34	2.30	2.25	2.21	2.18
10	2.92	2.73	2.61	2.52	2.32	2.24	2.20	2.16	2.11	2.08
15	2.70	2.49	2.36	2.27	2.06	1.97	1.92	1.87	1.82	1.79
20	2.59	2.38	2.25	2.16	1.94	1.84	1.79	1.74	1.68	1.64
30	2.49	2.28	2.14	2.05	1.82	1.72	1.67	1.61	1.54	1.50
60	2.39	2.18	2.04	1.95	1.71	1.60	1.54	1.48	1.40	1.35
120	2.35	2.13	1.99	1.90	1.65	1.55	1.48	1.41	1.32	1.26

INDEX